T0344806

LIMITS OF DETECTION
IN CHEMICAL ANALYSIS

CHEMICAL ANALYSIS

A Series of Monographs on Analytical Chemistry
and Its Applications

VOLUME 185

A complete list of the titles in this series appears at the end of this volume.

LIMITS OF DETECTION IN CHEMICAL ANALYSIS

EDWARD VOIGTMAN

Registered Offices
John Wiley & Sons, Inc., 111 River Street, Hoboken, NJ 07030, USA

Editorial Office
111 River Street, Hoboken, NJ 07030, USA

For details of our global editorial offices, customer services, and more information about Wiley products visit us at www.wiley.com.

Wiley also publishes its books in a variety of electronic formats and by print-on-demand. Some content that appears in standard print versions of this book may not be available in other formats.

Library of Congress Cataloging-in-Publication Data

Names: Voigtman, Edward, 1949– author.
Title: Limits of detection in chemical analysis / Edward Voigtman.
Description: Hoboken, NJ : John Wiley & Sons, 2017. | Series: Chemical
    Analysis Series | Includes bibliographical references and index.
Identifiers: LCCN 2017000471 (print) | LCCN 2017005375 (ebook) | ISBN
    9781119188971 (cloth) | ISBN 9781119188995 (pdf) | ISBN 9781119188988 (epub)
Subjects: LCSH: Chemistry, Analytic.
Classification: LCC QD75.22 .V65 2017 (print) | LCC QD75.22 (ebook) | DDC
    543/.19–dc23
LC record available at https://lccn.loc.gov/2017000471

Cover image: Courtesy of the Author
Cover design by Wiley

Set in 10/12pt, TimesLTStd by SPi Global, Chennai, India.

Printed in the United States of America

10 9 8 7 6 5 4 3 2 1

CONTENTS

PREFACE

INTRODUCTION

The limit of detection, equivalently known as the detection limit, is among the most important concepts in chemical analysis, lying at its very heart. The name itself is somewhat misleading, seemingly implying that detection (of the thing of measurement interest, whatever it may be) is possible *down to* a limiting value, below which detection cannot be assumed. This is the "traditional" detection limit definition, but there is another viable definition: so long as the thing of measurement interest is present at, or above, the limit of detection, it has low *a priori*–specified probability of *escaping* detection. The polite sparring between advocates of these two definitions has gone on for decades, but the ongoing ascendancy of the latter definition will become evident in progressing through the text.

First, though, does it even matter? The past several decades have seen truly remarkable advances in our ability to manipulate and study the properties of single atoms, ions, molecules, nanoparticles, quantum dots, and so on. In view of this, what is the current relevance of limits of detection in chemical analysis? If it is possible to work effectively with single chemical entities, does this not imply that limits of detection are superfluous?

The short answer is no: detection limits are important figures of merit for the myriad "ordinary" chemical instruments in daily use in thousands of laboratories around the world. They are one of the initial performance characteristics examined when considering the potential applicability of a specific chemical measurement system, and its associated measurement protocol, to an analytical measurement task.

As figures of merit, detection limits are rather like fuel efficiency ratings for passenger automobiles: very useful for comparison or optimization purposes, but not to the exclusion of all other pertinent factors, such as purchase price or ease of operation. Crucially, figures of merit are only valid and transferable, that is, useful for quantitative comparison purposes, insofar as they are computed in accordance with properly defined and standardized protocols.

Highly sophisticated chemical instruments, capable of detecting single chemical entities, are much more like expensive race cars: their regimes of applicability are very different from those of their ubiquitous workaday counterparts. Systems in this rarefied category are often unique, or nearly so, and they have been optimized with meticulous attention devoted to every salient aspect of their performance. Consequently, detection limits (and race car fuel efficiencies, for that matter) are almost entirely irrelevant relative to other desired performance characteristics.

WHO SHOULD READ THIS BOOK AND WHAT IS IT ABOUT?

This book is for anyone who wants to learn about limits of detection in chemical analysis. The focus is on the fundamental theoretical aspects of limits of detection and how they may be properly computed for ordinary univariate chemical measurement systems. Attention is restricted to the *simplest* possible such systems, with the *simplest* possible types of measurement errors, because these are extraordinarily useful as model systems and as "zero-order" approximations of real systems. Demonstrating how to calculate statistically valid detection limits is this text's entire *raison d'être*. In particular, Currie's detection limits schema [1] is explored in detail, clearly demonstrating how it is instantiated correctly and explaining how it actually works.

Along the way, many figures are provided to illustrate important concepts and to summarize key mathematical results. Simulation models, videos, screencasts, and animations are also provided, at a companion website, to enhance understanding. All of the software models used in the text are freely available at the website: see Appendix B for the URL.

WHAT ABOUT THE STANDARDS ORGANIZATIONS?

Over a span of many years, various standards organizations have generally provided valuable guidance on both theoretical and practical aspects of measurement methodologies and metrology. The result is an immense, and ever-growing, set of modern standards. On the whole, these organizations have been highly successful in fulfilling their responsibilities.

However, the official detection limit standards, with a notable exception (see Chapter 17), yield biased detection limits. In contrast, the correct instantiation of Currie's schema is actually simpler than the irrelevant standards it replaces, is free from bias, and may be validly extended to models with heteroscedastic noise.

NOMENCLATURE

The nomenclature used in this text is entirely conventional and mostly familiar to anyone having an acquaintance with elementary parametric statistics. As an aid, Appendix A provides a very brief review of necessary statistical definitions, concepts, and results. Whenever possible, Greek letters are reserved for population parameters and true values, while Roman letters are used for almost all other constants, variables, and variates. The focus is on understanding fundamental concepts and practical applications, using the long-established principles and terminology of frequentist statistics.

As for definitions, they are much like tourniquets: they should be tight enough to do the job, but not so tight as to detrimentally constrict, so there will be no reticence about either loosening or evolving definitions, as may be useful. This means that long-used terminology will generally be employed, with due care taken to communicate as clearly and unambiguously as possible.

BACKGROUND KNOWLEDGE REQUIRED

Aside from a cursory acquaintance with elementary statistics, only simple algebra is required: a few results obtained using elementary calculus are presented, but there is no need for readers to do calculus in order to master the content of the book. Of utmost importance is simply the ability to think and reason logically.

WHAT WILL NOT BE FOUND IN THIS BOOK?

There are no lengthy tables of detection limits and no references to publications that merely report "the world's lowest detection limits," because these never stand the test of time. There will not be an extensively detailed history of detection limit research. Rather, Chapter 1 presents an extremely brief version, focusing on the most germane events and issues that have occurred since Kaiser's seminal publication in *1947* [2]. Many relevant references are listed in the text and in the Bibliography, but, despite my best efforts, it is certain that some papers will have been missed. For this, I offer my sincere apology: no slight is intended and it is hoped that nothing crucial has escaped detection.

Important topics and areas, such as multivariate calibration, regression on calibration data having errors on both independent and dependent variables, systems having only low levels of shot noise, practically all of chemometrics and formal metrology *per se*, proficiency testing, tolerance intervals, and *many* others, are not discussed at all. Aside from the obvious facts that books have finite page limits and authors have finite "field competency" limits, two factors are of paramount importance.

First, anyone who actually reads and works through the hundreds of detection limit publications, even ignoring websites entirely, quickly realizes that misconceptions, mistakes and oversimplifications abound. Furthermore, even the nomenclature is an embarrassing farrago of terminologies and symbols and the various standards organizations, not for want of trying, have not resolved this issue. Second, without a rigorous theoretical basis upon which to build, more advanced developments may be compromised or even invalidated. Accordingly, the present book resolutely focuses on the fundamental, foundational aspects of analytical detection limit theory, in the hope that this will ultimately be of most general utility and lasting value.

Edward Voigtman
Easthampton, MA, USA
November 2, 2016

REFERENCES

1. L.A. Currie, "Limits for qualitative and quantitative determination: application to radio-chemistry", *Anal. Chem.* **40** (1968) 586-593.
2. H. Kaiser, "Die Berechnung der Nachweisempfindlichkeit", *Spectrochim. Acta* **3** (1947) 40–67.

ACKNOWLEDGMENT

First and foremost, I thank my wife, Janiece Leach, for her extraordinary support throughout my career and during the many hours spent working on this text. Without her love and unshakable faith in me, I never would have finished it. She has my eternal love, respect, and gratitude!

I am grateful to my many colleagues in the Department of Chemistry at the University of Massachusetts – Amherst. Professors Julian Tyson, Ramon Barnes, David Curran, Peter Uden, Ray D'Alonzo, and Alfred Wynne were especially helpful and supportive during my 29 years as a faculty member and I owe them debts of gratitude too large to ever repay. Professor Tyson, in particular, is thanked for arranging for me to meet with Prof. Vitha (*vide infra*).

My former PhD students (Mitch Johnson, Uma Kale, and Dan Montville) really kept me on my toes, as did three excellent undergraduates (Kevin Abraham, Jill Carlson, and Artur Wysoczanski). All deserve my sincere gratitude for their efforts.

At the University of Florida, Gainesville, I thank Dr. Ben Smith, Prof. Nico Omenetto (Editor, *Spectrochimica Acta Part B*) and Prof. Emeritus Jim Wincfordner. I learned an enormous amount from these three wise men, over a period of many years, and am honored to have them as valued friends and scientific colleagues.

Finally, a special thanks to everyone at John Wiley & Sons, Inc., particularly Prof. Mark F. Vitha (Editor of the Chemical Analysis Series), Bob Esposito (Associate Publisher), Michael Leventhal (who works with Bob Esposito), Divya Narayanan (Project Manager), and Anumita Gupta (Production Editor) and the entire production team for their efforts and assistance in making this book a success. Without their help and consummate professionalism, this text could not have happened.

As for errors, any such are entirely mine and an erratum list will be posted to the text's companion website.

ABOUT THE COMPANION WEBSITE

This book is accompanied by a companion website:
www.wiley.com/go/Voigtman/Limits_of_Detection_in_Chemical_Analysis

The website includes
- *LightStone*® Software for Windows-based PC
- Simulation models used in the monograph
- Videos
- Screencasts
- Reference files
- Additional simulation models

1

BACKGROUND

1.1 INTRODUCTION

For some purposes, qualitative detection of a substance of interest may be sufficient; for example, is there melamine adulterant in milk [1] or ^{210}Po in an ex-spy [2]? In many cases, quantitation of the substance of interest, generally referred to as the analyte, is either desired or required. Three simple examples are as follows:

- What is the total organic content (TOC) in a drinking water specimen?
- What is the Cr^{3+} number density in a ruby laser rod?
- What is the pinene concentration in an air specimen collected in a pine forest?

In each of these cases, what matters is quantitative, since it is already known that all drinking water contains some organic content, every ruby has (and gets its color from) its Cr^{3+} content, and pinene contributes to the fragrance of a pine forest. What matters are the specific numerical concentrations, quantities, or amounts, that is, the quantitative analyte contents. In the drinking water example, unacceptable TOC levels should, politics and costs aside, trigger subsequent decisions and corrective actions.

In science fiction, it is common for instruments to be capable of scanning entire alien planets, and perfectly inventorying everything in them, perhaps in preparation for errorlessly teleporting up everything valuable. The real world is different: the laws of nature are always obeyed and experimental measurements are afflicted with measurement uncertainties. The immediate consequence of the latter is that only estimates of underlying true values may be obtained and these estimates, except by generally unknowable coincidence, do not equal the true values. These limits may be reduced further only if the relevant measurement uncertainties are reduced.

Limits of Detection in Chemical Analysis, First Edition. Edward Voigtman.
© 2017 John Wiley & Sons, Inc. Published 2017 by John Wiley & Sons, Inc.
Companion Website: www.wiley.com/go/Voigtman/Limits_of_Detection_in_Chemical_Analysis

The elementary theory of detection limits in chemical measurement systems evolved over a century, through the dedicated efforts of many individuals. Based on their collective work, the next several chapters are devoted to the methodic and rigorous development of the concepts of decision level and limit of detection, with emphasis placed on understanding what they are and the purposes they serve. In Chapter 19, the concept of limit of quantitation is finally introduced, thereby completing the classic detection triptych.

1.2   A SHORT LIST OF DETECTION LIMIT REFERENCES

The refereed scientific literature contains hundreds of publications dealing with limits of detection. These attest to the fact that there was not a consensus understanding of chemical analysis detection power for many years. Although the large majority of early papers dealing with detection limits have few further insights to yield, those wishing to judge this for themselves have never had it easier, thanks to Internet availability of many publications and translation software for use with papers in some otherwise inaccessible languages. Accordingly, a lengthy list of early publications will not be presented, but references are provided to a few papers that contain such lists.

The review paper by Belter et al. [3] focuses on an historical overview of analytical detection and quantification capabilities and cites work back to 1911. Currie's 1987 book chapter [4] is also a valuable source of early references, as is his classic 1968 publication [5] and his paper in 1999 [6]. Additional papers containing especially relevant references include those of Gabriels [7], Kaiser [8–10], Boumans [11], Linnet and Kondratovich [12], Eksperiandova et al. [13], Mocak et al. [14], EPA document EPA-821-B-04-005 [15], and the Eurachem/CITAC Quantifying Uncertainty in Analytical Measurement (QUAM) Guide CG 4 (3rd Ed.) [16]. As well, there are maintained websites where useful lists of relevant publications may be found [17]. Additional references are provided throughout this text and in the Bibliography.

1.3   AN EXTREMELY BRIEF HISTORY OF LIMITS OF DETECTION

In 1947, Kaiser published what might be considered the first paper to deal explicitly with detection limit concepts as they apply to chemical analysis methods [18]. Others followed, including Altshuler and Pasternak [19], but it was the landmark 1968 publication by Currie [5] that marked the true beginning of the modern era of analytical chemistry detection limit theory. Subsequently, Currie tirelessly advocated for the basic precepts articulated in his heavily cited paper. His detection limit schema was based on classical Neyman–Pearson hypothesis testing principles, using standard frequentist statistical methodology. As a consequence, both false positives and false negatives must be taken into consideration. Neither Currie nor Kaiser was the first to recognize the value of considering both types of error: the prior development of receiver operating characteristics, briefly discussed in Chapter 6, clearly proves this. But Currie was the first highly regarded analytical chemist to clearly identify and discuss the issue, and bring it to the attention of practicing analysts.

Implicit in Currie's paper was his belief that a chemical measurement system possessed an underlying true limit of detection, temporarily denoted as L_D, and that this was the fundamental figure of merit that needed to be estimated with minimal, or even negligible, uncertainty. As shown in Chapter 15, it is easy to construct a simple experimental chemical measurement system that possesses an L_D, though this does not prove that every such system must possess an L_D. It does, however, definitively rule out general nonexistence of an L_D [20].

1.4 AN OBSTRUCTION

It might have been expected that Currie, or one of his contemporaries, would have rather quickly arrived at the results to be presented in subsequent chapters, for example, Chapters 7–14. Unfortunately, as every scientist knows only too well, scientific progress is far from a clean, linear progression. It actually evolves by creeps and jerks, many mistakes are made, and there is often no obvious way forward. Worse still, correcting mistakes may be a lengthy process even when the facts are incontrovertible. As it happened, a major problem arose only 2 years after Currie's paper: Hubaux and Vos [21] published an influential paper that effectively sidetracked Currie's program.

Hubaux and Vos' method obtained detection limit estimates by employing standard calibration curves, processed via ordinary least squares, with the customary prediction limit hyperbolas. The statistical methodology they employed was both familiar and entirely conventional. Not surprisingly, this led to a long period where very little progress was made because, with the notable exception of Currie, the experts at the time thought the matter was largely settled. A perfect example was provided in 1978 by Boumans, who confidently declared, as the lead sentence in his detection limit tutorial [22]: "Are there any problems left to be solved in defining, determining and interpreting detection limits?" His immediate answer was "Fundamentally most problems have been adequately discussed in the literature." It is now known that his answer was incorrect and that the Hubaux and Vos method, discussed in Appendix E, was an inadvertent obstruction. But this was not at all obvious at the time and there appeared to be no reason to doubt Boumans' expert opinion.

1.5 AN EVEN BIGGER OBSTRUCTION

Early in 1988, Currie published a lengthy book chapter [4] in which he came close to solving the problem of how to instantiate his 1968 detection limits schema. Tellingly, the Hubaux and Vos paper and method, which had absolutely no need for either true underlying detection limits or explicit hypothesis testing, was neither mentioned in Currie's book chapter nor listed in the references at its end. However, a more important event preceded Currie's book chapter by a few months: Clayton *et al.* published their highly influential detection limits paper [23]. Their work was based on the "statistical power" of the *t* test and used critical values of the noncentrality parameter, δ, of the noncentral *t* distribution. These critical values are specific numerical values of the δ parameter, and, for brevity below, they are generically denoted by $\delta_{critical}$.

Currie adopted the δ_{critical} method and promoted it as the underlying basis of the analytical detection limits methodology currently sanctioned by various standards organizations, for example, the International Standards Organization (ISO) and the International Union of Pure and Applied Chemistry (IUPAC). Currie apparently felt that the δ_{critical} method was the key to correctly bringing to fruition his 1968 detection limits schema, but an insurmountable problem eventually surfaced: in 2008, the δ_{critical} method was proven to be irrelevant to detection limits [24, 25].

This quite unexpected finding is discussed in detail in Chapter 17, which, among other things, shows how to perform simple *Mathematica*® simulations to verify the result. Unlike German standard DIN 32645 [26], the ISO and IUPAC official detection limit protocols are directly based upon the δ_{critical} method, thereby rendering them irrelevant. Accordingly, it is the primary purpose of this book to demonstrate the rigorous theory upon which Currie's detection limit schema is correctly founded and to demonstrate, via both properly designed experiments and comprehensive computer simulations, that it works as predicted.

1.6   WHAT WENT WRONG?

As will be seen in progressing through the chapters, several unfortunate mistakes were made between 1987 and 2008. First, what many found irresistibly appealing about the δ_{critical} method was that it appeared to circumvent the problem of true detection limits, which are errorless, being unobtainable from experimental data. In contrast, with reference to the Hubaux and Vos method, Currie stated [27, p. 163]

> The major drawback with the method is that it produces 'detection limits' that are *random variables*, as acknowledged in the original publication of the method ([27], p. 850), different results being obtained for each realization of the calibration curve, even though the underlying measurement process is fixed. N.B.: The italics and single quotation marks are those of Currie [27] and [27] in the quotation is [21] here.

In this, Currie was incorrect: the Hubaux and Vos method cannot legitimately be criticized for producing experimental detection limits, that is, estimates, from experimental data. The method can, however, be found to be disadvantageous for other reasons, as per Appendix E.

The second mistake was the assumption that the δ_{critical} method's true theoretical detection limit was unique. In fact, there is *another* true underlying detection limit, known long before 1987, that is always fundamentally superior. Currie knew this at least as far back as 1968: with suitable definition of "k," it is simply his eqn 5b [5]. Failure to recognize that there could be two true underlying detection limits was ultimately due to the baseless assumption that there was only one possible hypothesis applicable to the testing of false negatives. Indeed, Clayton *et al.* explicitly state two hypotheses applicable to testing false positives, yet, with regard to false negatives, they state [23, p. 2507].

Other techniques involving similar assumptions, such as those cited previously, purport to provide fixed type I and type II error rates but avoid the use of noncentral t probabilities. As pointed out by Burrows, the derivation of such methods must include some type of logical or mathematical fallacy;

In the quotation, both Burrows and Clayton *et al.* were incorrect: the current problem of irrelevant officially sanctioned detection limit methodology could have been entirely avoided if the two relevant false negative hypotheses had been carefully formulated, articulated, and then tested. This is demonstrated in Chapter 17.

The third mistake involved the failure to properly test the $\delta_{critical}$ method. Indeed, it is still quite common to encounter just the opposite, that is, experimental "validation" via overly complicated experiments having only marginal demonstrated compliance with the assumptions upon which the theory was predicated. Even the experiments reported by Clayton *et al.* [23] were not in compliance with their proposed method, as discussed in Chapter 17.

At minimum, any posited detection limit theory should be rigorously tested via both real experiments, which are properly designed and instantiated, and via comprehensive computer simulations. Yet, in the two decades between 1987 and 2008, detection limit computer simulations were rarely performed. In Chapter 17, computer simulations demonstrate that the $\delta_{critical}$ method is a red herring and an opportunity is provided to verify this result for oneself.

The above trio of missteps seriously obstructed progress in correctly instantiating Currie's 1968 detection limits schema. Given the typically long time constant required for science to self-correct, it is unknown when matters may be set straight, but it is hoped that this book will accelerate the process. Otherwise, Azoulay *et al.* [28] may be proven correct.

1.7 CHAPTER HIGHLIGHTS

This chapter began by briefly stating the obvious case for making quantitative measurements. Then a few of the relevant older literature references were provided, followed by an extremely concise history of analytical detection limit progress, stagnation, and obstructions, with a short discussion of how the obstructions arose.

REFERENCES

1. J. Macartney (22 September 2008), "China baby milk scandal spreads as sick toll rises to 13,000", The Times (London), http://www.timesonline.co.uk/tol/news/world/asia/article4800458.ece.

2. A.J. Patterson, "Ushering in the era of nuclear terrorism", *Crit. Care Med.* **35** (2007) 953–954.

3. M. Belter, A. Sajnog, D. Barałkiewicz, "Over a century of detection and quantification capabilities in analytical chemistry – historical overview and trends", *Talanta* **129** (2014) 606–616.

4. L.A. Currie, "Detection: overview of historical, societal, and technical issues", Chap. 1 in *Detection in Analytical Chemistry*, ACS Symposium Series, ACS, Washington, DC, 1987 (published 1988), 1–62.

5. L.A. Currie, "Limits for qualitative and quantitative determination – application to radiochemistry", *Anal. Chem.* **40** (1968) 586–593.

6. L.A. Currie, "Detection and quantification limits: origins and historical overview", *Anal. Chim. Acta* **391** (1999) 127-134.

7. R. Gabriels, "A general method for calculating the detection limit in chemical analysis", *Anal. Chem.* **42** (1970) 1439–1440.

8. H. Kaiser, "Zum problem der nachweisgrenze", *Fresenius Z. Anal. Chem.* **209** (1965) 1–18.

9. H. Kaiser, "Quantitation in elemental analysis", *Anal. Chem.* **42(2)** (1970) 24A–41A.

10. H. Kaiser, "Part II quantitation in elemental analysis", *Anal. Chem.* **42(4)** (1970) 26A–59A.

11. P.W.J.M. Boumans, "Detection limits and spectral interferences in atomic emission spectrometry", *Anal. Chem.* **66** (1994) 459A–467A.

12. K. Linnet, M. Kondratovich, "Partly nonparametric approach for determining the limit of detection", *Clin. Chem.* **50** (2004) 732–740.

13. L.P. Eksperiandova, K.N. Belikov, S.V. Khimchenko, T.A. Blank, "Once again about determination and detection limits", *J. Anal. Chem.* **65** (2010) 223–228.

14. J. Mocak, A. M. Bond, S. Mitchell, G. Scollary, "A statistical overview of standard (IUPAC and ACS) and new procedures for determining the limits of detection and quantification: application to voltammetric and stripping techniques", *Pure Appl. Chem.* **69** (1997) 297–328.

15. Revised Assessment of Detection and Quantitation Approaches, U.S. Environmental Protection Agency, Washington, DC 20460, EPA-821-B-04-005, October, 2004.

16. Eurachem/CITAC Guide CG 4, *Quantifying Uncertainty in Analytical Measurement*, 3rd Ed., joint EURACHEM/CITAC Working Group 2012.

17. R. Boqué, http://www.chemometry.com/Expertise/references/referencesLOD.html (Accessed on 30 September 2016).

18. H. Kaiser, "Die Berechnung der Nachweisempfindlichkeit", *Spectrochim. Acta* **3** (1947) 40–67.

19. B. Altshuler, B. Pasternak, "Statistical measures of the lower limit of detection of a radioactivity counter", *Health Phys.* **9** (1963) 293–298.

20. D. Coleman, L. Vanatta, "Part 28 – statistically derived detection limits", *Am. Lab.* **39(20)** (2007), 24–27.

21. A. Hubaux, G. Vos, "Decision and detection limits for linear calibration curves", *Anal. Chem.* **42** (1970) 849–855.

22. P.W.J.M. Boumans, "A tutorial review of some elementary concepts in the statistical evaluation of trace element measurements", *Spectrochim. Acta Part B* **33B** (1978) 625–634.

23. C.A. Clayton, J.W. Hines, P.D. Elkins, "Detection limits with specified assurance probabilities", *Anal. Chem.* **59** (1987) 2506–2514.

24. E. Voigtman, "Limits of detection and decision. Part 2", *Spectrochim. Acta, B* **63** (2008) 129–141.

25. E. Voigtman, "Limits of detection and decision. Part 3", *Spectrochim. Acta, B* **63** (2008) 142–153.

26. L. Brüggemann, P. Morgenstern, R. Wennrich, "Comparison of international standards concerning the capability of detection for analytical methods", *Accred. Qual. Assur.* **15** (2010) 99–104.

27. L.A. Currie, "Detection: international update, and some emerging di-lemmas involving calibration, the blank, and multiple detection decisions", *Chemom. Intell. Lab. Syst.* **37** (1997) 151–181.

28. P. Azoulay, C. Fons-Rosen, J.S. Graff Zivin, "Does science advance one funeral at a time?", Working Paper 21788, National Bureau of Economic Research, 2015, http://www.nber.org/papers/w21788 (Accessed on 30 September 2016).

2

CHEMICAL MEASUREMENT SYSTEMS AND THEIR ERRORS

2.1 INTRODUCTION

The purpose of making a quantitative measurement is to estimate, with minimal uncertainty, the true numerical value (with units) of a property or variable of interest. For example, it may be desired to determine the fluoride ion concentration, in parts per million, in a well water sample. If the determined value were found to be above a recognized health safety limit, then this would be a major factor in the decision of whether or not to perform remediation work. To judge the efficacy of any such endeavor, further quantitative measurements would be performed, leading to subsequent decisions. Conversely, if the fluoride concentration was initially found to be acceptable, the prudent decision might be to monitor the situation by performing more determinations on a periodic basis. Making the requisite measurements requires proper operation of a suitable quantitative measurement system.

2.2 CHEMICAL MEASUREMENT SYSTEMS

A quantitative measurement system is an experimental measurement system having an output response from which it is possible to acquire or calculate a numerical value (with units), for at least one of the system's input variables, which are called measurands. Every quantitative measurement system has at least one measurand, defined as that input subjected to the measurement process. For simplicity in what follows, it is assumed that there is only a single measurand of interest: the system is univariate, not multivariate. The measurand's numerical value is denoted by lowercase x.

It is further assumed that, qualitatively, the measurand is a chemical analyte, that is, a chemical substance for which quantitative content information, such as its concentration, quantity, or amount, is sought. The analyte is present, at $x \geq 0$, in a specimen

Limits of Detection in Chemical Analysis, First Edition. Edward Voigtman.
© 2017 John Wiley & Sons, Inc. Published 2017 by John Wiley & Sons, Inc.
Companion Website: www.wiley.com/go/Voigtman/Limits_of_Detection_in_Chemical_Analysis

under test (SUT). Under these assumptions, the system is a univariate quantitative chemical measurement system, henceforth denoted by the acronym CMS. In the example above, fluoride ion is both the qualitative measurand, that is, the analyte, and its true (unknown) concentration in the well water SUT is the quantitative measurand that is the input to the CMS. The CMS itself might be a fluoride ion-selective electrode and associated high-input impedance digital millivoltmeter. The output response of the CMS, or some function of it, for example, an average over 1 s, would be the measurement, with attendant measurement uncertainty. The estimated fluoride ion concentration, with its uncertainty, would then be back-calculated from the measurement and its uncertainty.

The analytical blank, usually simply referred to as the blank, is an input SUT that contains no analyte, that is, $x = 0$, but is otherwise as similar in composition, as is possible or may be feasible, to analyte-containing SUTs that are subjected to the same measurement process. Particularly with SUTs that are solids, blanks may not exist in any useful sense, but effective workarounds often exist, for example, standard additions or isotopic dilution.

It is important to note two initial simplifications. First, the analyte content of an SUT is assumed to be time independent: $x \neq x(t)$. This does not mean that only one value of analyte content is allowable: it simply means that analytes that significantly change content value within an SUT, over the time required for measurements, are not considered. Second, it is assumed that, for a given SUT as input to the CMS, the CMS response is constant except for the additive noise due to the CMS.

As a practical matter, the assumption that the analyte content is constant in an SUT is thoroughly violated in any real CMS that intrinsically affects a separation or reaction of chemical species in the SUT. Thus, all chromatographic and flow injection analysis (FIA) systems are in this important category. Furthermore, even when the analyte content is constant in the SUT, a real CMS may produce a response that is time dependent even in the complete absence of noise. A simple example is afforded by graphite furnace atomic absorption spectroscopy (GF-AAS) instruments, where the SUT is a few microliters of analyte-containing solution placed into the graphite furnace and the response may be a rather complicated function of time, other variables, and prior conditions [1].

Both of these initial simplifications are removed later, but it is beneficial to start as simply as possible, in order that the underlying valid principles of analytical detection limit theory may be explicated as clearly as possible. Once this is done, it is seen that the theory may be extended, in relatively straightforward manner, to the transient signals arising in chromatography, GF-AAS, FIA, spectroscopy, and so on. Accordingly, consideration of these issues is deferred to Chapters 20–23.

2.3 THE IDEAL CMS

Figure 2.1 shows a simplified model of an ideal CMS that has a single output response variable, denoted by $r(x)$, with lowercase x denoting the analyte content input variable. The physically meaningful range of possible x values is $x \geq 0$, but $x < 0$ is allowed mathematically.

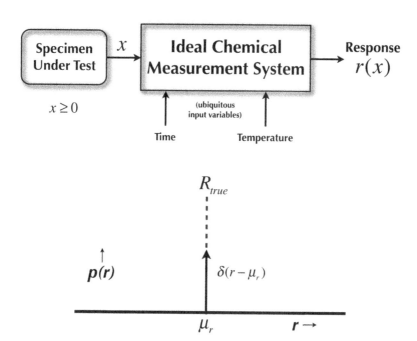

Figure 2.1 An ideal CMS model with one measurand, x, and one response variable, $r(x)$. The corresponding probability density function (see text) is shown at bottom.

In addition to its only deliberate input variable, this model also has the ubiquitous "nuisance" inputs of time (t) and temperature (T). In what follows, it is assumed that the CMS does not, itself, change with time in any significant manner during its usage periods. It is further assumed that temperature is controlled adequately. These last two assumptions mean that the defining parameters of the CMS, whatever they may be, are constants. Any other nuisance input variables are also ignored, for reasons of simplicity and clarity. In general, this set of simplifying assumptions is acceptable for an ideal CMS but are not entirely satisfactory for real CMSs.

Since absolute error is defined as the signed difference between an obtained result and an underlying true value, only an ideal, that is, errorless, measurement system would be able to yield measurement results that would be true values. Hence, if x were an otherwise arbitrary constant, denoted by X, then the output of the ideal CMS would always be the constant $r(X)$. It is useful to distinguish between variables, such as $r(x)$, and constants, such as $r(X)$. Notationally, this is achieved by using uppercase letters in place of the respective lowercase letters for variables. Thus, with X being an arbitrary constant instantiation of x, $r(X)$ is denoted by $R(X)$.

A convenient use of subscripts is to indicate specific constant values of variables. For example, $X_1, X_2, \ldots, X_i, \ldots, X_N$, designate N specific constant values of x and these might represent N calibration standards from which a calibration curve may be constructed. In addition, subscripts need not be single letters. Thus, if a true value of x

is denoted by X_{true}, then the corresponding $r(X_{\text{true}})$ is denoted by R_{true}. Similarly, it is convenient in later chapters to use subscripts to label important computed constants, such as theoretical decision levels and detection limits. Generally, subscripts will be as concise as is feasible.

The analytical blank is especially important in that it is constant, that is, $x = 0$, and has the smallest physically meaningful value of x. It is denoted by X_0, and, in general, the subscript "0" is reserved for the blank whenever possible. In accordance with the first simplifying assumption above, any viable calibration standard has constant analyte content, that is known with negligible uncertainty, and the analytical blank is simply the unique standard with zero analyte content, that is, $x \equiv X_0 \equiv 0 \equiv x_0$.

2.4   CMS OUTPUT DISTRIBUTIONS

In general, the output response variable of any CMS is characterized by a probability density function (PDF), described by at least two population parameters: a location parameter, also called a centrality parameter, and a scale parameter, also called a dispersion parameter. As a possible aid, Appendix A provides a very short overview of PDFs and their most pertinent properties. For the ideal CMS in Fig. 2.1, the output is $r(X_{\text{true}}) \equiv R(X_{\text{true}}) \equiv R_{\text{true}} \equiv \mu_r$, when the input is $x \equiv X_{\text{true}}$. Hence, an ideal CMS has $r(x)$ characterized by a PDF having unit area, zero width, and a location parameter, μ_r, exactly equal to the true response value, R_{true}, that is, its PDF is a so-called "delta function" distribution, located at $\mu_r \equiv R_{\text{true}}$. Therefore, any output response of the ideal CMS is a sample from the following PDF:

$$p(r) = \delta(r - \mu_r) = \delta(r - R_{\text{true}}) \tag{2.1}$$

where, for clarity, $r(x)$ is denoted by r in eqn 2.1 and R_{true} is constant for a given constant analyte content input X_{true}. This situation is graphically depicted at the bottom of Fig. 2.1.

If X_{true} changes, R_{true} and μ_r will change accordingly, remaining equal, and the ideal CMS will always yield errorless results. The functional relationship between response and analyte content may be depicted graphically or, if the response function is known explicitly, it may be expressed analytically. Generally, it is required that $r(x)$ is a monotonic function of x, that is, the response does not have regions with zero slopes or have slope reversals. Without loss of generality, it is assumed that response functions are always increasing functions, that is, the slope at any point is always greater than zero. Three plausible monotonic response functions are shown in Fig. 2.2.

2.5   RESPONSE FUNCTION POSSIBILITIES

There exist many additional monotonic response functions beyond the 3 shown in Fig. 2.2. However, the most common and important one, by far, is #3 in Fig. 2.2: the linear response function. It is ubiquitous for many reasons, including the following:

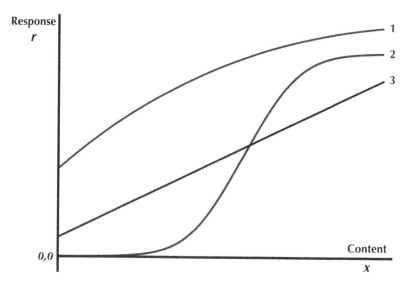

Figure 2.2   Three plausible monotonic response functions for an ideal CMS.

- It is as simple as possible, both conceptually and mathematically, since $r(x) = \alpha + \beta x$, where α is the true offset and β is the true slope.
- It is common practice to design CMSs to be as linear, over the widest usable "linear dynamic range," as is economically and practically feasible.
- If systematic error exists in the system, or arises at some point, it is most readily detected as such, for example, if deviations from theoretical linearity occur in an empirically obtained response function, they may manifest as "nonstraightness" of the empirical data plot.
- The constant β, defined as the "sensitivity" of the linear CMS, is simply the constant slope of the linear response function.

Additional reasons for "linearity model bias" might be listed, but there is little to gain from doing so. Indeed, the linear response function is so frequently encountered and actively desired, that nonlinear response functions often appear to serve as examples of poor CMS designs. It is useful to note that this model bias may have an unwanted side effect, that is, it may lead to erroneous extrapolations about the required behavior of every possible response function. To offset model bias slightly, a nonlinear monotonic response function, as shown in Fig. 2.3, is considered initially. Later, the "traditional" linear response function, that is, $r(x) \equiv \alpha + \beta x$, is used almost exclusively due to its undeniable importance in both theory and practice.

The response function shown in Fig. 2.3 provides a unique 1–1 correspondence between x and $r(x)$. Thus, if $r(x) \equiv$ intercept $+ y(x)$, then $x = y^{-1}[r(x) - $ intercept$]$. The r axis intercept is arbitrarily shown as being greater than zero, but zero and negative

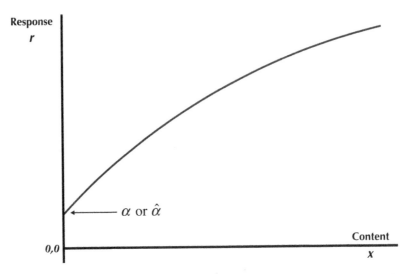

Figure 2.3 A plausible nonlinear monotonic response function for an ideal CMS.

intercepts are equally as valid in practice. Occasional exceptions and shot noise situations aside, the intercept is often of little fundamental importance; it is simply a nuisance offset that must be removed before $y(x)$ may be inverted. Typically, the goal is to use inverse calibration to quantify, with minimal uncertainty, the value of x in an input specimen having $x = X_{unknown}$. Notationally, the true value of the r axis intercept is denoted by α and estimates of it are generically denoted by $\hat{\alpha}$. Thus, the response function is

$$r(x) = \alpha + y(x) \tag{2.2}$$

where $\alpha \equiv r(0) \equiv R(0) \equiv R_0$. Subtraction of α from both sides of eqn 2.2 results in a net response function:

$$y(x) = r(x) - \alpha \tag{2.3}$$

This process is commonly referred to as blank subtraction. Other terms in use include blank correction, blank referencing, background correction, offset removal, DC correction, pedestal removal, and so on.

In practical application, the net response is almost always more convenient than the "raw" response. However, there is a price: if α is unknown, it must be estimated by $\hat{\alpha}$. This seemingly trivial requirement has nonnegligible consequences; $\hat{\alpha}$ has error associated with it, while α is errorless. The consequences of error in the blank subtraction process, and the additional noise imparted to the net response, are examined in later chapters. First, however, it is necessary to discuss the possible types of error that might be imparted.

2.6   NONIDEAL CMSs

In the real world, both systematic errors and random errors exist in CMSs. The former are deterministic, in the simplest cases, and may be categorized as either essential or inessential. Treatment of systematic errors is often quite cursory, even to the point of being wrongly dismissive of them as being somehow always inessential or unimportant. This may be understood in part due to their diverse possible manifestations in real CMSs, which frustrates attempts at comprehensive analyses of their causes, effects, and mitigation.

2.7   SYSTEMATIC ERROR TYPES

The ideal CMS was free from systematic error, which is defined as the signed difference between the location parameter of the relevant PDF and the true value. The simplest systematic errors in measurement systems are determinate errors, that is, they are of knowable origin, in principle, and they can be classified as either inessential or fundamental. It is easy to provide examples of commonly occurring inessential systematic errors:

- An analytical balance, which always gives results that are too high by a constant amount.
- Use of an incorrect units conversion factor.
- Use of an incorrect, incomplete or ill-formed model.
- A sloppily prepared standard solution or pH buffer solution.
- Use of the wrong value of a constant in a calculation, for example, using $\pi \cong 3.41$.

Unfortunately, only the simplest systematic errors are constant. Denker [2, pp. 81–84] discusses this in great detail and the following example is inspired by one of his: suppose that 20 students in a first year chemistry laboratory section were given the task of each preparing a liter of 0.100 M NaOH solution. If 10 aliquots of the nominal 0.100 M solutions were to be used in subsequent experiments, it is clear that taking all 10 from a single volumetric flask would not be the same as taking no more than one aliquot from each of 10 randomly selected flasks. In other words, the time average would not equal the ensemble average. Furthermore, there are 184 756 ways to randomly choose the 10 unique aliquots from 20 possible flasks, so the subsequent experimental results would depend upon *exactly which* 10 solutions were sampled, even if there was no random measurement error.

This simple example demonstrates that randomness may be introduced via the sampling process prior to the CMS, leading to the distressing realization that, in general, systematic error and random error are overlapping categories. Even worse, systematic errors may change in an essentially unlimited variety of ways: deterministically or randomly, continuously or discretely, rapidly or over years, and so on.

In view of the above, *there is no general way to deal with every possible way that systematic error may arise or evolve over time.* Denker's advice [2, p. 81] is as good as any: "if you kept all the data, you might be able to go back and recalibrate the data without having to repeat the experiment." The operative word is "might."

Presumably, then, a properly thought-out and executed calibration procedure should expose inessential systematic errors, thereby leading to either modification of the CMS, so as to avoid the errors in the first place, or, if they cannot be entirely avoided, to correct for their effects. In view of this, and in the absence of experimental evidence to the contrary, systematic errors are almost always assumed, for pragmatic reasons, to be either constant or to change only slowly and in relatively simple manner, for example, linearly or periodically. Then frequent calibration, modulation, ratioing, and so on are common ways to reduce or eliminate systematic error, with the specifics being dependent upon the experiment. Henceforth, *it will be assumed that there is no inessential systematic error.*

2.7.1 What Is Fundamental Systematic Error?

Much of the important work of scientists lies in trying to discover fundamental systematic errors, since these manifest as quantitative disagreements between the best theoretical predictions available, from models that are thought to be well-formed, correct and complete, and meticulously obtained experimental results. Thus, if any significant quantitative disagreement between best theory (with best models) and best experiment is found, theory is necessarily either incorrect, incomplete or ill-formed in some essential way and, as a consequence, must either be fixed or be replaced with a better theory. An example of this is replacement of Newtonian physics with relativity. Obviously, the word "significant" is crucial: how much disagreement must occur in order to be considered significant? Note also that errors in the values of physical constants, for example, the curiously ill-known gravitational constant "big G," are considered to be fundamental systematic errors, so these are nontrivial in every sense.

2.7.2 Why Is an Ideal Measurement System Physically Impossible?

An ideal CMS is physically impossible for at least the following two reasons:

- It can never be certain that fundamental systematic error is zero and *remains so.*
- All measurement systems have nonzero *random noise.*

The first reason ideal CMSs cannot exist is very important; even if it is thought that a CMS is completely free from all systematic errors, there is no guarantee that unknown sources of fundamental systematic error might not be present or arise subsequently. The prudent course of action is to avoid saying that systematic error is zero and instead be able to say that no evidence of systematic error, at some specified confidence level, has been demonstrated. Then, of course, work must be done to keep it that way.

As for the second reason, every experimentalist knows that random noise (also known as indeterminate error, measurement uncertainty, random error, or simply as noise) is their constant adversary whenever measurements are performed. More is said about this, beginning in Chapter 3.

2.8 REAL CMSs, PART 1

As noted above, real CMSs possess both systematic error and random noise. If the random noise was zero, but the systematic error was nonzero, the real CMS might be modeled as shown in Fig. 2.4. The systematic error ε_s, also known as bias, is defined by

$$\varepsilon_s \equiv \mu_r - R_{\text{true}} \tag{2.4}$$

with μ_r being the PDF's constant location parameter. The corresponding PDF is

$$p(r) \equiv \delta(r - (\varepsilon_s + R_{\text{true}})) \tag{2.5}$$

where μ_r is shifted away from R_{true} due to the bias. This is shown at the bottom of Fig. 2.4.

Systematic error may arise in an extraordinary number of ways and, even if meticulous care has been taken to eliminate or reduce it in a given CMS, it may develop seemingly out of nowhere. It may be time- or temperature-dependent, may arise from component aging or malfunctions, or it may be due to a change in an unsuspected,

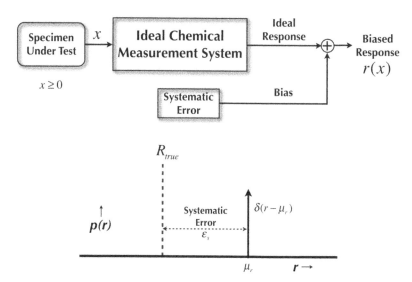

Figure 2.4 Model of a plausible, hypothetical CMS with zero random noise, with PDF at bottom.

and therefore uncontrolled, input variable. Simply put, there is essentially an unlimited number of opportunities for systematic error to occur and reducing its deleterious effects to a minimum is a constant task in real experiments performed with real CMSs. The assumption that systematic error is constant, as depicted in Fig. 2.4, is really a goal, not a given.

2.8.1 A Simple Example

Despite the many ways that systematic error may arise in an experiment, there are often obvious places where it occurs. An important illustration is provided in the following example. Assume that $r(x)$ is related to x via the following linear response function:

$$r(x) = \alpha + \beta x \qquad (2.6)$$

where α is the true offset, β the true slope, and there is no explicit random error in eqn 2.6, that is, it is not $r(x) \equiv \alpha + \beta x + \text{noise}$. The noiseless straight line equation violates linear superposition and is not, therefore, linear in a strict mathematical sense. However, subtraction of α from both sides results in a true linear net response function:

$$y(x) = \beta x \qquad (2.7)$$

where $y(x)$ is defined as in eqn 2.3. If R_{X_u} is the errorless value obtained when the errorless value X_u is substituted into eqn 2.6, it is trivial to calculate the exact value of X_u, via eqn 2.8:

$$X_u = \frac{R_{X_u} - \alpha}{\beta} \equiv \frac{Y_{X_u}}{\beta} \qquad (2.8)$$

assuming that both α and β are known. If an incorrect value of either α or β were to be used in eqn 2.8, then the computed value of X_u would be incorrect due to inessential systematic error. In principle, this systematic error is correctable simply by using correct parameter values. However, it often happens that α or β are inaccurately known or have somehow changed. This may happen for a wide variety of reasons and every experienced analyst can readily supply examples, some even humorous after the passage of sufficient time.

Continuing the above example, it may be that estimates are available for α or β, denoted by $\hat{\alpha}$ or $\hat{\beta}$, respectively. However, there are nontrivial consequences to using them: all estimates, regardless of how they are obtained, have error associated with them because they are samples from distributions that are *not* "delta functions." Consequently, using $\hat{\alpha}$ or $\hat{\beta}$ introduces their respective random error and it becomes impossible to calculate the errorless value X_u. Rather, X_u is estimated by the random variate x_u, equivalently denoted by $\hat{X}_u$, given by one of the following expressions:

$$x_u = \frac{R_{X_u} - \hat{\alpha}}{\beta} \qquad (2.9)$$

$$x_u = \frac{R_{X_u} - \alpha}{\hat{\beta}} \qquad (2.10)$$

or

$$x_u = \frac{R_{X_u} - \hat{\alpha}}{\hat{\beta}} \qquad (2.11)$$

In each case, x_u is a random sample from a distribution of possible values and it is referred to as a point estimate test statistic of X_u. The best case scenario is that it is unbiased, that is, its expectation value, $E[x_u]$, equals X_u. From eqn A.25 in Appendix A, this means

$$X_u = E[x_u] \equiv \int_{-\infty}^{\infty} x_u p(x_u) dx_u \qquad (2.12)$$

where $p(x_u)$ is the PDF of x_u and assuming the integral exists. Note that $p(x_u)$ depends upon which of eqns 2.9–2.11 is applicable and upon the PDFs of $\hat{\alpha}$ or $\hat{\beta}$, as thereby relevant. In any case, statistics becomes essential, since randomness has entered the picture.

2.9 RANDOM ERROR

Every real CMS has inescapable random (that is, indeterminate) error on its output response, which is therefore properly referred to as a variate rather than a variable. This error can only be dealt with in a statistical manner. If we were to instantaneously sample the output response variate repeatedly, it would be found that the numerical sample values obtained would generally not be numerically identical. In fact, it would be found that the numerical sample values were distributed in approximate accordance with the PDF of the output response variate. This assumes, of course, that the numerical sampled values were not artificially compromised in their apparent accuracy, that is, they were not deliberately or inadvertently rounded off or otherwise truncated excessively, thereby creating the false impression that they were numerically identical.

Inadvertent round-off or truncation is generally due either to the CMS not being used properly, for example, a sensitivity or display setting is incorrect, or to the CMS being fundamentally incapable of achieving the necessary sensitivity, for example, trying to use an inexpensive digital voltmeter, capable of reading to the nearest millivolt, to measure submicrovolt potential differences. This is the type of error, that is, "flat baseline," that generations of students are taught, in their early laboratory work, to recognize and avoid.

As discussed above, a real CMS has output responses that are random variates, characterized by PDFs with nonzero scale parameter. Most commonly, real CMSs are carefully constructed with the goal of minimizing all known sources of systematic error, although it is not true that systematic error is invariably detrimental. In fact, there are simple cases where random noise can be minimized only if inessential

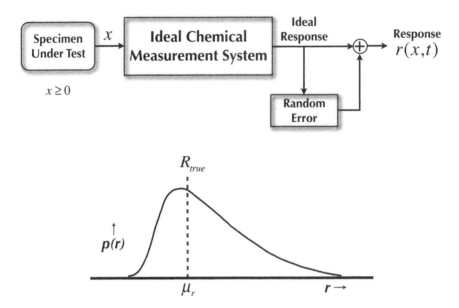

Figure 2.5   Model of a plausible real CMS with zero systematic error, with PDF at bottom.

systematic error is nonzero. In any event, having zero systematic error means that the location parameter of the PDF of the output response variate is either the PDF's population mean, if it exists, or its median.

Thus, if the systematic error was zero, but the random noise was nonzero, the real CMS might be modeled as shown in Fig. 2.5. For this type of model CMS, the response PDF is that of the random error (that is, noise), but with its location parameter, μ_r, being the true value, R_{true}, of what would otherwise be a noiseless output response. The particular PDF depicted in Fig. 2.5 happens to be asymmetric, though in many other cases, the PDF is symmetric, for example, Gaussian distributed. Beginning in Chapter 5, attention begins to focus upon the Gaussian distribution, for reasons explained there.

If the input to the CMS changes, the PDF might exhibit a change in its scale parameter, that is, be heteroscedastic ("different scattering"), but consideration of this possibility is deferred to Chapter 18. Until then, it is assumed that the noise is homoscedastic ("same scattering"), that is, the PDF's scale parameter is independent of analyte content. A common alternative way of stating this is to say that the random noise variance is constant. In Fig. 2.5, this would mean that the random error would be independent of the ideal response.

In Fig. 2.5, systematic error is assumed to be zero, thereby ensuring that $\mu_r \equiv R_{\text{true}}$, assuming μ_r exists, and that any change in R_{true} would automatically be equaled by μ_r. This is important for the following reason: suppose independent and identically distributed (*i.i.d.*) samples are taken of the response PDF and that these samples are then used, in the customary way, to compute a sample mean, denoted by $\bar{r}_{\text{responses}}$.

Then $\bar{r}_{\text{responses}}$ is the point estimate of μ_r. Since $\mu_r \equiv R_{\text{true}}$, $\bar{r}_{\text{responses}}$ is then also the point estimate of R_{true}. It is clearly useful if μ_r may be estimated in an unbiased way, because μ_r is the proxy for R_{true}. For the analytical blank, $x \equiv X_0 \equiv 0$, so that $R_0 \equiv r(X_0) \equiv \alpha$ and $\bar{r}_0$, the sample mean of M_0 blank replicate measurements, is a point estimate of α, that is, $\bar{r}_0 = \hat{\alpha}$.

2.10   REAL CMSs, PART 2

Real CMSs have both random error and systematic error, with the latter generally minimized by both instrument design and adherence to a rigorous, and meticulously enforced, measurement protocol. Hence, a plausible schematic diagram of such a CMS is shown in Fig. 2.6.

Under the assumption that systematic error has deliberately been reduced to zero, the model in Fig. 2.6 may be well approximated by that in Fig. 2.5. In other words, *systematic error may be neglected from further consideration and this will be the assumption in subsequent chapters.* By way of rationalization, this decision is customary, pragmatic, and somewhat regrettable, but there is little practical alternative because of the enormous variety of possible systematic errors.

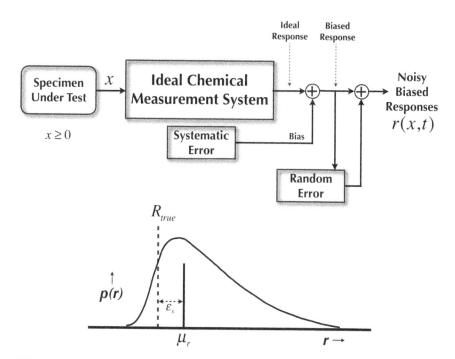

Figure 2.6   Model of a plausible real CMS with both types of error, with PDF at bottom.

2.11   MEASUREMENTS AND PDFs

Having briefly discussed CMSs, their output response variables or variates, PDFs and types of error, the focus now shifts to the measurement results themselves. Since a nonideal CMS's output response is characterized by a PDF, it might be supposed that quantitative measurements are simply samples, taken as close to instantaneously as possible, of the output response of a CMS. These are defined as elemental measurements. More generally, a compound measurement is a user-defined function of the output response of the CMS. For example, the difference between two CMS output responses is a compound measurement. Similarly for the average of the CMS's output over a time period, for example, an inexpensive digital voltmeter typically averages over 0.4 s to eliminate possible 50 or 60 Hz line frequency interferences.

Given the vast number of possibilities for functions, it is necessary to impose a reasonable constraint: the user-defined function should not deliberately subvert the utility of limits of detection as meaningful, transferable figures of merit. This is elaborated upon in Chapter 16, after the relevant detection limit concepts have been presented in detail.

2.11.1   Several Examples of Compound Measurements

Simple examples of compound measurements abound. An ordinary ruler illustrates the latter situation for the case of a measurement system that is nonchemical: the measurand is simply whatever the ruler is placed upon and the compound measurement, that is, a length, is then the positive numerical difference between an appropriate pair of marks on the ruler. Another common example involves finding the area under a peak, for example, in chromatography, and defining this area as the measurement result. This, then, is simply the integral, over a specified time interval, of the output response of a CMS. Weighing by difference is a ubiquitous example: the estimated mass of the desired specimen is the difference in masses between that of the specimen plus weighing vessel and that of the empty weighing vessel. A slightly more complicated example involves averaging an output response over a specified time interval, yielding a numerical result, then changing the analyte content (often to zero) and averaging the output response a second time, yielding a second numerical result. The quantitative measurement is then defined as the difference between the average values, with appropriate units.

Far more sophisticated measurement processing exists and is routinely used, as the situation warrants. However, no matter how complicated the process by which output responses are made to yield the ultimate quantitative compound measurement, one thing is always true: *every quantitative measurement, whether elemental or compound, is a random sample from some PDF*. Accordingly, measurements are instantiations of random variates, that is, they are numerical estimates of the measurand, with uncertainties that are fundamentally due to the PDFs of which they are random samples.

To illustrate this notion, consider the preparation of an aqueous In^{3+} calibration standard solution, having 500 ppm In^{3+} as its desired concentration. With due care, the prepared solution will have a constant concentration that should be close to the targeted value, but the true solution concentration will be unknown. By carefully taking into account the various constituent measurements involved, and their uncertainties, it is possible to arrive at a best estimate of the prepared solution's concentration, along with an estimate of the "combined standard uncertainty" [3]. For illustrative purposes, suppose the In^{3+} concentration was calculated to be 501.4 ppm, with uncertainty of 0.7 ppm. Then 501.4 ppm, a compound measurement, is the numerical estimate of the unknown true In^{3+} solution concentration. Note that some of the constituent measurements may be compound measurements themselves, for example, a weighing by difference. A detailed example along these lines is provided in Appendix B. It corrects several simulation mistakes made in example A2 in Ref. [3].

In general, the combined standard uncertainty estimate is based on distributional assumptions for the constituent uncertainties and is often calculated using a propagation of errors (POEs) approximation [3]. As a result, it is *not* the population standard deviation of the PDF, most likely unknown, of which the compound measurement is a random sample. Furthermore, as a calculated function of *assumed* population parameters and *actual* measurement results, the uncertainty estimate is problematic in regard to construction of statistically valid confidence intervals. The issue is usually covered over by multiplying the uncertainty by a "coverage factor," k, arbitrarily defined as 2 or 3 [3]. This results in a so-called "expanded combined standard uncertainty" [3], but the basic issue remains: "How may statistically valid confidence intervals be constructed?"

2.12   STATISTICS TO THE RESCUE

Not surprisingly, the proper analysis of noisy quantitative measurements requires the use of statistics. Among many other things, analysts need to know how to estimate the population parameters of a PDF from sample data from that PDF and how to compute statistically valid confidence intervals that do not rely upon *ad hoc* "coverage factors." They need to know how to compare sets of measurement results to ascertain the likelihood that they are samples from the same or equivalent distributions. Even in the In^{3+} concentration example, is 0.7 ppm concentration uncertainty negligible in the context of the experiments in which the solution would be used? Answering such questions requires information about the relevant sampled PDFs.

If the processing required to obtain the desired quantitative measurements is not too complicated, it may be possible to compute, analytically, the PDF from which the quantitative measurements are samples. Only in the simplest case, where the quantitative measurements are elemental measurements, is the PDF characterizing the quantitative measurements exactly the same as that characterizing the CMS's output response. Even then, an analytic PDF may be out of reach.

In the large majority of cases, where processing of the CMS's output response is needed in order to obtain the quantitative compound measurement, it is simply infeasible to attempt an analytic calculation of the relevant PDF. However, it is often possible to obtain an empirical realization of the PDF that accurately characterizes the set of possible quantitative measurement results. This empirical realization of the PDF is called a histogram and it is obtained by "gaming" the CMS, making repeated independent measurements and empirically determining how the results are distributed. Thus, a histogram is basically a finite data approximation of a PDF, and, conversely, a PDF is simply the infinite data limit of representative histograms.

The gaming strategy involved in obtaining the histogram, the famous Monte Carlo method invented by Ulam, is especially effective when it is possible to make many quantitative measurements. It is used extensively in later chapters, in both real experimental studies involving bootstrapping (for example, see Chapter 15) and in computer simulations. The Monte Carlo method, properly employed, also makes it possible to obtain statistically valid results, such as confidence intervals, without the use of arbitrary "coverage factors." See the QUAM example A2, in Appendix B, for a detailed illustrative example.

2.13 CHAPTER HIGHLIGHTS

This chapter was mainly devoted to a discussion of necessary measurement concepts and definitions. Among the most important are the CMS, measurand, analyte, blank, systematic error, random error, elementary and complex measurements, PDFs, and uncertainty. Several illustrative qualitative examples were provided.

REFERENCES

1. J.B. Dawson, R.J. Duffield, P.R. King, M. Hajizadeh-Saffar, G.W. Fisher, "Signal processing in electrothermal atomization atomic absorption spectroscopy (ETA-AAS)", *Spectrochim. Acta* **43B** (1988) 1133–1140.
2. J. Denker, "Uncertainty as applied to measurements and calculations", https://www.av8n .com/physics/uncertainty.htm, (2003) 1–100.
3. EURACHEM/CITAC Guide, *"Quantifying Uncertainty in Analytical Measurement"*, 3rd Ed., 2012.

3

THE RESPONSE, NET RESPONSE, AND CONTENT DOMAINS

3.1 INTRODUCTION

In Chapter 2, we introduced a plausible model of a real chemical measurement system (CMS) having random error but no systematic error. As a consequence of the random error, which is time dependent, the output of the CMS is a random variate, that is, a random sample from a PDF of some sort. The general situation is as in Fig. 3.1.

In general, the output response of the CMS is the random variate given by

$$r(x, t) \equiv r(x) + \text{noise}(t) \tag{3.1}$$

where $r(x)$ is the time-independent response function and the noise is time dependent, with zero mean, that is, zero location parameter. If the CMS has a constant analyte content X_i as its input, its response is still a noisy continuous variate, that is, $r(X_i, t)$. In the absence of noise, the output response would be the constant R_i, where $R_i \equiv R(X_i) \equiv r(X_i)$. Thus, the simplest relationship between $r(X_i, t)$ and R_i is

$$r(X_i, t) \equiv R_i + \text{noise}(t) \tag{3.2}$$

so that R_i would be a specific constant location parameter of the noise's PDF. If $x = X_i \equiv X_{true}$, then $r(X_i) = r(X_{true}) \equiv R_{true}$, with R_{true} being the noise PDF's constant location parameter. This is shown in Fig. 2.5.

Samples of $r(X_i, t)$ are typically collected, via digitization, and then various statistics may be computed. The reason for this is simple: it is desired to determine, as accurately as possible, $r(x)$ and its defining parameters, but what is available are samples of $r(x, t)$. Therefore, $r(x)$ is inextricably embedded in the confounding additive

Limits of Detection in Chemical Analysis, First Edition. Edward Voigtman.
© 2017 John Wiley & Sons, Inc. Published 2017 by John Wiley & Sons, Inc.
Companion Website: www.wiley.com/go/Voigtman/Limits_of_Detection_in_Chemical_Analysis

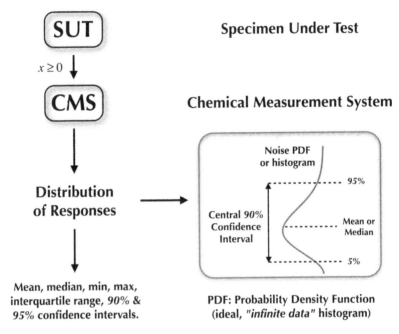

Figure 3.1   A plausible CMS's responses and computed statistics.

noise: it has become the noise PDF's variable location parameter. Hence, it is desired to minimize the CMS's output noise variance, thereby narrowing the noise's PDF, and use statistics to reduce, as much as possible, the effects of remaining measurement uncertainty.

The computed statistics mentioned above include at least some of the following: sample mean, sample standard deviation, median, minimum, maximum, interquartile range, and both 90% and 95% central confidence intervals (CIs). If sufficiently many collected data are available, they may be binned into a histogram, which is simply an empirical, finite data instantiation of the response variate's true theoretical PDF. With sufficiently large numbers of collected response data, for example, many thousands or more data, constructed histograms may be quite accurate approximations of theoretical PDFs.

CIs are typically constructed from smaller sets of data, for example, two to several hundred or so. In general, when the collected number of data is small, the central CIs are often of most utility, assuming the parent PDF is at least roughly known. Figure 3.2 has an inset showing a theoretical noise PDF or experimental histogram, plus central 90% confidence limit bars and the location parameter: the mean, if possible, or the median, otherwise. The location parameter of the distribution of responses is of fundamental importance and it matters whether it is the true value or an unbiased estimate of it.

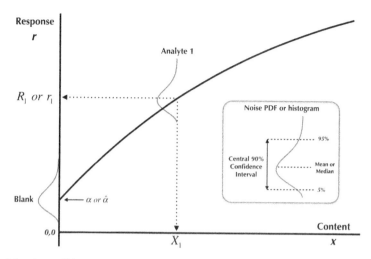

Figure 3.2 A possible response function with response distributions for the analytical blank, at $x = X_0 = 0$, and an analyte at X_1. For notational simplicity, $r(x, t)$ is denoted by r.

It is convenient to make several additional initial assumptions, all but the first one to be discarded in the following chapter:

- The noise is homoscedastic.
- The CMS's input is either the analytical blank (that is, $x = 0$) or analyte at X_1, where $X_1 \gg 0$.
- The CMS has the response function shown in Fig. 2.3.
- For the blank, the noise PDF is asymmetrically distributed, with location parameter, α, and scale parameter, σ_0.

With these assumptions, the resulting response behavior is depicted in Fig. 3.2.

3.2 WHAT IS THE BLANK'S RESPONSE DOMAIN LOCATION?

There are several important aspects of Fig. 3.2. First, the analytical blank's response distribution, regardless of whether it is a theoretical PDF or an experimental histogram, has a location value, α or $\hat{\alpha}$, that is not necessarily zero. As noted in Chapter 2, the true value of the blank, denoted by α, is the r axis intercept of the true response function. If α is unknown, as will almost always be the case in real experiments, then it must be estimated, preferably in an unbiased way, and its estimate will be denoted by $\hat{\alpha}$. One estimation possibility is to make N replicate determinations of the analytical blank and use their sample mean, $\hat{\alpha} \equiv \bar{r}_0$, as noted in Chapter 2. This also applies to a histogram of experimental blank responses. Another possibility is to perform a regression fit to a valid model of the response function and use its estimated intercept, a, as $\hat{\alpha}$. In any event, $\hat{\alpha}$ has error while α is errorless, and this difference must be taken into account.

3.3   FALSE POSITIVES AND FALSE NEGATIVES

A second observation, in regard to Fig. 3.2, is that the response distributions for the analytical blank, and for analyte 1, at X_1, have negligible overlap. Hence, if the only two possible X inputs to the CMS were analyte 1 and the blank, it would be very easy to use their respective responses to decide, with negligible error, the value of X. However, this opportunity vanishes if the response distributions have nonnegligible overlap, as shown in Fig. 3.3; it is now impossible for every response variate, for example, r' in the figure, to always be correctly attributed to its distribution of origin. This leads directly to the concept of errors of Type 1 and Type 2, that is, false positives and false negatives, respectively.

The following two chapters present a more detailed discussion of false positives and false negatives and introduce the traditional limit of detection. First, however,

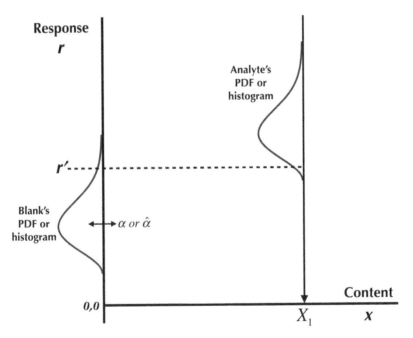

Figure 3.3   Response distributions for the analytical blank, at $x = X_0 = 0$, and an analyte at X_1, with nonnegligible overlap. For notational simplicity, $r(x, t)$ is denoted by r.

it is necessary to discuss the procedure for obtaining net responses from responses, and then determine how net responses are used to compute content values or their estimates.

3.4 NET RESPONSE DOMAIN

The net response domain is defined as the blank subtracted response domain, that is, the net response is the response minus either the blank's true value, α, or its unbiased best estimate, $\hat{\alpha}$. This gives two versions of the net response domain function, depending on whether α or $\hat{\alpha}$ is used.

If α is known, it should be used because it is errorless. Then $y(x)$ is defined as the net response with ideal blank subtraction:

$$y(x) \equiv r(x) - \alpha \qquad (3.3)$$

where $r(x)$ is the time-independent response function of the CMS. For a real CMS, it is almost always the case that α is unknown. Therefore, α must be estimated by $\hat{\alpha}$ and $y(x)$ is alternatively defined as the net response with nonideal blank subtraction:

$$y(x) \equiv r(x) - \hat{\alpha} \qquad (3.4)$$

Either of these two equations will be used as applicable, with care taken to reveal all relevant consequences.

3.5 BLANK SUBTRACTION

Figure 3.4 shows the net response domain equivalent of Fig. 3.3, with $y(x)$ as in eqn 3.3, for PDFs, or eqn 3.4, for histograms. Other combinations are discussed as follows. As shown in Fig. 3.4, the net blank's distribution is centered on zero in the net response domain and the same would apply to the central 90% CI. This is either exactly true or it may be an approximation, only true, on average, for an infinite ensemble of histograms. Since either α or $\hat{\alpha}$ may be employed for blank subtraction, this leads to several cases that must be considered carefully in order to accurately account for the noise effects. These are given in Table 3.1, with notes and further discussion following.

Case 1 is the most important, by far, because it is directly involved when considering how a single future blank (or the sample mean of M future $i.i.d.$ blank replicates) compares with a decision level. As noted in Table 3.1, the convolved PDF becomes centered exactly on zero, but the random error is increased and the shape of the PDF might also change, depending upon the specific error distributions involved. Of particular importance is the fact that there will only be a scale parameter increase, but no shape change, if the blank's PDF is Gaussian and $\hat{\alpha}$ has Gaussian distributed error. In

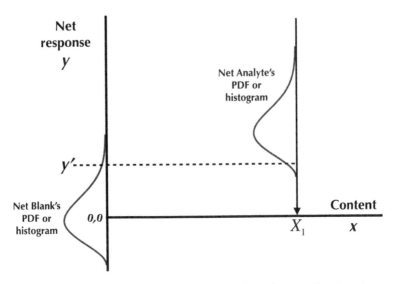

Figure 3.4 Net response distributions for the analytical blank, at $x = X_0 = 0$, and an analyte at X_1, with nonnegligible overlap. For notational simplicity, $y(x, t)$ is denoted by y.

Table 3.1 Effects of Blank Subtraction

Case	Action	Effect	Comments
1	Subtract $\hat{\alpha}$ from the blank's theoretical PDF	Centers PDF exactly on zero	Adds noise from $\hat{\alpha}$; possibility of shape change in PDF (see text)
2	Subtract α from the blank's theoretical PDF	Centers PDF exactly on zero	No added noise or shape changes
3	Subtract $\hat{\alpha}$ from the blank's experimental histogram	Centers histogram exactly on zero	No added noise or shape changes
4	Subtract α from the blank's experimental histogram	On average, centers histogram on zero	No added noise or shape changes (see text)
5	Subtract $\hat{\alpha}$ from CI having $\hat{\alpha}$ as "best estimate"	Centers CI exactly on zero	No added noise and no change in range of interval

The location parameter of the blank's response domain PDF is α and $\hat{\alpha}$ is its unbiased estimate.
For histograms, $\hat{\alpha}$ is the sample mean of the histogram data, not an independent alternative estimate, for example, *not* the sample mean of additional, independent blank replicates.
For all CIs, $\hat{\alpha}$ is the "best estimate" unless α is known, in which case, α is the "best estimate."

other words, the convolved PDF is Gaussian and centered *exactly* on zero, but has a *larger* scale parameter than the original Gaussian PDF of the blank.

In Case 4, a suitable statistical test should be performed to determine if $\hat{\alpha}$ is statistically equivalent to α. For Gaussian distributed error, a standard t test of $\hat{\alpha}$ versus α would be satisfactory. If $\hat{\alpha}$ and α were found to be statistically inequivalent, investigation would be warranted because systematic error might be present, for example, the response function might plateau at small values of x.

If $\hat{\alpha}$ and α were found to be statistically equivalent, subtraction of α from the blank's histogram yields the net blank histogram, centered on $\hat{\alpha} - \alpha$. The net blank histogram is centered on zero, on average, because $E[\hat{\alpha} - \alpha] = 0$. Even for a single net blank histogram, $\hat{\alpha} - \alpha$ tends to be small due to the many histogram data involved in estimating $\hat{\alpha}$.

If the blank's distribution function is known, along with α, then curve fitting is a better option. For example, if the blank is known to be Gaussian distributed, with known α, then a Gaussian curve fit could be performed on the net blank histogram. The result would be $N:\alpha, \hat{\sigma}_0$, where the fitting parameter, $\hat{\sigma}_0$, estimates the unknown σ_0. Then subtraction of α from $N:\alpha, \hat{\sigma}_0$ yields the net blank's estimated distribution function, $N:0, \hat{\sigma}_0$, which is exactly centered on zero. In any event, this case is relatively rare because α is usually unknown.

3.6 WHY BOTHER WITH NET RESPONSES?

Why bother with the net response domain? The short answer is that analyte content information is desired and it is necessary to pass through the net response domain in order to get to the content domain. A more informative answer is the following: a future true blank is simply a random sample from the blank's theoretical PDF, which has location parameter α. Assuming the other parameters of the blank's theoretical PDF were known, the future true blank would be judged, in the response domain, against a constant decision level, as discussed in detail in Chapter 4. But the constant decision level is errorless and is a fixed interval above α. If α is unknown, and must be estimated by $\hat{\alpha}$, then additional error is introduced: the location of the decision level is no longer known. Indeed, the decision level becomes a random variate due to the use of $\hat{\alpha}$. In short, future true blanks, or sample means of M i.i.d. such, are judged with ultimate reference to α, if it is known, or to $\hat{\alpha}$, if α is unknown. Further discussion of this matter, and how it must be taken into account, is contained in Chapter 8 and Appendix C.

3.7 CONTENT DOMAIN AND TWO FALLACIES

Note that no use has yet been made of either the response function, $r(x)$, or a net response function, $y(x)$. In fact, there is no requirement that either be explicitly used: it happens that *implicit* use is sufficient, as will be demonstrated in Chapter 6. For the moment, however, suppose that the schematic distributions in Fig. 3.4 were processed

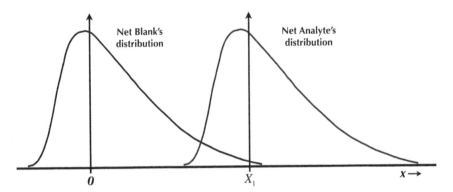

Figure 3.5 Back-calculated content domain distributions for the net blank, at $x = X_0 = 0$, and the net analyte at X_1, with nonnegligible overlap.

via inversion of a plausible linear $y(x)$ function, that is, $y(x) \equiv \beta x = r(x) - \alpha$, with α and β known (or estimated) and no prohibition of negative x values. Then Fig. 3.5 shows qualitatively how the net distributions, which are *fictitious, inferred* distributions, would appear in the content domain.

The back-calculated content domain distributions still overlap nonnegligibly, as expected, and half of the area under the net blank's distribution is in the "unphysical" negative x region. The fictitious net blank distribution does *not* imply that negative analyte content exists and this has long been recognized [1–3], though some have found it disconcerting [4–6]. The negative half of the net blank distribution is simply a harmless consequence of "input referencing" the observed output noise of the real CMS. In other words, if it was physically possible for x to have negative values, then an input distribution of x values, such as that shown in Fig. 3.5, would result in the actual observed output distribution of the blank's responses, *even though the CMS was noiseless*.

To illustrate this important point, Fig. 3.6 (upper) shows a schematic real CMS having linear response function. When the input is an analytical blank, that is, $x \equiv X_0 \equiv 0$, there will be a distribution of responses due to the additive noise; the real CMS is modeled as an ideal CMS plus additive noise. But Fig. 3.6 (lower) shows an exactly equivalent way to model the real CMS: the response distribution is processed by subtraction of α, and then division by β, to yield the back-calculated distribution of fictitious blank variates. This fictitious distribution of x variates, as inputs to an ideal CMS, would then result in the identical observed distribution of responses. Indeed, in this example, all that really happened was to apply a linear transformation and then its inverse, thereby nullifying the two operations.

A related conceptual error involves using a value other than zero as the presumed location of the blank's fictitious content domain distribution, for example, $\bar{x}_{blank}$ [7], (Blank mean) [8], or c_B [9]. From Table 3.1, excepting the relatively rare Case 4, it is evident that back-calculation from the net response domain to the content domain results in the blank's fictitious distribution (or convolved PDF) being exactly centered on zero. In Case 4, for a *single* histogram, but *without* curve

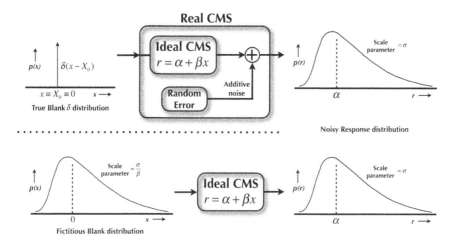

Figure 3.6 Illustration of how output noise may be input referenced, resulting in a fictitious distribution of blank inputs, even when the input is identically zero, that is, a true blank.

fitting, the back-calculated blank distribution in the content domain is centered on $(\hat{\alpha} - \alpha)/(\beta \text{ or } \hat{\beta})$. This has zero expectation value and is usually close to zero due to the many histogram data involved in estimating $\hat{\alpha}$.

Note that if a content domain blank distribution is shown as being centered on, for example, $\bar{x}_{blank}$, then this *must* be as per Case 4 and the distribution is a back-calculated histogram, *not* a PDF. Furthermore, $\bar{x}_{blank}$ is *not* a measured value because *it is impossible to make measurements in the content domain*. Rather, $\bar{x}_{blank}$ is a compound measurement, back-calculated via

$$\bar{x}_{blank} = \frac{(\hat{\alpha} - \alpha)}{(\beta \text{ or } \hat{\beta})} \tag{3.5}$$

where $\hat{\alpha}$ is, itself, a compound measurement, that is, the sample mean of the response histogram data. Similarly, $\hat{\beta}$ would be a compound measurement, if it must be used in eqn 3.5.

In highly dilute specimens under test (SUTs), X_1 is slightly greater than zero. Then, with reference to Fig. 3.6 (upper), the CMS output distribution is shifted up slightly, by the amount βX_1, and the back-calculated x_1 input distribution is centered on X_1. Hence, the fictitious x_1 input distribution extends below zero, though slightly less so than the blank's fictitious x distribution, and this does *not* require any corrective action: see Chapter 7 for full details.

3.8   CAN AN ABSOLUTE STANDARD TRULY EXIST?

Esoteric contrivances aside, the short answer is no: there is always some uncertainty in the analyte content value, even if this is just round-off error. However, the nominal content values for *standards*, even highly dilute ones, are random samples from distributions that are so narrow that they are well-approximated as "delta" functions centered on known values. In other words, the analyte content uncertainty, even propagated via the response function, is negligible in comparison with the additive random error of the CMS. An SUT with nonnegligible analyte content uncertainty is unacceptable for use as a calibration standard: its true content value would not be known well enough, thereby adversely affecting estimation of the response function.

The assumption that a standard has negligible analyte content uncertainty is due to its presumably careful preparation, with subsequent verification. Depending on the circumstances, the latter may involve testing versus certified reference materials or primary standards, t tests, ANOVA, sensitivity analysis, and so on. As well, freshly prepared standards are almost invariably preferred because "old" standards may have altered with the passage of time, even if properly stored.

Finally, with regard to the 0.100 M NaOH solution preparation example discussed in the previous chapter, it is obvious that none of the solutions could be used as standards unless they were subsequently standardized, for example, by titration versus primary standard potassium hydrogen phthalate (KHP). If standardization were omitted, and all of the solutions were simply assumed to be 0.100 M, then these would be random samples from a distribution that was *not* narrow: the content domain uncertainty would *not* be negligible. The direct consequence would be increased noise in the response and net response domains, that is, the total noise would consist of the additive random noise of the CMS plus propagated "analyte content" noise due to the invalid nominal concentration assumption.

3.9 CHAPTER HIGHLIGHTS

The chapter began the discussion of the response, net response, and content domains, and how they are related. The essential concepts of false positives and false negatives were introduced, the ramifications of blank subtraction were discussed, and a pair of fallacies, particularly concerning the analytical blank's fictitious distribution of content, were laid to rest.

REFERENCES

1. M. Thompson, "Do we really need detection limits?", *Analyst* **123** (1998) 405–407.
2. Analytical Methods Committee of the Royal Society of Chemistry, "What should be done with results below the detection limit? Mentioning the unmentionable," 5 (April 2001) 2 pp. (Fig. 2).
3. M. Thompson, S.L.R. Ellison, "Towards an uncertainty paradigm of detection capability", *Anal. Methods* **5** (2013) 5857–5861.
4. M. Korun, P.M. Modec, "Interpretation of the measurement results near the detection limit in gamma-ray spectroscopy using Bayesian statistics", *Accred. Qual. Assur.* **15** (2010) 515–520.

5. M. Korun, B. Zorko, "Reporting measurement results of activities near the natural limit: note and extension of the article 'Interpretation of the measurement results near the detection limit in gamma-ray spectroscopy using Bayesian statistics'", *Accred. Qual. Assur.* **18** (2013) 175–179.

6. P.-Th. Wilrich, "Note on the correction of negative measured values if the measurand is nonnegative", *Accred. Qual. Assur.* **19** (2014) 81–85.

7. J. Fonollosa, A. Vergara, R. Huerta, S. Marco, "Estimation of the limit of detection using information theory measures", *Anal. Chim. Acta* **810** (2014) 1–9.

8. N.P. Hyslop, W.H. White, "An empirical approach to estimating detection limits using collocated data", *Environ. Sci. Technol.* **42** (2008) 5235–5240.

9. J. Jiménez-Chacón, M. Alvarez-Prieto, "An approach to detection capabilities estimation of analytical procedures based on measurement uncertainty", *Accred. Qual. Assur.* **15** (2010) 19–28.

4

TRADITIONAL LIMITS
OF DETECTION

4.1 INTRODUCTION

In Chapter 3, blank and analyte distributions were tacitly assumed to be those relevant to chemical measurement system (CMS) responses, that is, histograms were comprised of elementary measurements, rather than compound ones, and probability density functions (PDFs) were those of the CMS itself. This was simply an expository convenience, since no modifications are fundamentally necessary in regard to PDFs and histograms of compound measurements. There are, however, implications concerning the extent to which postprocessing of CMS responses is reasonable or justifiable; see Chapter 16. This aside, the same tacit assumption will continue to be made, for purposes of clarity.

4.2 THE DECISION LEVEL

Consider Fig. 4.1, which differs from Fig. 3.2, in which it explicitly shows a possible decision level, r_C. As before, it is immediately evident that the response distributions have negligible overlap. Therefore, if the only two possible inputs to the CMS were analyte 1, at X_1, and the blank, it would be possible to use their respective responses to decide, with negligible error, the value of X. All that would be necessary would be to define r_C appropriately, as done in Fig. 4.1, and then judge CMS responses against it: any response below r_C would imply $x = X_0 = 0$ while any response above r_C would imply $x = X_1$. The probability of having a response exactly at r_C would be negligible in the case at hand, but, in general, a decision rule could be defined to handle such rare cases.

Limits of Detection in Chemical Analysis, First Edition. Edward Voigtman.
© 2017 John Wiley & Sons, Inc. Published 2017 by John Wiley & Sons, Inc.
Companion Website: www.wiley.com/go/Voigtman/Limits_of_Detection_in_Chemical_Analysis

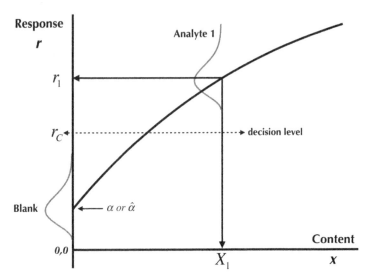

Figure 4.1 A possible response function with response PDFs or histograms for the analytical blank, at $x = X_0 = 0$, and an analyte at X_1. A possible decision level, r_C, is depicted. If the distributions are PDFs, then r_C is replaced by the theoretical decision level, R_C.

4.3 FALSE POSITIVES AGAIN

Now suppose that r_C were to be lowered until it was equal to the upper limit of the central 90% CI for the analytical blank. This is shown in Fig. 4.2. Then any response exceeding r_C would be judged to be due to nonzero analyte content. However, there would be a 5% chance that a true blank had exceeded r_C. In this case, the decision would be incorrect and the result would be called a false positive or a Type I error. The 5% probability arises because the central 90% CI for the blank is flanked by upper and lower tails having equal 5% probabilities. The probability of false positives is denoted by p and the long-standing *de facto* canonical value for p is 0.05, that is, 5% probability [1–3].

The shape of the blank's PDF is almost entirely irrelevant to the matter of false positives. In principle, essentially any unimodal distribution is acceptable as the PDF of the responses of the CMS. The uniform (that is, rectangular) PDF is also acceptable. An extreme "U"-shaped PDF would likely be problematic, but these are almost never encountered in real CMSs.

In Fig. 4.1, analyte 1 was far above the blank and there was effectively no possibility of mischaracterizing either analyte or blank responses. If the analyte content is lowered, its response distribution also lowers, as shown in Fig. 4.3, and complications will arise if X_1 is lowered enough so that the analyte's distribution has significant area below r_C; some analyte responses would be incorrectly categorized as blanks. Such errors are called false negatives, or Type II errors, and their probability is denoted by q. As for p, the long-standing *de facto* canonical value for q is 0.05, that is, 5% probability [1–3]. In Fig. 4.3, $q = 0.05$.

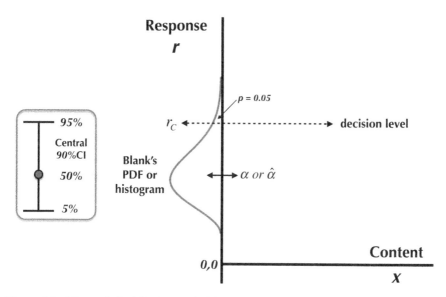

Figure 4.2 The analytical blank, the decision level, r_C, and false positives. If the blank's distribution is a PDF, then r_C is replaced by the theoretical decision level, R_C.

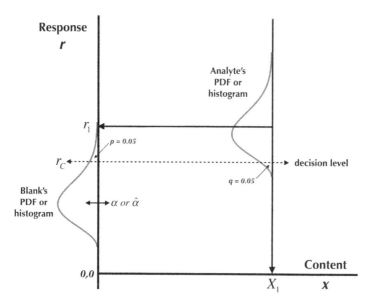

Figure 4.3 Only 5% of analyte responses due to X_1 are below the decision level, r_C. If the blank's distribution is a PDF, then r_C is replaced by R_C.

4.4 DO FALSE NEGATIVES REALLY MATTER?

If false negatives are of no concern, then the decision level may serve also as a detection limit. This, in fact, is exactly the case with traditional detection limits, and it was even the case with the highly publicized official "5σ" detection of the long-sought Higgs boson at the Large Hadron Collider (LHC), a discovery that led directly to the 2013 Nobel Prize in Physics for Professors Higgs and Englert. What mattered was the near certainty that the Higgs boson had been detected at long last, that is, that the probability of a "false discovery" was less than about 300 ppb. Thus, even if large numbers of Higgs bosons might have had to be brought into fleeting existence in order to reduce p sufficiently, this was entirely acceptable: detection efficiency was unimportant, except to the large teams of experimentalists and technicians constantly toiling at, on and in the LHC. In a CMS, however, having the decision level serve also as the detection limit means that 50% of analyte responses at the detection limit will be false negatives, that is, $q = 0.5$, as elaborated upon immediately below and discussed further in the following two chapters.

4.5 FALSE NEGATIVES AGAIN

Suppose the analyte content was such that its response distribution had exactly 50% of its area below r_C and the other 50% above r_C. This is shown in Fig. 4.4, where

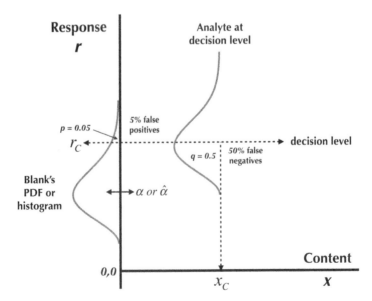

Figure 4.4 Response domain analyte responses at the decision level, r_C. If the blank's distribution is a PDF, then r_C is replaced by the theoretical decision level, R_C.

the analyte content is identified as x_C, that is, the point x_C, r_C is a point on the unseen (not plotted) response function.

The analyte content value x_C is called the experimental content domain decision level and r_C is the corresponding experimental response domain decision level. If these are also used as traditional detection limits, then $q = 0.5$. Many authors, at least as far back as 1968 [4], have found this to be unacceptable: there is only a (fair) coin flip chance of detecting analyte at the traditional detection limit. Yet, it was almost universal practice, and is still extremely common, to find traditional detection limits that are simply decision levels, or oversimplified variants of them.

4.6 DECISION LEVEL DETERMINATION WITHOUT A CALIBRATION CURVE

The above issue is revisited in more detail in the following two chapters, but first note that it provides an opportunity, if histograms of responses may be obtained, to estimate the theoretical content domain decision level, X_C. The first step is to obtain a histogram of blank responses, specify p, and thereby determine r_C, as illustrated in Fig. 4.4. Then, assuming the response function is known to be monotonic, the procedure is simple: iteratively input different analyte contents, that is, $X_1, X_2, X_3, \ldots$, until the last one produces 50% false negatives with respect to r_C. The last one is then x_C, that is, the experimental estimate of X_C. Obviously, this procedure requires more effort than just making a few measurements per analyte content input, but is actually quite easy with some modern CMSs. The advantage it has is that there is no requirement to construct a calibration curve, provided that the response function is, in fact, monotonic. Indeed, *only the last X value matters*: the ones preceding it were simply means to an end and could be entirely ignored once x_C had been found.

Practicality aside, the process reveals an amusing irony: by carefully quantifying false negatives, iteratively changing X until 50% false negatives are achieved, a traditional detection limit is obtained, that is, one that "completely ignores" false negatives. This is a nice example of how, on occasion, a disadvantage may be turned to an advantage.

4.7 NET RESPONSE DOMAIN AGAIN

Figure 4.5 shows the net response domain equivalent of Fig. 4.4, with $y(x) \equiv r(x) - \alpha$ or $y(x) \equiv r(x) - \hat{\alpha}$, as relevant. Therefore, the experimental net response domain decision level, y_C, is simply r_C minus either α or $\hat{\alpha}$, again as relevant. The net response domain distributions are as per the relevant row of Table 3.1.

The various decision level symbols are summarized in Table 4.1. If false negatives are ignored, then $q = 0.5$ and these decision levels do double duty as traditional detection limits. In this case, the "C" subscripts are replaced by "DC" subscripts, for purposes of clarity.

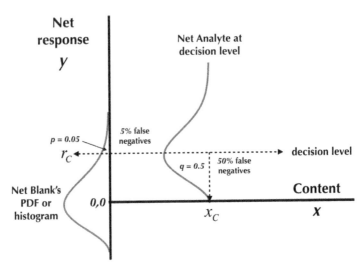

Figure 4.5 Net response domain analyte responses at the decision level. If the blank's distribution is a PDF, then y_C is replaced by the theoretical decision level, Y_C.

Table 4.1 Decision Level Symbols in the Three Domains (q Not Ignored)

Domain	Theoretical (True)	Experimental Estimate
Response	R_C	r_C
Net response	Y_C	y_C
Content	X_C	x_C

4.8 AN OVERSIMPLIFIED DERIVATION OF THE TRADITIONAL DETECTION LIMIT, X_{DC}

Assume that α, β, and σ_0 are known and that the blank distribution in Fig. 4.4 and the net blank distribution in Fig. 4.5 are both PDFs. Further assume that the unseen response function is linear, that is, $r(x) \equiv \alpha + \beta x \equiv \alpha + y(x)$, with $\beta > 0$. Since q is implicitly 0.5, $X_{DC} \equiv X_C$, $R_{DC} \equiv R_C = r_C$, and $Y_{DC} \equiv Y_C = y_C$. Then $r(X_{DC}) = R_{DC}$ and $y(X_{DC}) = Y_{DC} = \beta X_{DC}$, so $X_{DC} = Y_{DC}/\beta = (R_{DC} - \alpha)/\beta$. The constant R_{DC} is simply chosen to provide the desired *a priori* specified probability of false positives, for example, $p \equiv 0.05$. Since $R_{DC} > \alpha$, their difference is a positive constant that may be parsed in units of the scale parameter, σ_0, of the blank's PDF. Thus,

$$R_{DC} - \alpha = c_p \sigma_0 = Y_{DC} = \beta X_{DC} \tag{4.1}$$

where c_p is a positive constant chosen so that the blank's PDF has 100p% of its area above R_{DC}. With $q = 0.5$, X_{DC} is the traditional theoretical content domain detection limit, given by eqn 4.2:

$$X_{DC} = \frac{c_p \sigma_0}{\beta} \qquad (4.2)$$

Finally, if the blank's PDF was a Gaussian distribution, that is, $N{:}\alpha,\sigma_0$, rather than the asymmetric PDF shown in Figs 4.4 and 4.5, then c_p would simply be the critical z value, z_p. For $p \equiv 0.05$, $z_p \cong 1.644854 \simeq 1.645$. Therefore

$$X_{DC} = \frac{z_p \sigma_0}{\beta} \simeq \frac{1.645 \sigma_0}{\beta} \qquad (4.3)$$

Replacing z_p by a larger number results in a p value below 0.05, for example, if $z_p \equiv 3$, then $p \cong 0.135\%$.

4.9 OVERSIMPLIFICATIONS CAUSE PROBLEMS

The derivation of eqn 4.3 depended upon an unsettling number of significant simplifying assumptions. It was intended only as a preliminary introduction to the traditional detection limit notion and will be entirely supplanted by superior developments in later chapters. As will become clear, failure to respect the numerous simplifying assumptions leading to eqn 4.3 has been a major problem afflicting both detection limit theory and practice. Indeed, it is common to see eqn 4.3 modified by replacing z_p with a constant that is typically defined as 3, and often σ_0 and β are replaced by estimates, without replacing z_p by the appropriate t_p. As well, the effects of replacing α with $\hat{\alpha}$ are also commonly ignored. Then it is erroneously asserted that $p = 0.135\%$. In fact, such misuse of eqn 4.3 almost invariably results in p being neither properly specified nor controlled, which is a high price to pay given that false negatives are also entirely ignored, that is, $q = 0.5$. Chapter 5 begins the process of fixing these needlessly inflicted mistakes.

4.10 CHAPTER HIGHLIGHTS

Last chapter's preliminary discussion of false positives and false negatives was elaborated upon in this chapter. A preliminary argument was advanced for why neglecting false negatives may be problematic, and it was shown that there was no fundamental requirement for using a response function, since Currie's detection limit schema may be implemented without one. An oversimplified derivation of the traditional detection limit was also given, together with a discussion of why it was inadequate.

REFERENCES

1. L.A. Currie, "Nomenclature in evaluation of analytical methods including detection and quantification capabilities (IUPAC Recommendations 1995)", *Pure Appl. Chem.* **67** (1995) 1699–1723.

2. L.A. Currie, "Detection: international update, and some emerging di-lemmas involving calibration, the blank, and multiple detection decisions", *Chemom. Intell. Lab. Syst.* **37** (1997) 151–181.

3. M. Thompson, S.L.R. Ellison, "Towards an uncertainty paradigm of detection capability", *Anal. Methods* **5** (2013) 5857–5861.

4. L.A. Currie, "Limits for qualitative and quantitative determination – application to radio-chemistry", *Anal. Chem.* **40** (1968) 586–593.

5

MODERN LIMITS OF DETECTION

5.1 INTRODUCTION

As discussed in Chapter 4, decision levels may also be used as detection limits, provided that an analyte, present at the detection limit, has only 50% probability of being detected. Many analysts still use and report these traditional detection limits, x_{DC}, apparently quite content with the obsolete 1975 IUPAC definition [1, 2], even though it was supplanted in 1995 [3]. Enormous numbers of publications already contain these detection limits, which may be calculated with minimal thought or effort. They also require no modification for heteroscedastic noises because only the blank is involved in the definition of x_{DC}. Lastly, for the most commonly encountered assumptions, that is, $p = q$ and homoscedastic noise with symmetric PDF, the decision level is exactly half the detection limit. Then why bother with a factor of 2?

There may well be no definitive argument that can be made either in favor of implicitly defining q as 0.5 or in favor of explicitly defining q at a value below 0.5, for example, the canonical $q = 0.05$. Since there is no more than p probability that a blank response will be mistaken for an analyte response, p could simply be lowered, if desired. Indeed, for the Higgs boson detection and the first direct gravitational wave detection [4], the detection criterion was 5σ, that is, p less than about 300 ppb. Obviously, in these two historically momentous experimental endeavors, the detections were profoundly important confirmations of long-standing predictions of the Standard Model and General Relativity, respectively. Both cases were essentially qualitative: detection, regardless of efficiency, was of enormous significance, while nondetection would simply have been the cause to think more deeply, then upgrade and keep trying.

Detection of analytes is categorically different: mere qualitative detection does not necessarily have any particular importance or constitute a prize-winning discovery.

Limits of Detection in Chemical Analysis, First Edition. Edward Voigtman.
© 2017 John Wiley & Sons, Inc. Published 2017 by John Wiley & Sons, Inc.
Companion Website: www.wiley.com/go/Voigtman/Limits_of_Detection_in_Chemical_Analysis

Indeed, if an SUT contains analyte content at the traditional limit of detection, then detection *could* occur, with 50% probability. As seen below, the cure for poor detection efficiency is the use of distinct decision levels and detection limits. However, even aside from poor detection efficiencies, traditional detection limits are usually computed with minimal thought or effort. For example, if x_{DC} is defined as

$$x_{DC} \equiv \frac{k \times \text{standard deviation of the blank}}{\text{slope of calibration curve}} \tag{5.1}$$

where k is a "coverage factor," then a variety of questions ensue, including the following:

- What is the purpose of k?
- Does k determine p?
- Is k equal to a constant, *for example*, 3?
- If k is not a constant, is it a function or composite of some sort?
- Does k take into account the noise on the blank's sample mean?
- Is the standard deviation a population parameter, σ_{blank}, or a sample test statistic, s_{blank}?
- If there are several possible candidates for s_{blank}, which one should be used?
- Where do degrees of freedom enter the picture, if at all?
- Is the slope the true slope, β, or an estimate, b?

In the large majority of cases, these questions are either ignored or cavalierly dismissed, and k is simply defined as 3. Consequently, published x_{DC} values usually neither properly specify p nor control it. Furthermore, it is rarely possible to correct such values because the necessary information is not available in the associated publications. This significantly degrades the utility of such values for purposes of comparison or optimization.

5.2 CURRIE DETECTION LIMITS

In contrast to the above, an ever-increasing number of analysts, and all of the standards organizations, favor the basic approach introduced by Currie in his landmark 1968 publication [5]. To be sure, there was significant prior work in the field [6–8], but Currie eloquently presented the rationale, to chemical analysts, for the necessary consideration of both false positives and false negatives. His paper resonated with analysts who found it unacceptable, either for philosophical or pragmatic reasons, to have a detection limit with only coin flip detection efficiency. As seen below, if an SUT contains analyte content at the Currie limit of detection, then detection *should* occur, with $100(1 - q)\%$ probability.

Currie's paper also made clear, without explicitly stating so, that he felt there was an underlying true limit of detection, L_D, for a given instrumental method. The accurate and unbiased estimation of this fundamental detection limit was his ultimate goal and the reason for this is readily evident: a well-trained analyst, using a properly functioning instrument and following a valid protocol, ideally should be able to repeatedly and reliably estimate, with low uncertainty, the L_D value. This would then fairly characterize an important aspect of the instrumental method and serve as a verifiable, transferable figure of merit. Conversely, if repeated determinations of the detection limit exhibited large uncertainties, then detection limits would be of little practical utility.

Currie's paper resolutely focused on p and q: all that matters fundamentally is being able to specify and control both p and q, and reproducibly demonstrating it. Toward this end, consider Fig. 5.1. This figure shows the response distributions for analyte present at content values of x_C, the experimental content domain Currie decision level, and at x_D, the experimental content domain Currie detection limit. The canonical probability definitions at the detection limit are $p \equiv 0.05$ and $q \equiv 0.05$. Figure 5.2 shows the equivalent depiction in the net response domain.

The various detection limit symbols are summarized in Table 5.1. These differ from the decision level symbols in the previous chapter in having a "D" subscript and note that L_D will never be used again: R_D, Y_D, and X_D are what L_D actually is in each domain.

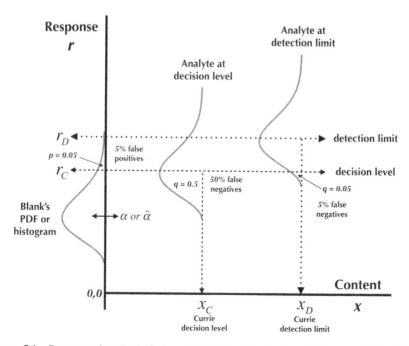

Figure 5.1 Response domain depiction of Currie decision levels and detection limits. If the distributions are PDFs, then r_C and r_D are replaced by R_C and R_D, respectively: see Table 5.1.

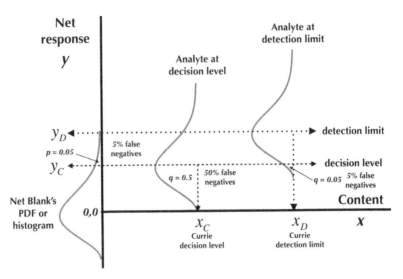

Figure 5.2 Net response domain depiction of Currie decision levels and detection limits. If the distributions are PDFs, then y_C and y_D are replaced by Y_C and Y_D, respectively: see Table 5.1.

Table 5.1   Detection Limit Symbols in the Three Domains

Domain	Theoretical (True)	Experimental Estimate
Response	R_D	r_D
Net response	Y_D	y_D
Content	X_D	x_D

5.3   WHY WERE p AND q EACH ARBITRARILY DEFINED AS 0.05?

Limits of detection are only useful as transferable figures of merit if they are standardized. In this regard, there is no difficulty with the *arbitrary*, canonical assumption that $p \equiv 0.05 \equiv q$. Indeed, there is no prohibition on reporting nonstandard detection limits in addition to reporting standard detection limits and it is certainly true that there are cases where this is the best course of action. But *standard* detection limits and decision levels must always be in compliance with the canonical standard definitions of p and q, as *ipso facto* set by the relevant standards organizations [9]. The specific value of 0.05 is of no importance whatsoever except for the minor fact that it has been widely used for decades and, far more importantly, there is neither a persuasive nor compelling rationale for changing to another arbitrary value. This is entirely a matter for the standards organizations, but perhaps it is best to let this particular triviality lie.

5.4 DETECTION LIMIT DETERMINATION WITHOUT CALIBRATION CURVES

From Figs 5.1 and 5.2, it is immediately evident how Currie's program may be instantiated. In the preceding chapter, it was shown that an x_C value, which estimates X_C, may be obtained via an iterative procedure, without knowing anything more about the response function than that it was monotonic. Assuming it is possible to obtain a histogram of blank responses, and for the defined value of p, the distribution of blank responses (Fig. 5.1) provides an r_C estimate and the distribution of net blank responses (Fig. 5.2) provides a y_C estimate. In place of histograms, reasonably accurate central 90% CIs may be used to estimate r_C or y_C, respectively. If the blank and net blank distributions were actual PDFs, then the corresponding decision levels would be R_C or Y_C, respectively.

Obviously, x_D must exceed x_C. Finding x_D would again involve iteratively performing measurements on analytes having content above x_C. Iteration would halt when the last sample tested was found to yield 5% false negatives, when its responses or net responses were judged against their respective decision levels. Then $x_D = X_{last}$, that is, the estimate of X_D.

Despite its simplicity, the above outlined experiment has a number of very important implications. First, it is in full compliance with the Currie schema requirement of specification and control of both p and q, even though all that is known of the true response function, or net response function, is that it is monotonic. The noise distribution is almost entirely unspecified other than being reasonably well behaved. Indeed, although the hypothetical distributions in Figs 5.1 and 5.2 are homoscedastic, this is not essential, and they are shown as asymmetric simply to underscore the point that Gaussian distributions are not fundamentally required.

Second, there is no requirement that a calibration curve be used: all that is required is the ability to make sufficiently many replicate measurements to construct reasonably accurate histograms (or central 90% CIs) for the blank and either analyte responses or analyte net responses. Third, there is no need for any regression technique, even if the noise is heteroscedastic, and there is also no need for prediction intervals or prediction bands. Finally, the decision levels and detection limits may be estimated entirely nonparametrically. Hence, use of a parametric method, or construction of a calibration curve and use of regression, should achieve some *deliberate* purpose. In other words, methodology inessential to implementation of Currie's program should *not* be used unless there are significant net advantages in doing so.

5.5 A NONPARAMETRIC DETECTION LIMIT BRACKETING EXPERIMENT

In 2011, Voigtman and Abraham [10] demonstrated how to obtain many millions of experimental detection limits from a simple laser-excited molecular fluorescence instrument. It is instructive to use some of the data from that set of experiments to illustrate the initial steps in the iterative process described above. All of the

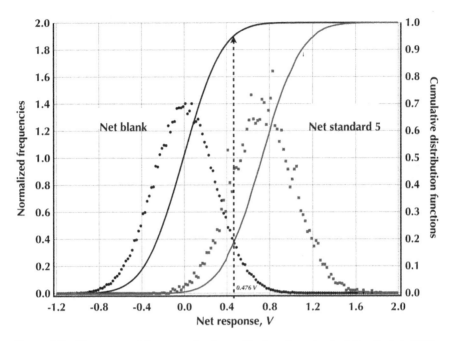

Figure 5.3   Net response histograms for the net blank and net standard 5, with their CDFs.

experimental details mentioned below are contained in the above reference, which is reprinted in Chapter 15. As an aside, the experimentally determined calibration curves were found to be linear, to an excellent approximation, and the noise was demonstrated to be Gaussian distributed and approximately white in power spectral density. But this information is initially ignored below.

Figure 5.3 shows a plot of the histograms of the 143 360 binned net blank responses and 12 001 binned net responses for analyte standard 5 (0.2120 mg/100 mL, rhodamine 6G tetrafluoroborate in ethanol), together with their respective cumulative distribution functions (CDFs), that is, normalized integrated running areas.

From the CDF of the net blank, assuming $p = 0.05$, the experimental net response domain Currie decision level, y_C, is 0.476 V, that is, $0.476\,V = CDF^{-1}_{net\,blank}(0.95)$. This is shown as the vertical arrow in Fig. 5.3. Net standard 5 is clearly below the experimental content domain Currie detection limit, x_D, since its histogram has $q = 0.187 = CDF_{standard\,5}(0.476\,V)$, that is, 18.7% false negatives, which is well above the defined value of $q = 0.05$.

Figure 5.4 is a similar plot of the net blank histogram versus 12 001 binned net responses for analyte standard 6 (0.2968 mg/100 mL), together with their CDFs. Net standard 6 is above x_D, since $q = 0.027 = CDF_{standard\,6}(0.476\,V) < 0.05$. Hence, x_D must lie between 0.2120 and 0.2968 mg/100 mL, consistent with the reported value of 0.252 mg/100 mL [10]. Finding x_D experimentally would then be a matter of preparing analyte solutions with intermediate concentrations and testing them as

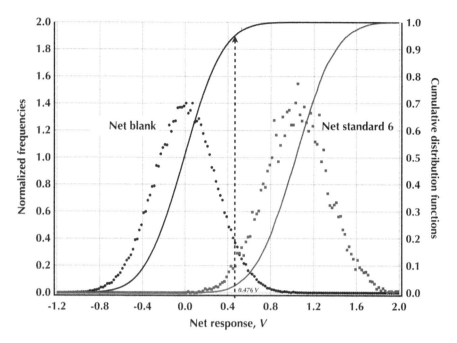

Figure 5.4 Net response histograms for the net blank and net standard 6, with their CDFs.

above, stopping when the last one, at $x_D = X_{last}$, resulted in $q = 0.05 = \text{CDF}_{x_D}(0.476\,\text{V})$. It would simply be a matter of perseverance.

5.6 IS THERE A PARAMETRIC IMPROVEMENT?

The nonparametric analysis presented above entirely ignored the fact that the experimental data distributions were known to be Gaussian because they were deliberately designed to be Gaussian. Freedom from parametric constraints can be helpful, particularly when the underlying functions are unknown. However, knowledge that the experimental data distributions are Gaussian in the present example offers the possibility of an improvement in efficiency, simply by performing Gaussian curve fits on the distributions. The fitted curves are shown in Fig. 5.5.

The Gaussian curve fit results are in Table 5.2, for the Gaussian function in eqn 5.2:

$$f(y) = \text{Offset} + A e^{-(y-y_0)^2/2\sigma^2} \tag{5.2}$$

where $f(y)$ is the normalized frequency, that is, the ordinate in Fig. 5.5. From the Gaussian fit of the net blank distribution, $y_C = 0.4774$ V for $p = 0.05$, only slightly different from the nonparametrically determined value of 0.476 V. The probabilities of false negatives were then computed to be $q = 0.1867$ (net standard 5) and $q = 0.0272$

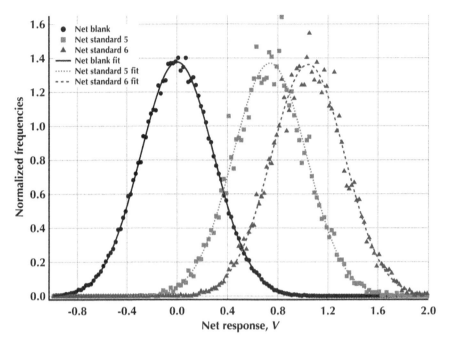

Figure 5.5   Gaussian fits for the net blank, net standard 5 and net standard 6.

Table 5.2   Gaussian Curve Fit Results ± 1 Standard Error

Sample	Offset	A	y_0	σ
Net blank	-0.0003 ± 0.0019	1.3779 ± 0.0041	0.0008 ± 0.0009	0.2897 ± 0.0011
Net standard 5	-0.0008 ± 0.0067	1.3702 ± 0.0150	0.7365 ± 0.0034	0.2918 ± 0.0040
Net standard 6	-0.0008 ± 0.0059	1.3630 ± 0.0130	1.0411 ± 0.0030	0.2933 ± 0.0035

(net standard 6), negligibly different from the nonparametrically determined values. In this case, then, no efficiency improvement was observed.

5.7   CRITICAL NEXUS

In Chapter 6, a very brief introduction to receiver operating characteristics (ROCs) is presented. Using both synthetic data and real experimental data, it is shown how they may be used to determine detection limits. Then, the next eight chapters deeply focus on the most well-behaved CMSs possible: those with homoscedastic additive Gaussian white noise (AGWN) and perfectly linear response functions.

5.8 CHAPTER HIGHLIGHTS

A stronger case was made for explicitly specifying an *a priori* maximum allowable probability of false negatives. The rationale for Currie's detection limit schema was presented and it was shown that detection limits in compliance with his schema did not require knowledge of the response function other than that it be known to be monotonic. Real experimental results corroborated this assertion and it was found that no parametric advantage existed for the particular real data examined.

REFERENCES

1. IUPAC, "Nomenclature, symbols, units and their usage in spectrochemical analysis – II. Data interpretation, analytical chemistry division", *Spectrochim. Acta, B* **33** (1978) 241–245.
2. G.L. Long, J.D. Winefordner, "Limit of detection: a closer look at the IUPAC definition", *Anal. Chem.* **55** (1983) 712A–724A.
3. L.A. Currie, for IUPAC, "Nomenclature in evaluation of analytical methods including detection and quantification capabilities" *Pure Appl. Chem.* **67** (1995) 1699–1723, IUPAC ©1995.
4. B.P. Abbott *et al.*, "Observation of gravitational waves from a binary black hole merger", *Phys. Rev. Lett.* **116** (2016) 061102, doi:10.1103/PhysRevLett.116.061102.
5. L.A. Currie, "Limits for qualitative and quantitative determination – application to radiochemistry", *Anal. Chem.* **40** (1968) 586–593.
6. B. Altshuler, B. Pasternak, "Statistical measures of the lower limit of detection of a radioactivity counter", *Health Phys.* **9** (1963) 293–298.
7. H. Kaiser, "Zum problem der nachweisgrenze", *Z. Anal. Chem.* **209** (1965) 1–18.
8. H. Kaiser, "Zur definition der nachweisgrenze, der garantiegrenze und der dabei benutzen begriffe: fragen und ergebnisse der diskussion", *Z. Anal. Chem.* **216** (1966) 80–94.
9. M. Thompson, S.L.R. Ellison, "Towards an uncertainty paradigm of detection capability", *Anal. Methods* **5** (2013) 5857–5861
10. E. Voigtman, K.T. Abraham, "Statistical behavior of ten million experimental detection limits", *Spectrochim. Acta, B* **66** (2011) 105–113.

6

RECEIVER OPERATING CHARACTERISTICS

6.1 INTRODUCTION

Receiver operating characteristic (ROC) curves [1] provide a powerful and informative way to view results such as those presented in Figs 5.3 and 5.4. As a nonparametric technique, ROCs afford a number of advantages, including general robustness and freedom from dependence upon population parameters that are often unknown. Indeed, Wysoczanski and Voigtman [1] demonstrated how easy it was to use ROCs in determining valid detection limits in a simple fluorescence CMS in which

- the response function was monotonic and nonlinear, but *deliberately* unknown;
- the noise probability density function (PDF) was highly asymmetric, thus non-Gaussian, but *deliberately* unknown; and
- both homoscedastic and heteroscedastic noise precision scenarios were involved.

There is an extensive literature on ROCs, so this chapter provides only a very brief introduction to their principles and construction. The paper by Fraga *et al.* [2] is recommended as a starting point for further information on ROCs in CMSs and, as well, there are numerous sources on information on the web, for example, Refs [3, 4].

6.2 ROC BASICS

To begin our consideration of ROCs, consider the nine pairs of response distributions for the blank (leftmost) and analyte (rightmost), as shown in Fig. 6.1. In Fig. 6.1, the response distributions may be either theoretical PDFs or experimental histograms and it makes no difference whether the responses are "raw" or referenced to the analytical

Limits of Detection in Chemical Analysis, First Edition. Edward Voigtman.
© 2017 John Wiley & Sons, Inc. Published 2017 by John Wiley & Sons, Inc.
Companion Website: www.wiley.com/go/Voigtman/Limits_of_Detection_in_Chemical_Analysis

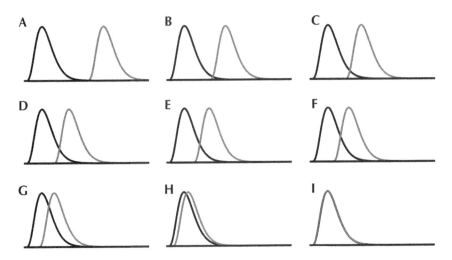

Figure 6.1 Blank and analyte response distributions illustrating progressively increasing degrees of overlap.

blank's mean. The distribution pair shown in Fig. 6.1a is almost perfectly resolved, that is, the distributions have negligible overlap. At the other extreme, the two distributions in Fig. 6.1i are completely overlapped; the blank distribution is plotted directly beneath the analyte distribution.

For simplicity, assume the distributions are PDFs, so that the response domain decision level is R_C. If R_C were to be set at the *leftmost* extreme of any of the nine plots shown in Fig. 6.1, virtually all blank responses would be misjudged as false positives, while virtually all analyte responses would be properly characterized. Thus, $p \cong 1$ and $q \cong 0$. On the other hand, if R_C were to be set at the *rightmost* extreme of any of the nine plots, virtually all analyte responses would be misjudged as false negatives, while virtually all blank responses would be properly characterized. Thus, $p \cong 0$ and $q \cong 1$. For any intermediate value of R_C, p, and q would have values between 0 and 1, so their joint "space of support" is a unit square.

A moment's thought reveals that the cumulative distribution functions (CDFs) of the respective blank and analyte PDFs provide the quantitative connection between R_C and the p,q pair. This follows because, as shown in eqn A.19, CDF(x) is simply the *lower tail area* of its associated PDF, that is, $p(x)$, with respect to argument x. False positives are *upper tail areas* of the blank's PDF, with respect to argument R_C. Hence, they are equal to one minus the complementary lower tail areas. False negatives are lower tail areas of the analyte's PDF, with respect to argument R_C. Thus, we have the following:

$$p \equiv (1 - blank's \ area \ below \ R_C) \equiv 1 - CDF_{blank}(R_C) = 1 - \int_{-\infty}^{R_C} p_{blank}(r)dr \quad (6.1)$$

and

$$q \equiv (analyte's\ area\ below\ R_C) \equiv CDF_{analyte}(R_C) = \int_{-\infty}^{R_C} p_{analyte}(r)dr \qquad (6.2)$$

Since $Y_C = R_C - \alpha$, a simple change of variables yields the equivalent net response domain equations:

$$p \equiv 1 - CDF_{blank}(Y_C) = 1 - \int_{-\infty}^{Y_C} p_{blank}(y)dy \qquad (6.3)$$

and

$$q \equiv CDF_{analyte}(Y_C) = \int_{-\infty}^{Y_C} p_{analyte}(y)dy \qquad (6.4)$$

It makes no difference which pair of equations is used, and it also makes no difference if α is consistently replaced by an estimate, $\hat{\alpha}$. The latter follows because use of *any one value* of $\hat{\alpha}$ would result in both distributions shifting by the same amount, in the same direction. Hence, there would be no net effect.

Note that in Fig. 6.1b–h, there is no R_C value that can simultaneously result in near-zero values of both p and q. The best that can be achieved is a compromise: R_C may be adjusted to reduce the probability of the least desirable type of error, thereby increasing the probability of the less undesirable type of error. This is a major advantage of ROCs. However, in conventional analytical detection limit theory, the default is to assume that false positives and false negatives are equally undesirable, that is, have "equal costs" associated with them. This is in agreement with the repeatedly encountered *a priori* canonical detection limit defaults, that is, $p \equiv 0.05$ and $q \equiv 0.05$.

The perfect overlap situation shown in Fig. 6.1i depicts analyte content indistinguishable from that of the analytical blank. Clearly, for the PDFs in Fig. 6.1i, it is necessarily true that $p + q = 1$. Furthermore, if R_C were to be located at the mean of the PDFs in Fig. 6.1i, then it would follow that each PDF would have half of its area below R_C and half above, that is, both p and q would equal 0.5. This is the "coin flip" scenario [5, Fig. 4]; no information is available.

6.3 CONSTRUCTING ROCs

An ROC plot is obtained by sweeping R_C (in the response domain) or Y_C (in the net response domain) from its extreme minimum value, where $p = 1$ and $q = 0$, to its extreme maximum value, where $p = 0$ and $q = 1$. For each possible R_C or Y_C value, there is a resulting p, q pair. A standard ROC plot is simply the succession of pairs of the form $p, (1 - q)$, that is, a curve in the unit square ROC support space. The quantity $(1 - q)$ is often called the "true positives probability" or "sensitivity", while the quantity $(1 - p)$ is often called the "true negatives probability" or "selectivity" [3]. Other names, for example, "detectivity," are also in use, but will be ignored herein, because they mostly just facilitate confusion.

In any CMS, what matters for detection purposes is minimization of both p and q, to the extent possible or feasible. Not surprisingly, a simple ROC variant involves plotting p, q pairs [6]. This has a significant advantage in terms of intuitiveness and the confusing names are no longer needed. A second advantage concerns data presentation: ideal detection is the lower-left corner, that is, the origin, of the ROC plot. This makes it qualitatively apparent, at a glance, what the data show. Accordingly, this is the only type of ROC plot used in the present work.

The full range ROC plots, for all nine of the detection situations depicted in Fig. 6.1, are shown in Fig. 6.2. The necessary plot data was generated via numerical evaluation of the simple model function that was used to generate the distributions shown in Fig. 6.1. As noted earlier, generating ROC curves is easy; every curve is a plot of the analyte's CDF versus one minus the CDF of the blank, each as a function of the relevant decision level: R_C or Y_C.

The asymmetric curvatures of the ROC curves in Fig. 6.2 are due to the asymmetry of the assumed distributions. The ROC curve for the distributions in Fig. 6.1i is a straight line, that is, $q = 1 - p$. It is called the "line of indecision," for obvious reasons. Also note that these curves have no noise on them, since they were generated by numerical evaluation of a noiseless model function. With real experimental data, the ROC curves will exhibit the effects of noise, to varying degrees.

Mathematically, p and q have upper limits of unity, as shown in Fig. 6.2. However, real CMSs have upper range limits of 0.5 for both p and q. As stated previously [1], "The reason is simple: $p > 0.5$ would mean that the decision level was set below

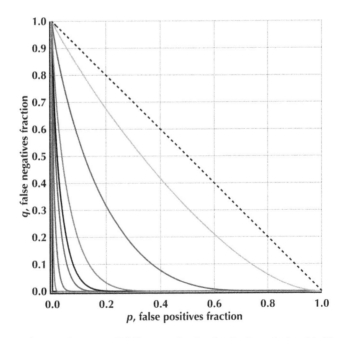

Figure 6.2   The full range ROC curves for the distributions depicted in Fig. 6.1.

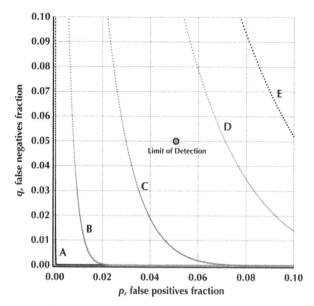

Figure 6.3   An expanded scale version of Fig. 6.2. The distributions in Fig. 6.1a generate the ROC "A" curve, and so on, in order.

the blank's mean, while $q > 0.5$ would mean that the decision level was set above the detection limit." Both situations are irrelevant, as Currie illustrated long ago [5, Fig. 4].

An expanded view of the lower-left corner of Fig. 6.2 is shown in Fig. 6.3. From Fig. 6.3, it is clear that the distributions depicted in Fig. 6.1a are almost perfectly separated because their ROC curve almost touches the origin. Hence, both p and q are readily made extremely small by appropriate choice of R_C or Y_C. The rest of the distribution pairs in Fig. 6.1 are not as well separated. However, the ROC curves for Fig. 6.1b and c pass well *below* the canonical limit of detection, so the analyte in those cases is *above* the limit of detection. In the remaining six cases in Fig. 6.1, the analyte is *below* the detection limit.

6.4   ROCs FOR FIGS 5.3 AND 5.4

At this point, it is beneficial to return to the experimental molecular fluorescence net response data presented in Figs 5.3 and 5.4 and use it to construct ROC plots. The results are shown in Fig. 6.4, with an expanded view in the inset at the upper right of the figure.

From Fig. 6.4, it is apparent that standard 6 is above the detection limit and standard 5 is below it. If another standard were prepared with analyte content between those of standards 5 and 6, its net response CDF, together with that of the net blank, would generate an ROC curve between those shown in Fig. 6.4. If the analyte was

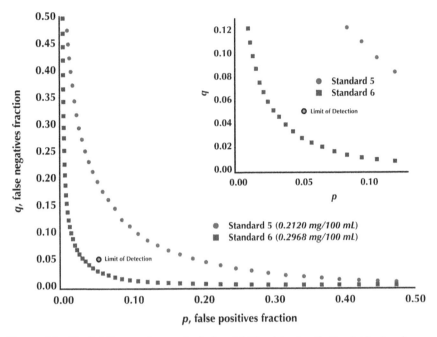

Figure 6.4   The ROC curves for standards 5 and 6. The detection limit point is also shown.

actually at the detection limit, where p and q are each 0.05, its ROC curve would pass through that specific point on the ROC plot. Note that the experimental ROC curves are relatively "smooth," since they are computed from integrated histograms of net responses, that is, experimental CDFs. The integration operation intrinsically performs smoothing and this is readily seen in the CDFs in Figs 5.3 and 5.4.

6.5   A FEW EXPERIMENTAL ROC RESULTS

With the ROC methodology, it is unnecessary to assume "nice" behavior for the dominant noise source, for example, a Gaussian distribution, and the response function may be nonlinear, so long as it is a monotonic function of the measurand. Furthermore, the noise may even be heteroscedastic and there is no need to posit a model. Consider then the block diagram of a ruby fluorescence instrument [1], constructed specifically in order to demonstrate that obtaining detection limits does not necessarily require great effort. It is shown in Fig. 6.5.

This fluorimeter has a very unusual twist: instead of having constant UV LED light intensity and variable measurand, roles are reversed: a ruby rod provides a constant fluorophore content while the UV LED's drive current is the measurand. In this way, the intensity of the ruby rod's 694.3 nm fluorescence is a monotonic, but nonlinear, function of the measurand. Furthermore, the fluorimeter was deliberately designed to

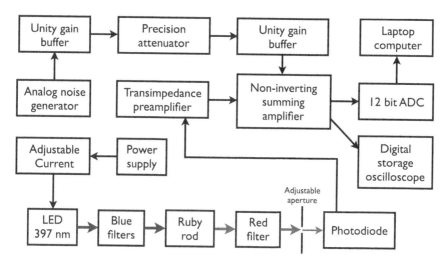

Figure 6.5 The block diagram of the ruby fluorescence instrument. Wysoczanski, 2014 [1]. Reproduced with permission of Elsevier.

have non-Gaussian noise PDF that could be either homoscedastic or heteroscedastic. *Importantly, none of the functional dependences were either known or modeled.* The noise's PDF histogram is shown in Fig. 6.6, while the nonlinear response, with ±1 standard deviation error bars, is shown in Fig. 6.7. For the heteroscedastic results, see Ref. [1].

Fluorescence data were collected (see Ref. [1] for all experimental details) and ROC curves were then computed using eqns 6.1 and 6.2. The full set of ROC curves, for the experiment performed with homoscedastic noise, is shown in Fig. 6.8, while Fig. 6.9 shows an expanded view of the results in Fig. 6.8.

From Fig. 6.9, it is clear that one ROC curve passes through the canonical detection limit point. It was computed from the data for measurand standard 26 [1, Table 2], which was 7.98 mA. As shown in Fig. 6.10, the histogram of responses for measurand at the detection limit (rightmost) has 5% of its area below the r_C decision level (vertical arrow) at 0.3337 V. Similarly, the blank's histogram (leftmost) has 5% of its area above the decision level.

A further set of experiments was then performed: by trial and error, measurand values were changed in increments of 0.01 mA and resulting histograms were obtained. It was found that a measurand value of 3.39 mA resulted in a histogram (center) having equal areas above and below r_C. Hence, in the measurand domain, the decision level was 3.39 mA and the detection limit was 7.98 mA. Finally, note that the detection limit is more than twice the decision level because the noise PDF tails to lower values. Had the noise tailed to higher values, such as might occur if the noise were log-normally distributed or the like, the detection limit would have been less than twice the decision level.

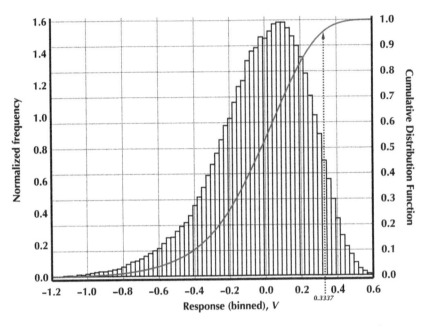

Figure 6.6 The histogram for the homoscedastic noise in the ruby fluorescence instrument. Note the decision level (vertical arrow) at 0.3337 V. The upper tail is 5% of the total area. Qualitatively compare with reverse of Fig. 2.8. Wysoczanski, 2014 [1]. Reproduced with permission of Elsevier.

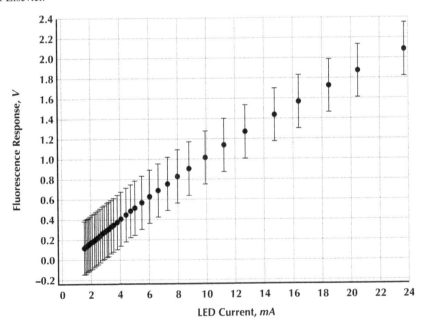

Figure 6.7 The nonlinear response, with homoscedastic noise, in the ruby fluorescence instrument. Qualitatively compare with Fig. 2.4. Wysoczanski, 2014 [1]. Reproduced with permission of Elsevier.

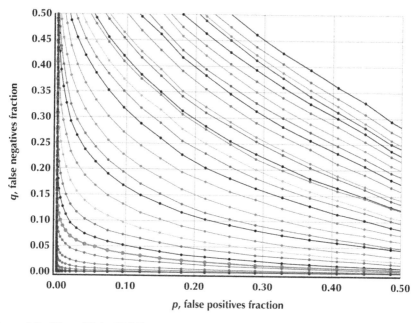

Figure 6.8 The published ROC curves pertinent to Figs 6.6 and 6.7. Wysoczanski, 2014 [1]. Reproduced with permission of Elsevier.

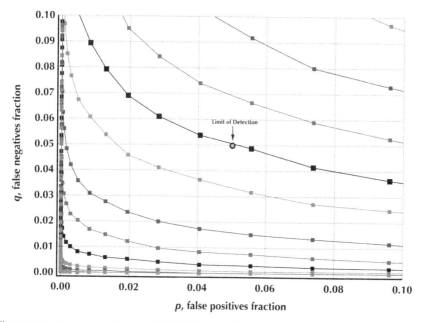

Figure 6.9 An expanded view of the ROC curves in Fig. 6.8. Wysoczanski, 2014 [1]. Reproduced with permission of Elsevier.

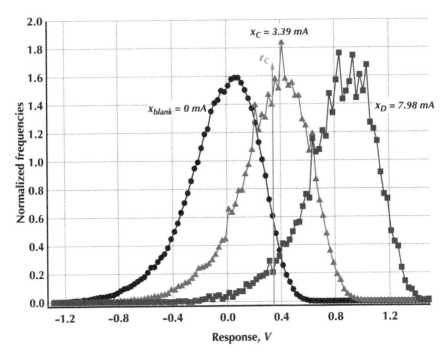

Figure 6.10 Experimental corroboration of the fact that, when measurand is present at the x_C decision level, half of its repeated responses must fall below the r_C decision level at 0.3337 V. Wysoczanski, 2014 [1]. Reproduced with permission of Elsevier.

6.6   SINCE ROCs MAY WORK WELL, WHY BOTHER WITH ANYTHING ELSE?

There was nothing difficult about the above experiment or the interpretation of the ROC results. However, the advantages that ROCs may provide, when relatively little information is available about a CMS's response function or the PDF of its dominant noise, are not without cost; it is not necessarily obvious what course of action should be followed in order to improve detection. Obviously, lowering the noise results in a lower limit of detection, but what else would be beneficial? Without parameters to consider, it is not obvious what might potentially be optimized or what would be the most effective optimization strategy.

Consequently, ROCs are an empirical method well-suited to dealing with complicated CMSs that are difficult to reduce to simpler subsystems, that is, ones that are likely to be easier to understand on a quantitative basis. This particularly applies when the noise is decidedly non-Gaussian and the customary Gaussian statistical precepts are inapplicable, for example, critical z or t values cannot be used. In the following few chapters, we study the most important parametric model, by far, and demonstrate the many advantages afforded by a model that is both rigorously defined and properly tested via quantitative Monte Carlo computer simulations.

6.7 CHAPTER HIGHLIGHTS

This chapter provided a brief introduction to ROCs and how they may be used to process the molecular fluorescence data examined in the previous chapter. The method is also used to obtain Currie detection limits for an experimental system in which the response function and noise are largely unknown. Lastly, a brief discussion of the pros and cons of ROCs was presented.

REFERENCES

1. A. Wysoczanski, E. Voigtman, "Receiver operating characteristic – curve limits of detection", *Spectrochim. Acta, B* **100** (2014) 70–77.
2. C.G. Fraga, A.M. Melville, B.W. Wright, "ROC-curve approach for determining the detection limit of a field chemical sensor", *Analyst* **132** (2007) 230–236.
3. https://www.spl.harvard.edu/archive/spl-pre2007/pages/ppl/zou/roc.html (Accessed on 03 October 2016).
4. http://www.anaesthetist.com/mnm/stats/roc/Findex.htm (Accessed on 03 October 2016).
5. L.A. Currie, "Limits for qualitative and quantitative determination – application to radiochemistry", *Anal. Chem.* **40** (1968) 586–593.
6. D. Coleman, L. Vannata, "Part 27 – receiver operating characteristic (ROC) curves", *Am. Lab.* **39(16)** (2007) 28–29.

7

STATISTICS OF AN IDEAL MODEL CMS

7.1 INTRODUCTION

The results presented in Chapter 6 demonstrate that Currie's program may be implemented nonparametrically, insofar as adherence to *a priori* specification of both p
and q is concerned, and ROC curves, if available, make it relatively easy to ascertain
whether analyte content is above or below the detection limit. However, nothing in
the previous discussions directly addressed the matter of accurate estimation of the
true detection limit, in any domain. In order to pursue this issue, it is advantageous
to define a parametric model CMS and evaluate its statistical behavior. Then, to the
extent a real instrumental system is well approximated by the model CMS, it may be
possible to estimate accurately the detection limit for the real system.

7.2 THE IDEAL CMS

For the model system, assume as previously [1–7] a CMS characterized by the linear
response model

$$r(x, t) \equiv \alpha + \beta x + \text{noise}(t) \tag{7.1}$$

where α is the true intercept, $\beta > 0$ is the true slope, x is the independent variable
in the content domain, and $r(x, t)$ is the dependent variable in the response domain.
Typically, x will be a concentration (as herein), quantity or amount, in appropriate
units, and $x \geq 0$. Note that $r(x, t)$ is time dependent exclusively because of the additive
Gaussian white noise (AGWN) with zero mean. Additional assumptions are that there
are no systematic errors and the noise is assumed to be homoscedastic, with constant
population standard deviation, σ_0.

From eqn 7.1, the time-independent response, $r(x)$, is simply

$$r(x) \equiv \alpha + \beta x \tag{7.2}$$

Limits of Detection in Chemical Analysis, First Edition. Edward Voigtman.
© 2017 John Wiley & Sons, Inc. Published 2017 by John Wiley & Sons, Inc.
Companion Website: www.wiley.com/go/Voigtman/Limits_of_Detection_in_Chemical_Analysis

Table 7.1   Summary of Model Definition, Parameters, Notation, and Assumptions

Item	Definition	Comment
CMS type	Univariate	One measurand only
Response function (noiseless)	$r(x) \equiv \alpha + \beta x$	Errorless *only* if x is noiseless and *both* α and β are used
True intercept	α	If unknown, estimated by unbiased $\hat{\alpha}$
True slope	β	If unknown, estimated by unbiased $\hat{\beta}$. Without loss of generality, $\beta > 0$
Measurand ("analyte") content	x	Physically allowed range in a CMS: $x \geq 0$
Full response function	$r(x, t) = r(x) + \text{noise}(t)$	Time dependence due *solely* to the noise
Noise probability density function (PDF)	Gaussian (that is, "Normal")	Additive, with white power spectral density (PSD). See "AGWN" below
Additive Gaussian white noise	AGWN	Commonly used shorthand. A white PSD is frequency independent
Population standard deviation of the blank	σ_0	Constant > 0
Noise precision model (NPM)	$\sigma(x) \equiv \sigma_0$	Homoscedastic
Ordinary least squares (OLS) calibration equation	$\hat{r}(x) = a + bx$	$\hat{\alpha} \equiv a$ and $\hat{\beta} \equiv b$
Number of standards in OLS	N	Multiple standards may have the same value: 1 "replicate" per standard
OLS intercept	a	An unbiased estimate of α
OLS slope	b	An unbiased estimate of β
Predicted OLS response	$\hat{r}(x)$	Not the same as $r(x) = \alpha + \beta x$
Sample standard deviation of the blank	s_0	Negatively biased point estimate of σ_0 as per Appendix A
Sample standard error about regression	s_r	The "*rms*" error in OLS. Sometimes awkwardly denoted by $s_{y/x}$
Number of *future* blank replicates	M	$M = 1$ is canonical and best choice for elementary measurements
Number of blank replicates	M_0	If blank replicates are used
Sample mean of M *future i.i.d.* blank replicates	$\bar{r}_{\text{future blanks}}$	For $M = 1$, $r_{\text{future blank}}$ or $r_{\text{one future blank}}$ is used. Similarly in the other domains
Sample mean of M_0 *i.i.d.* blank replicates	$\bar{r}_0$	A commonly used estimate of α

Table 7.1 *(Continued)*

Item	Definition	Comment
Specific *constant* value of x	$X_{\text{subscript}}$	Upper-case "X," with a subscript or label
Analytical blank's content	X_0	$x \equiv X_0 \equiv 0 \equiv x_0$
An x variable or variate	$x_{\text{subscript}}$	Lower-case "x," with a subscript or label
Degrees of freedom	ν	Depends on relevant details
Specific *constant* value of $r(x)$	$R(X_{\text{subscript}})$ or $R_{\text{subscript}}$	Upper-case "R"
Normal (that is, Gaussian) distribution shorthand notation	$N{:}\mu, \sigma$	Population mean $= \mu$ and population standard deviation $= \sigma$
If u is distributed as $N{:}\mu, \sigma$	$u \sim N{:}\mu, \sigma$	Commonly used shorthand
If u is a random sample of $N{:}\mu, \sigma$	$u \in N{:}\mu, \sigma$	Commonly used shorthand
Blank subtraction factor (that is, noise pooling factor)	$\eta^{1/2}$	See Appendix C for full details. If $M = 1$, then $\eta^{1/2} \geq 1$
Dummy variable, u	u	Useful if x would cause confusion

Thus, any specific measurement, r_i, with $x = x_i$, is distributed as $N{:}\alpha + \beta x_i, \sigma_0$. The average of N such *i.i.d.* measurements, denoted by $\bar{r}_i$, is distributed as $N{:}\alpha + \beta x_i, \sigma_0/N^{1/2}$. For the analytical blank, $x \equiv X_0 \equiv 0 \equiv x_0$, so $x_i = X_0$. Therefore, the sample mean of M_0 *i.i.d.* blank replicates, $\bar{r}_0$, is distributed as $N{:}\alpha, \sigma_0/M_0^{1/2}$. Similarly, the sample mean of M *future i.i.d.* blank replicates, $\bar{r}_{\text{future blanks}}$, is distributed as $N{:}\alpha, \sigma_0/M^{1/2}$.

Table 7.1 summarizes the model definition, parameters, notation, and assumptions, while Fig. 7.1 graphically shows $r(x)$ and the Gaussian noise PDF of the analytical blank. As is obvious from the above discussion, there are three controlling population parameters, listed in reverse order of importance: α, β, and σ_0. These are true values since systematic error is assumed to be zero. The three parameters may be either known or unknown, leading to 2^3 possibilities, each of which will be considered in turn. This chapter is devoted to the "most theoretical" case, that is, it is assumed that *all three parameters are known* and, therefore, no estimates of them are needed. The other seven possibilities are covered in the following seven chapters.

Figure 7.2 provides a more detailed view, than that in Fig. 7.1, of the Gaussian noise PDF of the analytical blank. As shown in Fig. 7.2, the constant value $\alpha + z_p\sigma_0$ divides the unit area of the PDF into an upper tail of area fraction p, and a remaining large lower area fraction of $1 - p$. With $p \equiv 0.05$, $z_p = z_{0.05} \cong 1.644854 \simeq 1.645$. Table A.1 provides other p, z_p pairs, and *Microsoft Excel*® provides an easy way to obtain a z_p value for any p: $z_p = -\text{NORMSINV}(p)$. In *Mathematica*®, the syntax is $z_p = \text{InverseCDF}[\text{NormalDistribution}[], 1 - p]$.

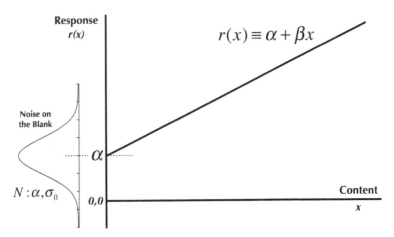

Figure 7.1 The time-independent response function, $r(x)$, and the Gaussian noise on the blank.

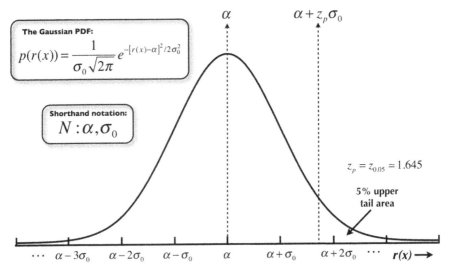

Figure 7.2 The Gaussian noise PDF of the analytical blank, showing the critical value z_p.

7.3 CURRIE DECISION LEVELS IN ALL THREE DOMAINS

The theoretical response domain Currie decision level, R_C, is defined as

$$R_C \equiv \alpha + z_p\sigma_0 \qquad (7.3)$$

Subtracting α from R_C results in the theoretical net response domain Currie decision level, Y_C:

$$Y_C \equiv R_C - \alpha = z_p\sigma_0 \qquad (7.4)$$

Then dividing Y_C by β results in the theoretical content domain Currie decision level, X_C:

$$X_C \equiv \frac{Y_C}{\beta} = \frac{z_p \sigma_0}{\beta} \tag{7.5}$$

Each of these decision levels is an errorless constant, with $Y_C > 0$ and $X_C > 0$.

A single future blank in the response domain, $r_{\text{future blank}}$, is distributed as $N{:}\alpha, \sigma_0$. Therefore, it has probability p of exceeding R_C and thereby being miscategorized as a false positive. In the net response domain, a single future net blank, that is, $r_{\text{future blank}} - \alpha$, is distributed as $N{:}0, \sigma_0$. This likewise results in probability p of it exceeding Y_C and thereby being a false positive.

The back-calculated "content noise" PDF for x_i is $N{:}x_i, \sigma_0/\beta$. This fictitious PDF is simply an "input referenced" noise distribution: it is the hypothetical PDF that x would need to have in order to produce the actual distribution of net responses, from an otherwise noiseless system, when $x = x_i$. In other words, since $y(x_i) = \beta x_i \sim N{:}\beta x_i, \sigma_0$, then $x_i \sim N{:}x_i, \sigma_0/\beta$. Hence, $x_{\text{future blank}} \sim N{:}0, \sigma_0/\beta$, since $x_i = X_0 = 0$ for a future blank.

7.4   CURRIE DETECTION LIMITS IN ALL THREE DOMAINS

Specification of limits of detection in the three domains requires specification of respective decision levels, as above, plus specification of an *a priori* maximum allowed probability of false negatives, q. As for p, the *de facto* canonical choice of q is 0.05. The general situation in the response domain, with $p \equiv 0.05 \equiv q$, is shown in Fig. 7.3.

From Fig. 7.3, it is readily seen that the theoretical response domain Currie detection limit, R_D, is simply

$$R_D \equiv \alpha + (z_p + z_q)\sigma_0 = R_C + z_q\sigma_0 \tag{7.6}$$

Then the theoretical net response domain Currie detection limit, Y_D, is

$$Y_D \equiv R_D - \alpha = (z_p + z_q)\sigma_0 = \frac{z_p + z_q}{z_p} Y_C \tag{7.7}$$

and the theoretical content domain Currie detection limit, X_D, is

$$X_D \equiv Y_D/\beta = (z_p + z_q)\sigma_0/\beta = \frac{z_p + z_q}{z_p} X_C \tag{7.8}$$

Each of these detection limits is an errorless constant, with $Y_D > 0$ and $X_D > 0$.

The decision level and detection limit equations obviously accommodate whatever p and q values may be desired. The only sensible reason to have p and q differ would be if false positives and false negatives were considered to be unequally deleterious.

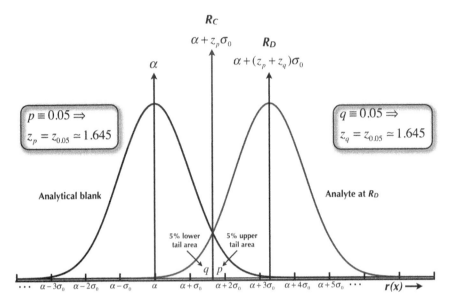

Figure 7.3 The Gaussian PDFs of the analytical blank and of the analyte, at the limit of detection.

In this case, the relevant p or q value could be defined lower to reduce the probability of the most deleterious errors. Since there is no *a priori* reason to have $p \neq q$, that is, the two error types have *a priori* "equal costs," the default will be $p \equiv 0.05 \equiv q$, unless specifically noted otherwise.

7.5   GRAPHICAL ILLUSTRATIONS OF eqns 7.3–7.8

Figures 7.4–7.6 graphically illustrate the Currie decision level and detection limit situations in all three domains, with the default p and q values. In each case, $p = q$ ensures that the Gaussian PDFs are symmetrically distributed about their respective decision levels. Note in Fig. 7.6 that the probability of a random blank exceeding X_D is only about 0.05%. By symmetry, a random sample from the analyte PDF centered on X_D would have the same negligibly small probability of having a negative value. Obviously, any negative content domain value is below X_C, resulting in a "not detected" decision.

In general, content domain distributions are either fictitious PDFs, as depicted in Fig. 7.6, or histograms of compound measurements. Either way, they are back-calculated from the response domain, via the net response domain, and they do not violate the physical impossibility of negative analyte content. On the contrary, negative values from content domain distributions are *not* measured values, because *measurements cannot be made in the content domain*, and they are *not* in need of any "correction" [8]. Simply stated, negative values, from the content domain

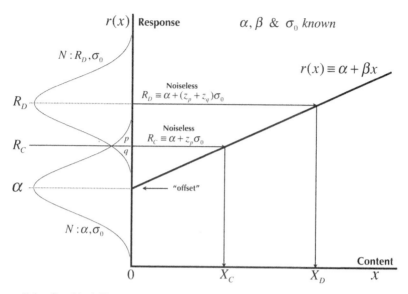

Figure 7.4 Graphical illustration of the Currie decision level and detection limit in the response domain.

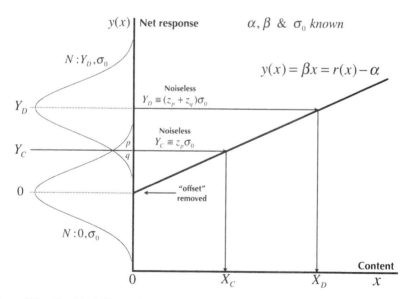

Figure 7.5 Graphical illustration of the Currie decision level and detection limit in the net response domain.

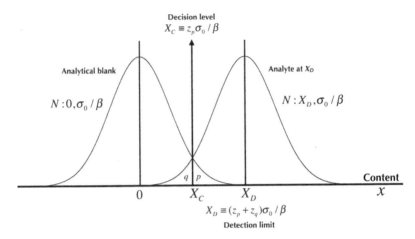

Figure 7.6 Graphical illustration of the Currie decision level and detection limit in the content domain. The Gaussian PDFs are fictitious, that is, back-calculated from the net response domain in Fig. 7.5.

distributions other than the blank, are the most extreme false negatives. For the blank's content domain distribution, negative values are the most extreme true negatives. The following example uses a small data set to illustrate that negative values are *not* problematic in *any* domain.

7.6 AN EXAMPLE: ARE NEGATIVE CONTENT DOMAIN VALUES LEGITIMATE?

Consider the following seven *i.i.d.* blank replicate measurements: 0.92, 1.49, 0.79, 0.58, 0.26, 1.63, and 1.44 nA. They are samples from N: 1, 0.6, rounded to two decimal places, with $\bar{r}_0 \cong 1.016$ nA and $s_0 \cong 0.517$ nA. Since $\alpha = 1$ nA, the net blank replicates, are -0.08, 0.49, -0.21, -0.42, -0.74, 0.63, and 0.44 nA, with $\bar{y}_0 \cong 0.016$ nA and s_0 unchanged. If β was exactly equal to 0.2 nA/μM, then the rounded back-calculated content "replicates," are -0.4, 2.5, -1.1, -2.1, -3.7, 3.2, and 2.2 μM, with $\bar{x}_0 \cong 0.09\,\mu$M and $s_0/\beta \cong 2.61\,\mu$M. Thus, in both the net response and content domains, four of seven values are negative. With $z_p = z_{0.05} \simeq 1.645$, eqns 7.3–7.5 yield $R_C \cong 1.987$ nA, $Y_C \cong 0.987$ nA, and $X_C \cong 4.935\,\mu$M, respectively. Therefore, in each domain, the seven values are legitimate and result in correct "not detected" decisions.

The data in the example do not allow quantitation of the zero analyte content of the blank because *no analyte was detected*. More generally, it is shown in Chapter 24 that if analyte content is in the vicinity of X_D, or lower, back-calculated $\hat{X}$ values are *unusable* for quantitation purposes. For example, if $x = X_D$, the central 95% CI for $\hat{X}_D$ is $X_D \pm z_{0.025}\sigma_0/\beta$. Following Coleman *et al.* [9], the relative measurement error (RME) is defined as the half-width of the central CI, divided by its center value. This is shown in Fig. 7.7, for $x \equiv X_D$.

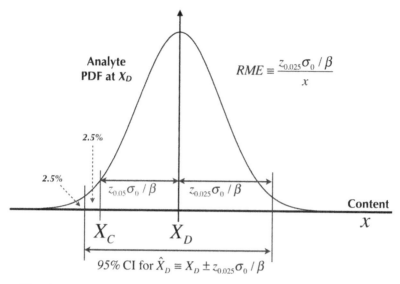

Figure 7.7   Illustration of the relative measurement error (RME) at the limit of detection, X_D. The Gaussian PDF is fictitious, as in Fig. 7.6.

Hence, with $z_{0.025} \cong 1.959964 \approx 1.960$ in eqn 7.8,

$$\text{RME at } X_D \equiv \frac{z_{0.025}\sigma_0/\beta}{X_D} \equiv \frac{z_{0.025}\sigma_0/\beta}{(z_p + z_q)\sigma_0/\beta} = \frac{z_{0.025}}{z_p + z_q} \cong 0.596 \quad \text{if } p \equiv 0.05 \equiv q$$

(7.9)

The 59.6% RME at $x = X_D$ is so large that $\hat{X}_D$ has zero significant figures of measurement precision, with greater than 95% confidence [9]. Coleman *et al.* [9] state as follows:

> Zero significant digits may initially seem pathetic or technically dubious, but it actually is the only logical definition of detection: a concentration so low that there is no information content in the reported measurement, other than a confident determination of the presence or absence of the analyte. 'Confidence' is defined here as acceptably low average risks of false detection and false non-detection.

Also in Chapter 24, it is shown that the semiquantitation regime is the range of x values for which $0.05 \leq \text{RME} < 0.5$, with 95% confidence. Consequently, $x = X_D$ is below the semiquantitation regime and is unusable for quantitation purposes.

Related to the fact that individual $\hat{X}$ values should not be censored or altered, the same applies to content domain distributions: they should *not* be "corrected" by

- discarding negative values;
- replacing negative values with zeroes;

- rectifying negative distribution regions by taking absolute values; or by
- *manipulating the distribution in any other such fashion.*

The reason is simple: doing so would degrade the information implicitly contained in the distribution, for example, the mean and standard deviation, and introduce bias. Unfortunately, the analyst does not always have control; some analytical instruments are designed to truncate the blank's content domain distribution at zero [10, 11], in a misguided attempt to be helpful.

7.7   TABULAR SUMMARY OF THE EQUATIONS

A convenient summary of the Currie decision level and detection limit equations is provided in Table 7.2. Testing the equations in Table 7.2 is of paramount importance. Realistically, there is no possibility that they could be incorrect: they are founded on long established, incontrovertible statistics. Nevertheless, all decision level and detection limit equations, in this and later chapters, have been extensively tested via Monte Carlo computer simulation.

In view of the enormous power and sophistication of modern computer hardware and software, it is strikingly odd that the use of quantitative computer simulations, to elucidate the statistical behavior of detection limits, has been so rarely practiced. Indeed, the univariate linear CMS model is as simple as possible, and the noise is likewise as benign as possible, so whatever the reasons may have been for not performing computer simulations, they were unacceptable. Furthermore, theoretical results *cannot* be validly tested using complicated experimental systems and protocols that do not verifiably satisfy the fundamental assumptions upon which theory is predicated. This means that, as important as real experiments are, computer simulations are also *essential.*

Table 7.2   Currie Decision and Detection Expressions

Parameter	Known	Unknown	Estimator
σ_0	✓		–
β	✓		–
α	✓		–

	Decision Level	Detection Limit
Response domain	$R_C = \alpha + z_p \sigma_0$	$R_D = \alpha + (z_p + z_q)\sigma_0$
Net response domain	$Y_C = z_p \sigma_0$	$Y_D = (z_p + z_q)\sigma_0$
Content domain	$X_C = z_p \sigma_0 / \beta$	$X_D = (z_p + z_q)\sigma_0 / \beta$

1. Optimum theoretical case because all quantities are errorless real numbers > 0.
2. If $M > 1$, multiply each σ_0 by $\eta^{1/2}$, where $\eta^{1/2} = 1/M^{1/2}$.

7.8 MONTE CARLO COMPUTER SIMULATIONS

The *ExtendSim*® simulation program, augmented with the author's free *LightStone*® libraries of component blocks, has been used extensively since 1990, resulting in scientific publications in which simulations have played the key role; see the references in Appendix B, which also provides a very brief tutorial on the use of *LightStone* and *ExtendSim*. Much more information, plus all of the simulation models discussed in this text, are freely available; see Appendix B for the *LightStone* website's URL.

In view of the simplicity of the model CMS and assumptions, there is no difficulty in performing quantitative Monte Carlo simulations to test eqns 7.3–7.8. To test the decision levels, all that is necessary is to generate large numbers of *i.i.d.* blanks, in each domain, and then keep track of how many blanks exceed their respective decision levels. In a simulation having 1 million steps, 1 million blanks are generated in each domain. With $p \equiv 0.05$, the expected number of false positives, in each domain, is simply 50 000, that is, 0.05×10^6. Running 10 simulations then yields both the sample mean and sample standard deviation of the obtained numbers of false positives. Ideally, the mean for 10 simulations should be close to 50 000, with small standard deviation.

Testing for false negatives proceeds similarly: 10 simulations of 1 million steps each are run. These are, in fact, the exact same simulations involved in testing the false positives probability. For each simulation step, a sample is randomly taken from each of the three analyte distributions centered at R_D, Y_D, and X_D. These are compared with their respective decision levels and those that fall below their decision levels are false negatives. With $q \equiv 0.05$, the expected number of false negatives, in each domain, is again 50 000.

It is useful to examine Figs 7.4–7.6 again. With each simulation step, random samples are taken from each distribution in all three figures. This results in three tests for false positives, one per domain, and three tests for false negatives, one per domain. For a single simulation of a million steps, a total of 6 million decisions are made. Running 10 simulations takes several minutes, results in a total of 60 million decisions, and reveals whether the theory has survived the testing or has failed.

To facilitate understanding the simulation program, Fig. 7.8 shows a schematic diagram of where the simulation's relevant components ("subroutines") are located. It also applies in the following seven chapters.

As shown in Fig. 7.8, roughly the left third of the *LightStone* and *ExtendSim* program deals with parameter specifications, generation of blanks, and documentation labels. The remainder is where eqns 7.3–7.5 are implemented as decision levels and analyte replicates are prepared by suitable recentering of blanks. For example, a random blank variate, from $N{:}\alpha, \sigma_0$, becomes a random analyte variate, from $N{:}R_D, \sigma_0$, if the constant $(z_p + z_q)\sigma_0$, which equals $R_D - \alpha$, is simply added to the blank variate. This should be immediately apparent in Fig. 7.9.

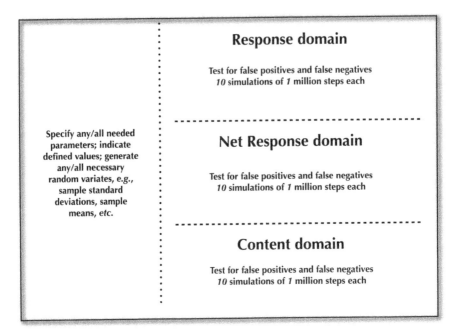

Figure 7.8   Schematic arrangement of the regions of the Monte Carlo simulation in Fig. 7.9.

7.9   SIMULATION CORROBORATION OF THE EQUATIONS IN TABLE 7.2

The results of running the 10 simulations are shown directly in Fig. 7.9, in the six data tables on the right side of the figure. For clarity, these are summarized in Table 7.3. Obviously, the numbers of false positives are exactly the same in each domain and similarly the numbers of false negatives are exactly the same in each domain. This is due to the fact that transformation from the response domain to the net response domain merely involves subtraction of α from R_C and both of the variates, that is, the blank variate and the analyte variate at R_D. This is just a shift in reference point, so the results do not change. Similarly, transformation from the net response domain to the content domain is done by dividing β into Y_C and both of the shifted variates. This is just a scale factor change, so again there is no relative effect.

Dividing by one million steps per simulation yields the following obtained probabilities: $5.005 \pm 0.033\%$ false positives and $4.998 \pm 0.028\%$ false negatives, in excellent agreement with the *a priori* specification of exactly 5% each. Thus, the theory has survived, but mere survival is only a necessary condition, not a sufficient one. The Monte Carlo simulation also proved its utility, and will continue to do so, as seen in subsequent chapters.

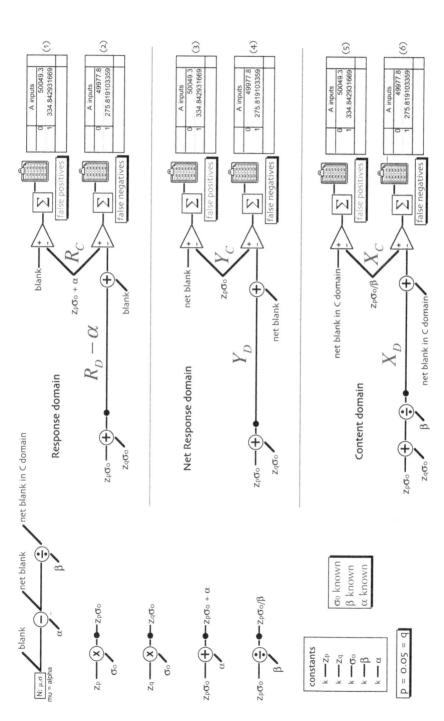

Figure 7.9 The Monte Carlo simulation, with $M=1$, that corroborates the equations in Table 7.2.

79

Table 7.3   Summary of Simulation Results for False Positives and False Negatives

Domain	False Positives ± 1 Standard Deviation	False Negatives ± 1 Standard Deviation
Response	$50\,049.3 \pm 334.8$	$49\,977.8 \pm 275.8$
Net response	$50\,049.3 \pm 334.8$	$49\,977.8 \pm 275.8$
Content	$50\,049.3 \pm 334.8$	$49\,977.8 \pm 275.8$

7.10   CENTRAL CONFIDENCE INTERVALS FOR PREDICTED x VALUES

Given that σ_0, β, and α are assumed to be known, it is easy to compute predicted x values and central CIs. Suppose $x = X_i$ is unknown and $\bar{r}_i$ is the mean of M i.i.d. measurements of X_i. Then $\bar{r}_i \sim N : \alpha + \beta X_i, \sigma_0 / M^{1/2}$ and $\bar{y}_i = (\bar{r}_i - \alpha) \sim N : \beta X_i, \sigma_0 / M^{1/2}$. The back-calculated estimate of X_i, denoted by $\bar{x}_i$ or $\hat{X}_i$, is given by $\hat{X}_i \equiv \bar{x}_i = \bar{y}_i / \beta$. Hence, $\hat{X}_i \sim N : X_i, \sigma_0 / M^{1/2} \beta$ and central CIs for X_i are readily calculated, for example, the 95% CI for X_i is

$$95\% \text{ CI for } X_i = \hat{X}_i \pm z_{0.025} \sigma_0 / M^{1/2} \beta \tag{7.10}$$

where $z_{0.025} \cong 1.959964 \simeq 1.960$. Inverting eqn 7.10, as per Appendix A, the 95% CI for $\hat{X}_i$ is

$$95\% \text{ CI for } \hat{X}_i = X_i \pm z_{0.025} \sigma_0 / M^{1/2} \beta \tag{7.11}$$

For the blank, $x \equiv X_0 \equiv 0 \equiv X_i$, so $\hat{X}_i \equiv \hat{X}_0$, $M \equiv M_0$ and the 95% CI for $\hat{X}_0$ is

$$95\% \text{ CI for } \hat{X}_0 = X_0 \pm z_{0.025} \sigma_0 / M_0^{1/2} \beta = \pm z_{0.025} \sigma_0 / M_0^{1/2} \beta \tag{7.12}$$

Applying eqn 7.12 to the negative content values example, the 95% CI for $\bar{x}_0$ is $\pm 2.22 \,\mu M$, so this particular 95% CI contains $\bar{x}_0 \cong 0.09 \,\mu M$. Similarly, applying eqn 7.10 to the same data, the 95% CI for the true analyte content of the blank, that is, 0, is $0.09 \pm 2.22 \,\mu M$.

There is no problem with central CIs having negative values: if $\bar{x}_0$ had been negative, the 95% CI would simply have been $-0.09 \pm 2.22 \,\mu M$. The *actual* problem with calculated central CIs is that of ensuring that they are negligibly biased. To illustrate, Fig. 7.7 shows that X_C is within the central 95% CI for $\hat{X}_D$ when $x = X_D$. Since $p = 0.05 = q$ and $M \equiv 1$, 5% of the Gaussian PDF's area is below X_C, in two equal portions: 2.5% below the 95% CI's lower limit and 2.5% between that lower limit and X_C. Therefore, at $x = X_D$, 95% of back-calculated $\hat{X}_D$ values would be usable for construction of CIs, as in eqn 7.10. However, the CIs would be biased high, by about 10.8%, due to rejection of the 5% of $\hat{X}_D$ values below X_C. As $x = X_i$ increases above X_D, bias declines because fewer $\hat{X}_i$ values are rejected. At $x = X_i = 2X_D$, only about 0.05% of $\hat{X}_i$ values are rejected, resulting in about 0.2% bias.

In view of the above, a prudent course of action might be to simply eschew CIs, when analyte content is less than about twice the detection limit, and focus on measurement uncertainty. In any event, when analyte content is below the semiquantitation regime, as elucidated in Chapter 24, CIs should be interpreted with caution.

7.11 CHAPTER HIGHLIGHTS

This chapter, and the following seven chapters, provides a detailed discussion of Currie's detection limit schema, with all of the relevant equations presented and tested via Monte Carlo simulations. Table 7.1 summarizes all of the necessary model assumptions and definitions, and a set of carefully drawn illustrations was provided. It was assumed that the three model parameters were known, but an important caution is in order; aside from computer simulations and purely theoretical considerations, it will *almost never* be true that σ_0, β, and α are known. Therefore, it should *not* be assumed that a parameter is known unless there is a provable basis for the assumption. Failure to heed this advice almost invariably leads to needless errors. If in doubt, assume the least and see Chapter 14.

REFERENCES

1. E. Voigtman, "Limits of detection and decision. Part 1", *Spectrochim. Acta, B* **63** (2008) 115–128.
2. E. Voigtman, "Limits of detection and decision. Part 2", *Spectrochim. Acta, B* **63** (2008) 129–141.
3. E. Voigtman, "Limits of detection and decision. Part 3", *Spectrochim. Acta, B* **63** (2008) 142–153.
4. E. Voigtman, "Limits of detection and decision. Part 4", *Spectrochim. Acta, B* **63** (2008) 154–165.
5. E. Voigtman, K.T. Abraham, "Statistical behavior of ten million experimental detection limits", *Spectrochim. Acta, B* **66** (2011) 105–113.
6. E. Voigtman, K.T. Abraham, "True detection limits in an experimental linearly heteroscedastic system. Part 1", *Spectrochim. Acta, B* **66** (2011) 822–827.
7. E. Voigtman, K.T. Abraham, "True detection limits in an experimental linearly heteroscedastic system. Part 2", *Spectrochim. Acta, B* **66** (2011) 828–833.
8. P.-Th. Wilrich, "Note on the correction of negative measured values if the measurand is nonnegative", *Accred. Qual. Assur.* **19** (2014) 81–85.
9. D. Coleman, J. Auses, N. Grams, "Regulation – From an industry perspective *or* Relationships between detection limits, quantitation limits, and significant digits", *Chemom. Intell. Lab. Syst.* **37** (1997) 71–80.
10. K. Linnet, M. Kondratovich, "Partly nonparametric approach for determining the limit of detection", *Clin. Chem.* **50** (2004) 732–740.
11. D.A. Armbruster, T. Pry, "Limit of blank, limit of detection and limit of quantitation", *Clin. Biochem. Rev.* **29** (2008) **Suppl (i)** S49–S52.

8

IF ONLY THE TRUE INTERCEPT IS UNKNOWN

8.1 INTRODUCTION

In Chapter 7, the CMS, and its associated definitions and constructs, was fully defined. There, the best case scenario was explored: α, β, and σ_0 had *known* values. Each of these parameters may be either known or unknown, so the remaining seven possibilities are examined, at the rate of one per chapter. All the assumptions made in Chapter 7 continue to hold in this, and the following six chapters, unless stated otherwise.

8.2 ASSUMPTIONS

Assume σ_0 and β are known, but α is unknown. Then $r(x) = \alpha + \beta x$ is estimated by

$$\hat{r}(x) = \hat{\alpha} + \beta x \tag{8.1}$$

where $\hat{\alpha}$ is the estimate of α. As detailed in Appendix C, there are two commonly used methods for obtaining unbiased, normally distributed $\hat{\alpha}$ estimates. Given that β is known, ordinary least squares (OLS) is not used. However, there is an obvious way to estimate α: pick a convenient constant x value, for example, X_1, average M_{X_1} replicate measurements of $r(X_1)$ to obtain $\bar{r}(X_1)$, and then subtract βX_1. Thus, $\hat{\alpha} = \bar{r}(X_1) - \beta X_1$ and the distribution of $\hat{\alpha}$ is $N : \alpha, \sigma_0/M_{X_1}^{1/2}$. The blank is certainly an allowable choice, that is, $x \equiv X_0 \equiv 0$, in which case $\hat{\alpha} = \bar{r}(X_0) \equiv \bar{r}_0$ and $\hat{\alpha} \sim N : \alpha, \sigma_0/M_0^{1/2}$. More generally, $\hat{\alpha} \sim N : \alpha, \sigma_a$, where σ_a is the population standard error of the blank.

8.3 NOISE EFFECT OF ESTIMATING THE TRUE INTERCEPT

In Chapter 7, α was assumed to be known and the Currie decision level, R_C, was higher than α by a fixed amount, that is, $R_C = \alpha + z_p \sigma_0$. Here, α is assumed unknown and must be estimated by $\hat{\alpha}$. This immediately leads to a referencing issue, as shown in Fig. 8.1.

Limits of Detection in Chemical Analysis, First Edition. Edward Voigtman.
© 2017 John Wiley & Sons, Inc. Published 2017 by John Wiley & Sons, Inc.
Companion Website: www.wiley.com/go/Voigtman/Limits_of_Detection_in_Chemical_Analysis

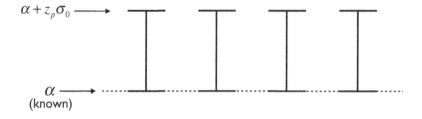

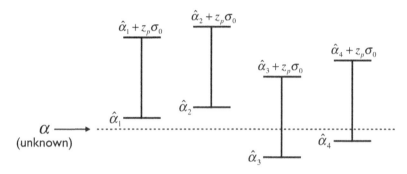

Figure 8.1   Noise effect when α is estimated by $\hat{\alpha}$.

The upper half of Fig. 8.1 shows the constant level that is R_C, when α is known. The lower half shows that, when α is estimated by $\hat{\alpha}$, the tops of the fixed length intervals are not at a constant level; they fluctuate depending upon whatever particular value of $\hat{\alpha}$ is used. In the infinite limit, the $\hat{\alpha}$ values average to α, as expected from the fact that $\hat{\alpha}$ is unbiased.

However, an analytical blank is always a random sample from $N{:}\alpha, \sigma_0$, regardless of whether α is known or not. This fact precludes the possibility of defining the Currie decision level as $r_C \equiv \hat{\alpha} + z_p\sigma_0$, because the independent Gaussian noises on the analytical blank, and $\hat{\alpha}$, add. As detailed in Appendix C, the result is a new Gaussian distribution having variance equal to $\eta\sigma_0^2$, that is, σ_d^2. Equivalently, the noise may be thought of as arising from a differencing, that is, the referencing to $\hat{\alpha}$ rather than to α. Either way, $\sigma_d = \eta^{1/2}\sigma_0$. This somewhat subtle issue is discussed in more detail further, but we first examine three simple simulations to see how the $\eta^{1/2}$ factor is used.

8.4   A SIMPLE SIMULATION IN THE RESPONSE AND NET RESPONSE DOMAINS

Figure 8.2 shows a simulation that allows for investigation of probabilities of false positives in both the response and net response domains. It is assumed that σ_0 is known, but α, aside from its obviously required use in generating properly distributed analytical blanks in the simulations, is assumed unknown. In Fig. 8.2, $\hat{\alpha}$ is the sample

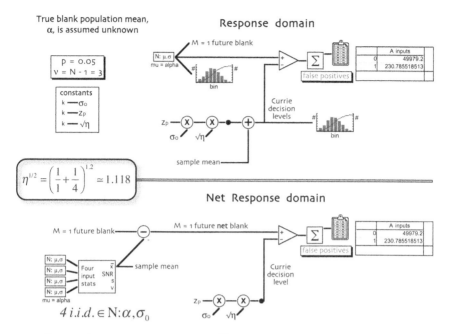

Figure 8.2 Example showing how $\eta^{1/2}$ is used. Note that $M = 1$ future blank.

mean of $M_0 \equiv 4$ independent and identically distributed (*i.i.d.*) analytical blanks. The value of $\eta^{1/2}$ is shown in the figure, calculated by eqn C.14 in Appendix C.

In the upper half of the figure, successive Gaussian distributed future blanks are compared against successive Gaussian distributed Currie decision levels, r_C, defined by

$$r_C \equiv \hat{\alpha} + z_p \eta^{1/2} \sigma_0 \qquad (8.2)$$

where $z_p \simeq 1.644854$. Ten simulations, of one million steps each, were performed, and the obtained probability of false positives was $4.998 \pm 0.023\%$. This is close to perfect. Exactly the same results were obtained in the lower half of the figure, in the net response domain.

Several observations are worth noting. First, in the net response domain, the Currie decision level, R_C, is an errorless constant, that is, $z_p \eta^{1/2} \sigma_0$. Thus, in the net response domain, successive values of the Gaussian distributed net response variates are compared with R_C. This small simplification is one of the reasons why the net response domain is usually preferred over the response domain. Note also that the net responses of the blank are distributed as $N{:}0, \sigma_d$, where $\sigma_d = \eta^{1/2} \sigma_0$, as above.

Second, it is the essential factor of $\eta^{1/2}$ that makes eqn 8.2 work. In Fig. 8.3, the $\eta^{1/2}$ factor was simply removed, which is equivalent to setting it to unity, that is, entirely neglecting the noise on $\hat{\alpha}$. Thus, this model tests the use of $r_C \equiv \hat{\alpha} + z_p \sigma_0$, after having ruled it out.

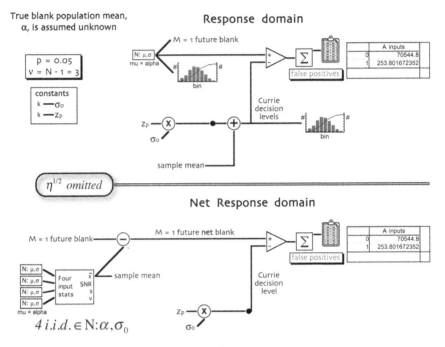

Figure 8.3   Example showing that omitting $\eta^{1/2}$ is incorrect. This is a very common mistake.

Clearly, neglecting the noise on $\hat{\alpha}$ leads to a 40% positive bias in the obtained probabilities of false positives: $7.054 \pm 0.025\%$. A further demonstration of the need to account for the noise on $\hat{\alpha}$ is provided in Fig. 8.4.

The simulation in Fig. 8.4 differs from that in Fig. 8.2 in only one regard: the sample mean is computed from four *i.i.d.* samples from a Gaussian distribution centered on α, but having 10 times the population standard deviation. Thus, $\hat{\alpha} \equiv \bar{r}_0 \sim N:\alpha, 10\sigma_0$. The sample mean is still unbiased, but its variance is increased by a factor of 100, thereby causing $\eta^{1/2}$, shown in Fig. 8.4, to become much larger than would be typical in properly designed real experiments. However, the obtained probability of false positives was $5.005 \pm 0.018\%$. Thus, $\eta^{1/2}$ works as expected.

8.5   RESPONSE DOMAIN EFFECTS OF REPLACING THE TRUE INTERCEPT BY AN ESTIMATE

In Chapter 7, Fig. 7.4 provided a graphical illustration of the Currie decision level and detection limit. If every instantiation of α in that figure was replaced with $\hat{\alpha}$, the result would be as shown in Fig. 8.5. Superficially, this figure makes sense: the errorless quantities R_C, R_D, X_C, and X_D have simply become the variates r_C, r_D, x_C, and x_D, respectively, and the distribution of the blank, $N:\alpha, \sigma_0$, became $N:\hat{\alpha}, \sigma_0$.

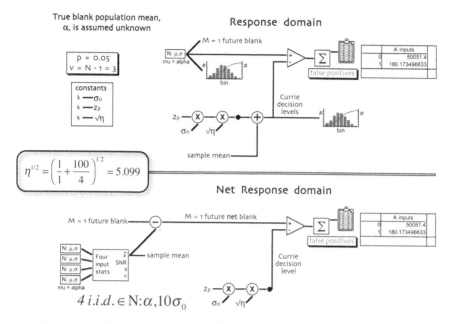

Figure 8.4 Example showing that $\eta^{1/2}$ even works under extreme conditions.

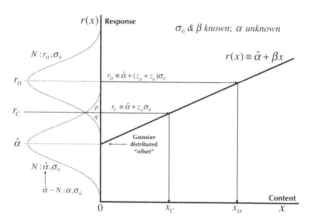

Figure 8.5 Very misleading graphical illustration of the Currie decision level and detection limit in the response domain.

The r_C and r_D equations in Fig. 8.5 are internally consistent with $N{:}\,\hat{\alpha}, \sigma_0$, that is, obtained probabilities of false positives and false negatives are as expected *a priori*.

However, Fig. 8.5 is very misleading: $N{:}\hat{\alpha}, \sigma_0$ is not the same as $N{:}\alpha, \sigma_0$. Rather, $N{:}\hat{\alpha}, \sigma_0$ is a *compound normal distribution*; it is a Gaussian distribution having, as its location "parameter," an independent Gaussian distribution. As demonstrated in the second example in Appendix B, $N{:}\hat{\alpha}, \sigma_0$ is simply a Gaussian distribution that is centered on α but is *wider* than that of the blank. In fact, $N{:}\hat{\alpha}, \sigma_0 = N{:}\alpha, \sigma_d = N{:}\alpha, \eta^{1/2}\sigma_0$. The reason that $N{:}\hat{\alpha}, \sigma_0$ works properly with the incorrect r_C and r_D equations in Fig. 8.5 is that r_C and r_D are *fixed* distances above *any* value of $\hat{\alpha}$ might be used. This is also as shown, for r_C, in the bottom half of Fig. 8.1 and the situation is precisely similar for r_D. Consequently, the graphical illustration in Fig. 8.5 is incorrect.

8.6   RESPONSE DOMAIN CURRIE DECISION LEVEL AND DETECTION LIMIT

The simple simulations shown in Figs 8.2 and 8.4 have already demonstrated that the experimental response domain Currie decision level, r_C, is defined as

$$r_C \equiv \hat{\alpha} + z_p \eta^{1/2}\sigma_0 \tag{8.3}$$

Then the corresponding experimental response domain Currie detection limit, r_D, is defined as

$$r_D \equiv \hat{\alpha} + (\eta^{1/2}z_p + z_q/M^{1/2})\sigma_0 = r_C + z_q\sigma_0/M^{1/2} \tag{8.4}$$

Note that the $\eta^{1/2}$ factor *only* multiplies z_p, *not* the sum of z_p and z_q. This is due to the fact that r_D is referenced to r_C, *not* α, and their difference, via eqn 8.4, is constant. Accordingly, r_D is the location "parameter" of a compound normal distribution having σ_0, *not* $\eta^{1/2}\sigma_0$, as its population standard deviation. Explicitly, the compound normal distribution is $N{:}r_D, \sigma_0$. Since $\hat{\alpha} \sim N{:}\alpha, \sigma_a$, $r_C \sim N{:}\alpha + z_p\eta^{1/2}\sigma_0, \sigma_a$, and $r_D \sim N{:}\alpha + (z_p\eta^{1/2} + z_q M^{-1/2})\sigma_0, \sigma_a$.

Beginning in Chapter 20, M will be greater than one because averaging is necessarily involved. Until then, however, the canonical and best choice for M is unity because only a *single* future measurement, either blank or analyte, would be compared with R_C or r_C.

8.7   NET RESPONSE DOMAIN CURRIE DECISION LEVEL AND DETECTION LIMIT

It was noted in the earlier section that the net response domain is usually preferred over the response domain. The basic reason for this is that all net responses are simply responses minus $\hat{\alpha}$, that is,

$$y(x) \equiv r(x) - \hat{\alpha} \tag{8.5}$$

Therefore, since the noise is homoscedastic, all net responses are distributed as $N{:}\beta x, \eta^{1/2}\sigma_0$, with net blanks, as a special case, distributed as $N{:}0, \eta^{1/2}\sigma_0$. Hence, the theoretical net response domain Currie decision level, Y_C, is defined as

$$Y_C \equiv z_p \eta^{1/2} \sigma_0 \qquad (8.6)$$

and the theoretical net response domain Currie detection limit, Y_D, is defined as

$$Y_D \equiv (z_p + z_q) \eta^{1/2} \sigma_0 = \frac{z_p + z_q}{z_p} Y_C \qquad (8.7)$$

Note that $Y_D \neq r_D - \hat{\alpha}$. Rather, Y_D is the location parameter of a Gaussian distribution having $\eta^{1/2} \sigma_0$ as its population standard deviation. Explicitly, the Gaussian distribution is $N:Y_D, \eta^{1/2} \sigma_0$.

Both Y_C and Y_D are errorless and their equations are nicely symmetric. With homoscedastic Gaussian noise, as assumed here, $Y_D = 2Y_C$, if $p = q$. This is exact, for all degrees of freedom, and is very important in practice, given the canonical specification of $p \equiv 0.05 \equiv q$.

8.8   CONTENT DOMAIN CURRIE DECISION LEVEL AND DETECTION LIMIT

The content domain expressions follow immediately from those in the net response domain. Thus, the theoretical content domain Currie decision level, X_C, is defined as

$$X_C \equiv Y_C / \beta = z_p \eta^{1/2} \sigma_0 / \beta \qquad (8.8)$$

and the theoretical content domain Currie detection limit, X_D, is defined as

$$X_D \equiv \frac{Y_D}{\beta} = \frac{(z_p + z_q) \eta^{1/2} \sigma_0}{\beta} = \frac{z_p + z_q}{z_p} X_C \qquad (8.9)$$

As for Y_C and Y_D, both X_C and X_D are errorless, their equations are nicely symmetric, and, with homoscedastic Gaussian noise, $X_D = 2X_C$, if $p = q$. This is exact, for all degrees of freedom.

8.9   GRAPHICAL ILLUSTRATIONS OF THE DECISION LEVEL AND DETECTION LIMIT EQUATIONS

Unfortunately, there is no way to make a sensible *single* graphical illustration of the response domain equations, that is, eqns 8.3 and 8.4, together with the analytical blank's distribution centered on the unknown α. In contrast, it is simple to illustrate the Currie decision level and detection limit situations in the net response and content domains. These are shown in Figs 8.6 and 8.7, respectively.

Both figures are properly scaled to illustrate the canonical $p \equiv 0.05 \equiv q$. From Fig. 8.6, it is easily seen that $\beta \equiv Y_C / X_C = Y_D / X_D$, as expected.

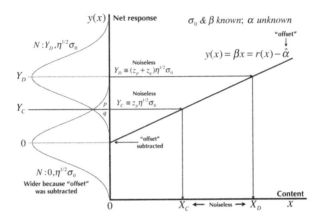

Figure 8.6 Graphical illustration of the Currie decision level and detection limit in the net response domain.

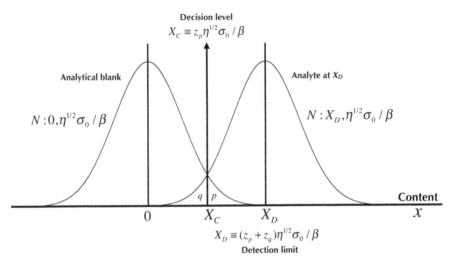

Figure 8.7 Graphical illustration of the Currie decision level and detection limit in the content domain. The Gaussian PDFs are fictitious, that is, back-calculated from the net response domain in Fig. 8.6.

8.10 TABULAR SUMMARY OF THE EQUATIONS

A convenient summary of the Currie decision level and detection limit equations is provided in Table 8.1.

Table 8.1   Currie Decision and Detection Expressions.

Parameter	Known	Unknown	Estimator
σ_0	✓		–
β	✓		–
α		✓	$\hat{\alpha}$

	Decision Level	Detection Limit
Response domain	$r_C = \hat{\alpha} + z_p \eta^{1/2} \sigma_0$	$r_D = \hat{\alpha} + (\eta^{1/2} z_p + z_q/M^{1/2})\sigma_0$
Net response domain	$Y_C = z_p \eta^{1/2} \sigma_0$	$Y_D = (z_p + z_q)\eta^{1/2}\sigma_0$
Content domain	$X_C = z_p \eta^{1/2} \sigma_0/\beta$	$X_D = (z_p + z_q)\eta^{1/2}\sigma_0/\beta$

1. No OLS calibration curve since β is known.
2. Make N measurements at any desired X_1 value (that is, homoscedasticity assumed). Then $\hat{\alpha} \equiv a = \bar{r}(X_1) - \beta X_1$.
3. $\eta^{1/2} \equiv (1 + N^{-1})^{1/2}$ if $M = 1$. This is a special case of $\eta^{1/2} \equiv (M^{-1} + N^{-1})^{1/2}$.

8.11   SIMULATION CORROBORATION OF THE EQUATIONS IN TABLE 8.1

The Monte Carlo simulation, with $M = 1$, is shown in Fig. 8.8. Given that β is known, OLS is not performed. Rather, for every simulation step, $N \equiv M_{X_1} = 7$ i.i.d. replicate responses at $x = X_1$ are averaged to yield $\bar{r}(X_1)$ values. Then the successive estimates of $\hat{\alpha}$, shown as "a" in Fig. 8.8, are $\hat{\alpha} \equiv a = \bar{r}(X_1) - \beta X_1$. Thus, $a \sim N{:}\alpha, \sigma_a$, where $\sigma_a \equiv \sigma_0/M_{X_1}^{1/2}$.

In the response domain, false negatives are tested by comparing i.i.d. samples from $N{:}r_D, \sigma_0$ with i.i.d. samples from the distribution of r_C, that is, $N{:}\alpha + z_p \eta^{1/2}\sigma_0, \sigma_a$. Note that $N{:}r_D, \sigma_0$ is constructed by simply adding r_D to the zero-centered "blank" in Fig. 8.8, that is, $N{:}r_D, \sigma_0 = r_D + N{:}0, \sigma_0$. In the net response domain, false negatives are tested by comparing i.i.d. samples from $N{:}Y_D, \eta^{1/2}\sigma_0$ with Y_C, and similarly in the content domain: i.i.d. samples from $N{:}X_D, \eta^{1/2}\sigma_0/\beta$ are compared with X_C. The noise pooling factor, $\eta^{1/2}$, was $(8/7)^{1/2}$.

The results of running the 10 simulations are shown directly in Fig. 8.8 and are summarized in Table 8.2.

The results demonstrate that the Currie decision level and detection limit equations in Table 8.1 are correct. Furthermore, they show that the net response domain, when α is unknown, has several significant advantages over the response domain:

- Y_C and Y_D are *errorless*, while r_C and r_D are Gaussian distributed.
- The net response domain equations are nicely symmetric, especially when $p = q$.
- Graphical illustration of the net response domain equations is trivially simple.

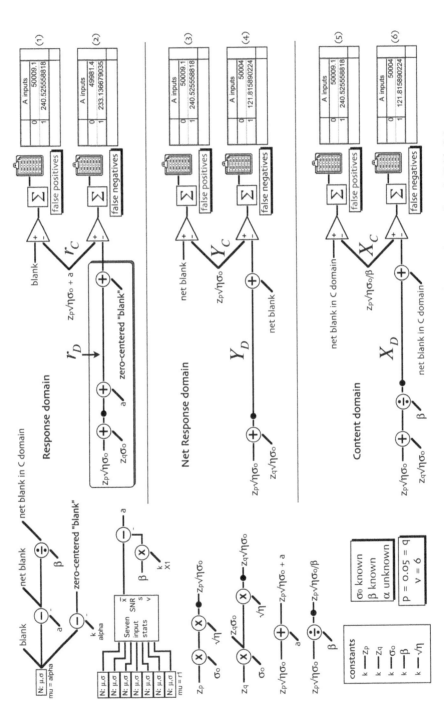

Figure 8.8 The Monte Carlo simulation, with $M = 1$, that corroborates the equations in Table 8.1.

Table 8.2 Summary of Simulation Results for False Positives and False Negatives.

Domain	False Positives ± 1 Standard Deviation	False Negatives ± 1 Standard Deviation
Response	$50\,009.1 \pm 240.5$	$49\,981.4 \pm 233.1$
Net response	$50\,009.1 \pm 240.5$	$50\,004.0 \pm 121.8$
Content	$50\,009.1 \pm 240.5$	$50\,004.0 \pm 121.8$

8.12 CHAPTER HIGHLIGHTS

In this chapter, σ_0 and β were assumed to be known, but α was assumed unknown, requiring that it be estimated. Both the means by which this may be done, and the consequences thereof, were discussed in detail. In particular, the essential use of the $\eta^{1/2}$ factor was demonstrated. The full set of decision level and detection limit equations was presented, tested via Monte Carlo simulations, and found to be correct.

9

IF ONLY THE TRUE SLOPE IS UNKNOWN

9.1 INTRODUCTION

All the assumptions made in Chapter 7 continue to hold, unless stated otherwise. Assume σ_0 and α are known, but β is unknown. Then $r(x) = \alpha + \beta x$ is estimated by

$$\hat{r}(x) = \alpha + bx \tag{9.1}$$

where b is the estimate of the slope. There are two commonly used methods for obtaining unbiased, normally distributed b estimates. The first is to simply perform ordinary least squares (OLS) on N calibration standards, denoted $X_1, X_2, \ldots, X_N$, forcing the regression line through α. Then $b \sim N{:}\beta, \sigma_b$, with σ_b being the population standard error of the slope:

$$\sigma_b = \sigma_0 \Big/ \left(\sum_{i=1}^{N} X_i^2 \right)^{1/2} \tag{9.2}$$

for forced regression through α.

Alternatively, given that α is known and two points define a straight line, pick the highest constant x value in the actual linear range, for example, X_H, average M_{X_H} replicate measurements of $r(X_H)$, and compute the slope:

$$b = \frac{\bar{r}(X_H) - \alpha}{X_H} \tag{9.3}$$

where $\bar{r}(X_H)$ is the sample mean of the M_{X_H} replicate measurements at $x = X_H$. Then $b \sim N{:}\beta, \sigma_0/X_H M_{X_H}^{1/2}$, showing that X_H should be as large as is feasible.

Limits of Detection in Chemical Analysis, First Edition. Edward Voigtman.
© 2017 John Wiley & Sons, Inc. Published 2017 by John Wiley & Sons, Inc.
Companion Website: www.wiley.com/go/Voigtman/Limits_of_Detection_in_Chemical_Analysis

9.2 POSSIBLE "DIVIDE BY ZERO" HAZARD

When β is known, the content domain Currie decision level and detection limit expressions are obtained by dividing β into the respective net response domain expressions, as in the two previous chapters. There is no possibility of dividing by zero because β is greater than zero and is constant. However, estimating β by b introduces a potential "divide by zero" hazard because b, as a normal variate, nominally has nonzero probability from $-\infty$ to ∞. This situation is illustrated in Fig. 9.1, where b values near zero would be problematic.

In Fig. 9.1, the zero value is shown as being reasonably far from β, that is, the center of the distribution of potential b variates. If β were close to 0, in units of σ_b, then there would be no viable response function, because the sensitivity, β, would simply be too small. Conversely, if β were many σ_b units away from 0, then the probability of a b variate being close to 0 would be negligibly small. The "gray area" in-between is dealt with via t testing of hypotheses.

9.3 THE t TEST FOR t_{slope}

Assume that b is estimated by one of the two methods described previously. Then there is an associated sample standard error of the slope, s_b, equal to either

$$s_0 / \left(\sum_{i=1}^{N} X_i^2 \right)^{1/2} \text{ or } s_0/X_H M_{X_H}^{1/2}.$$ In either case, $s_b > 0$ unless the fit is perfect. The

t test statistic for the slope, denoted by t_{slope}, is defined as b/s_b. If $t_{\text{slope}} > t_p$, where t_p is the critical t value for "risk" probability p, then the null hypothesis for the slope $(H_\emptyset : \beta = 0)$ is rejected, with confidence $1 - p$, and the one-tailed alternative hypothesis $(H_a : \beta > 0)$ is not rejected. In this case, the calibration curve is viable [1].

If $t_{\text{slope}} \leq t_p$, the null hypothesis is not rejected, because the risk is considered to be unacceptably high, so there is no viable calibration curve. Appendix A provides a discussion of the central and noncentral t distributions and Table A.2 therein contains commonly used critical t values. As well, critical t values are easily computed via

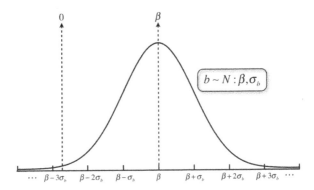

Figure 9.1   Possible "divide by zero" hazard if β is estimated by very small b.

software, for example, in *Microsoft Excel*®, $t_p = \text{TINV}(2p, v)$, where v is the degrees of freedom in the determination of s_0 above. If the fit should happen to be perfect, s_b is zero and t_{slope} is undefined. Then either $b = 0$, so there is no calibration curve, or $b > 0$, and there is negligible "divide by zero" hazard. In any event, perfect fits, occurring with *real* data, should *always* be suspiciously scrutinized.

This topic is examined further in the next chapter, in which the fundamental connection with the noncentral t distribution is made clear. But it is already possible to state that, so long as $t_{\text{slope}} > t_p$, with $p \leq 0.05$, there is no "divide by zero" hazard. In other words, there will be no "infinite" detection limits.

9.4   RESPONSE DOMAIN CURRIE DECISION LEVEL AND DETECTION LIMIT

The theoretical response domain Currie decision level, R_C, is defined as

$$R_C \equiv \alpha + z_p\sigma_0 \tag{9.4}$$

and the corresponding theoretical response domain Currie detection limit, R_D, is defined as

$$R_D \equiv \alpha + (z_p + z_q)\sigma_0 = R_C + z_q\sigma_0 \tag{9.5}$$

9.5   NET RESPONSE DOMAIN CURRIE DECISION LEVEL AND DETECTION LIMIT

The theoretical net response domain Currie decision level, Y_C, is defined as

$$Y_C \equiv z_p\sigma_0 = R_C - \alpha \tag{9.6}$$

and the theoretical net response domain Currie detection limit, Y_D, is defined as

$$Y_D \equiv (z_p + z_q)\sigma_0 = R_D - \alpha = \frac{(z_p + z_q)}{z_p}Y_C \tag{9.7}$$

Note that R_C, R_D, Y_C, and Y_D are exactly as in Chapter 7.

9.6   CONTENT DOMAIN CURRIE DECISION LEVEL AND DETECTION LIMIT

The content domain expressions follow immediately from those in the net response domain. Thus, the experimental content domain Currie decision level, x_C, is defined as

$$x_C \equiv Y_C/b = z_p\sigma_0/b \tag{9.8}$$

and the experimental content domain Currie detection limit, x_D, is defined as

$$x_D \equiv \frac{Y_D}{b} = \frac{(z_p + z_q)\sigma_0}{b} = \frac{(z_p + z_q)}{z_p}x_C \tag{9.9}$$

Both x_C and x_D are scaled reciprocals of Gaussian variates and are not mathematically equivalent to Gaussian variates.

9.7   GRAPHICAL ILLUSTRATIONS OF THE DECISION LEVEL AND DETECTION LIMIT EQUATIONS

A graphical illustration of the response domain equations, that is, eqns 9.4 and 9.5, together with the analytical blank's distribution, is shown in Fig. 9.2.

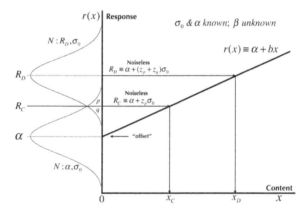

Figure 9.2   Graphical illustration of the Currie decision level and detection limit in the response domain.

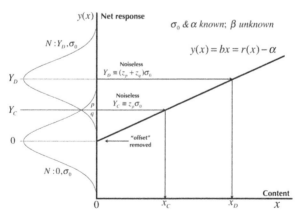

Figure 9.3   Graphical illustration of the Currie decision level and detection limit in the net response domain.

The response line in Fig. 9.2 is simply one *possible* straight line having α as its required intercept. Since b is normally distributed about β, it is unbiased. Therefore, the depicted response line is as likely to have $b > \beta$ as $b < \beta$.

In the net response domain, eqns 9.6 and 9.7 are illustrated in Fig. 9.3. The graphical illustrations in Figs 9.2 and 9.3 are properly scaled to illustrate the canonical $p \equiv 0.05 \equiv q$. As in Fig. 9.2, the response line in Fig. 9.3 is simply one *possible* straight line having α as its required intercept.

9.8 TABULAR SUMMARY OF THE EQUATIONS

A convenient summary of the Currie decision level and detection limit equations is provided in Table 9.1.

9.9 SIMULATION CORROBORATION OF THE EQUATIONS IN TABLE 9.1

The Monte Carlo simulation, with $M = 1$, is shown in Fig. 9.4. Since α is known, b is estimated by forced regression through α, using the following equation:

$$b = \frac{\left(\sum_{i=1}^{N} y_i x_i \right)}{\left(\sum_{i=1}^{N} x_i^2 \right)} \tag{9.10}$$

where $y_i = r_i - \alpha$ [2, eqn 2.50].

The results of running the 10 simulations are shown directly in Fig. 9.4 and are summarized in Table 9.2.

Table 9.1 Currie Decision and Detection Expressions

Parameter	Known	Unknown	Estimator
σ_0	✓		–
β		✓	b (from forced OLS)
α	✓		–

		Decision Level	Detection Limit
Response domain		$R_C = \alpha + z_p \sigma_0$	$R_D = \alpha + (z_p + z_q)\sigma_0$
Net response domain		$Y_C = z_p \sigma_0$	$Y_D = (z_p + z_q)\sigma_0$
Content domain		$x_C = z_p \sigma_0 / b$	$x_D = (z_p + z_q)\sigma_0 / b$

1. The value of b is from OLS forced through α.
2. If $M > 1$, multiply each σ_0 by $\eta^{1/2}$, where $\eta^{1/2} = 1/M^{1/2}$.

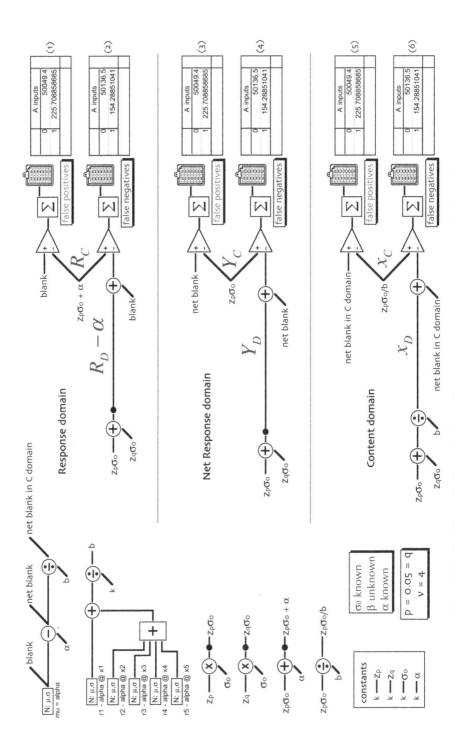

Figure 9.4 The Monte Carlo simulation, with $M = 1$, that corroborates the equations in Table 9.1.

Table 9.2 Summary of Simulation Results for False Positives and False Negatives

Domain	False Positives ± 1 Standard Deviation	False Negatives ± 1 Standard Deviation
Response	50 049.4 ± 225.7	50 136.5 ± 154.3
Net response	50 049.4 ± 225.7	50 136.5 ± 154.3
Content	50 049.4 ± 225.7	50 136.5 ± 154.3

The results in Table 9.2 demonstrate that the Currie decision level and detection limit equations in Table 9.1 are correct.

9.10 CHAPTER HIGHLIGHTS

In this chapter, σ_0 and α were assumed to be known, but β was assumed unknown, requiring that it be estimated. Both the means by which this may be done, and the consequences thereof, were discussed in detail. In particular, a potential "divide by zero" hazard was shown to be nonexistent. The full set of decision level and detection limit equations was presented, tested via Monte Carlo simulations, and found to be correct.

REFERENCES

1. E. Voigtman, "Limits of detection and decision. Part 1", *Spectrochim. Acta, B* **63** (2008) 115–128.
2. D.C. Montgomery, E.A. Peck, *Introduction to Linear Regression Analysis*, John Wiley and Sons, New York, 1982.

10

IF THE TRUE INTERCEPT AND TRUE SLOPE ARE BOTH UNKNOWN

10.1 INTRODUCTION

In this chapter, we examine what happens when both α and β are unknown. Not surprisingly, the results are a combination of those discussed in the previous two chapters. Assuming only σ_0 is known, $r(x) = \alpha + \beta x$ is estimated by

$$\hat{r}(x) = a + bx = \hat{\alpha} + \hat{\beta}x \tag{10.1}$$

where a (or $\hat{\alpha}$) is the estimate of α and b (or $\hat{\beta}$) is the estimate of β. By far, the most common method of estimation of a and b is ordinary least squares (OLS) on a set of N calibration standards: $X_1, X_2, \ldots, X_N$. Degenerate standards are allowed, that is, standards may be used more than once, but not as "replicates." The restriction of only one measurement per standard ensures that σ_r, the population standard error about regression, equals σ_0. Then s_r, the sample standard error about regression, may be used as s_0, with due attention to degrees of freedom. Under the conditions assumed for the model, as detailed in Chapter 7, a and b are the lowest variance linear unbiased estimators of α and β, respectively [1].

When OLS is performed, a variety of sample test statistics are generated. These almost invariably include, but usually are not limited to, the following:

- a, the intercept estimate
- b, the slope estimate
- s_a, the sample standard error of the intercept
- s_b, the sample standard error of the slope
- s_r, the sample standard error about regression
- r, the Pearson product-moment correlation coefficient
- t_{slope}, the t test statistic for the slope.

Limits of Detection in Chemical Analysis, First Edition. Edward Voigtman.
© 2017 John Wiley & Sons, Inc. Published 2017 by John Wiley & Sons, Inc.
Companion Website: www.wiley.com/go/Voigtman/Limits_of_Detection_in_Chemical_Analysis

Even a simple statistics software package may provide a great deal of additional useful statistics, for example, ANOVA and normality testing of the residuals. Thankfully, the vast majority of published work on OLS, and more advanced forms of regression, is entirely superfluous to present needs. Interested readers will have no difficulty in finding more information about regression techniques, in general, and OLS, in particular.

10.2   IMPORTANT DEFINITIONS, DISTRIBUTIONS, AND RELATIONSHIPS

Each of the OLS estimators listed above is a random variate. For example, b is distributed as $N{:}\beta, \sigma_b$, where σ_b is the population standard error of the slope. Table 10.1

Table 10.1   Summary of OLS Definitions, Distributions, and Relationships.

Item	Definition	Comment
N	Number of standards	One measurement per standard
X_i	ith Standard	Standards are assumed errorless
$\overline{X}$	Sample mean of the standards	$\overline{X} \equiv N^{-1} \sum_{i=1}^{N} X_i$
S_{XX}	Sum of squared differences from $\overline{X}$	$S_{XX} \equiv \sum_{i=1}^{N} (X_i - \overline{X})^2$
ν	Degrees of freedom	$\equiv N - 2$
σ_0	Population standard deviation of the noise	σ_0 is a positive constant, that is, homoscedastic. $\therefore \sigma_0 \neq \sigma(x)$
s_0	Sample standard deviation of the noise	Defined (eqn A.12) and distributed (eqn A.15) as in Appendix A
σ_r	Population standard error about regression	$\equiv \sigma_0$ (for 1 measurement/standard)
s_r	Sample standard error about regression	$\equiv \left[\nu^{-1} \sum_{i=1}^{N} [r_i(X_i) - a - bX_i]^2 \right]^{1/2} = s_0$
σ_a	Population standard error of the intercept	$\equiv \sigma_r [N^{-1} + \overline{X}^2 / S_{XX}]^{1/2}$
s_a	Sample standard error of the intercept	$\equiv s_r [N^{-1} + \overline{X}^2 / S_{XX}]^{1/2}$
σ_b	Population standard error of the slope	$\equiv \sigma_r / S_{XX}^{1/2}$
s_b	Sample standard error of the slope	$\equiv s_r / S_{XX}^{1/2}$
a	OLS intercept	$a \sim N{:}\alpha, \sigma_a$
b	OLS slope	$b \sim N{:}\beta, \sigma_b$
r	Pearson product-moment correlation coefficient	Many variants exist, for example, r^2 and R^2
δ	Noncentrality parameter of the noncentral t distribution	$\equiv \beta / \sigma_b$
t_{slope}	t Test statistic of the slope	$\equiv b/s_b$. Biased estimator of δ

provides a convenient summary of relevant OLS definitions, distributions, and relationships.

Note that s_0, s_r, s_a, and s_b are all χ variates, as per eqn A.15, or the more general eqn A.31, and all are negatively biased estimates of their respective population parameters. In contrast, the t_{slope} variate is a positively biased estimator of δ, as per eqn A.21.

10.3   THE NONCENTRAL t DISTRIBUTION BRIEFLY APPEARS

With the model CMS and ancillary assumptions made in Chapter 7, performing OLS on a set of N calibration standards yields the OLS estimators listed earlier. The t_{slope} variate is particularly important because it is noncentral t distributed, shown as follows:

$$t_{slope} \sim t(u|v, \delta) = t(u|N - 2, \beta/\sigma_b) \qquad (10.2)$$

where v is degrees of freedom, δ the noncentrality parameter, and the noncentral t distribution is described in Appendix A. The central t distribution, commonly called the "Student's" t distribution, is the $\delta \equiv 0$ special case of the noncentral t distribution. Since $\delta \equiv \beta/\sigma_b$, and $\sigma_b > 0$, it follows immediately that $\beta = 0$ if and only if $\delta = 0$. Similarly, $\beta > 0$ iff $\delta > 0$.

The t_{slope} variate, and its noncentral t distribution, is the underlying basis for hypothesis testing of the OLS slope. The null hypothesis, denoted by $H_\emptyset : \beta = 0$, is simply that $\beta = 0$. In this case, there is no linear term in the true response function, that is, x has no effect upon the response, assuming the model assumptions are valid. Given that β cannot be negative, as has been assumed from the beginning, the only alternative to $\beta = 0$ is $\beta > 0$, so the alternative hypothesis is $H_a: \beta > 0$.

The central t distribution is symmetric about zero, as in Fig. A.3. Hence, Fig. 10.1 shows only the lower positive region, with dummy independent variable u, together with the noncentral t distribution for one particular value of δ. For $p = 0.05$, the critical t value for $v = 4$ is approximately 2.131845. This means that the central t distribution has $100p\%$ of its area *above* $t_{critical} \equiv t_{p,v} = t_{0.05,4} \cong 2.131845$. With $\delta = 4.06728$, and $v = 4$, the noncentral t distribution similarly has 5% of its area *below* $t_{p,v} \cong 2.131845$. Thus, the δ value has been chosen so that, with respect to $t_{p,v}$, the noncentral t distribution has 0.05 area fraction below $t_{p,v}$.

To illustrate hypothesis testing of the slope in OLS, suppose $N = 6$ standards were used in the OLS analysis, so that $v = N - 2 = 4$, and that the obtained t_{slope} was equal to 3.5, as shown in Fig. 10.1. If $H_\emptyset$ was true, then $\beta = 0 = \delta$, so t_{slope} would be a random sample from the central t distribution. As Fig. 10.1 shows, this is certainly possible but relatively improbable. In fact, the probability of obtaining $t_{slope} \geq 3.5$ is easily computed: it is $1 - CDF(3.5) \cong 1.24\%$. If the null hypothesis were to be rejected, this is the risk of mistakenly rejecting a true null hypothesis.

If this risk is judged to be acceptably low, then $H_\emptyset$ may be rejected in favor of H_a. This means that it is thought to be more probable that t_{slope} was a random sample from

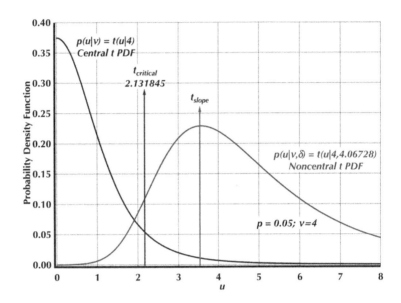

Figure 10.1   The central and noncentral t distributions for $v = 4$ and $\delta = 4.06728$.

a noncentral t distribution, with $\delta > 0$, but otherwise unknown δ. For this particular example, the confidence in H_a is approximately 98.76%, that is, CDF(3.5). This is simply one minus the risk. But δ is *unknown* and, in particular, it is almost certainly *not* equal to 4.06728.

10.4   WHAT PURPOSE WOULD BE SERVED BY KNOWING δ?

As Appendix D shows, both X_C and X_D would immediately be known if δ were known. Unfortunately, a single t_{slope} variate is a significantly positively biased estimator of δ, as given by eqn A.21 and as shown in Table 10.2. This eliminates using a single t_{slope} value to estimate δ, unless v is unrealistically high.

10.5   IS THERE A VIABLE WAY OF ESTIMATING δ?

There are at least two viable ways to estimate δ if $N' \gg 1$ independent t_{slope} variates are available. The first method involves binning the N' t_{slope} variates into a histogram and then curve fitting with the noncentral t distribution. Since $v = N - 2$, the only adjustable parameter is δ. The second method is even simpler; average N' t_{slope} variates and use their sample mean, $\bar{t}_{\text{slope}}$, as an approximation of $E[t_{\text{slope}}]$. Then $\hat{\delta} \simeq \bar{t}_{\text{slope}}/(E[t_{\text{slope}}]/\delta)$. Both methods work well when $N' \gg 1$ and note that v and N' are unrelated.

Table 10.2   Bias Factors for t_{slope} Variates.

v	$E[t_{\text{slope}}]/\delta$	v	$E[t_{\text{slope}}]/\delta$
2	1.772453851	12	1.068442477
3	1.381976598	15	1.053732355
4	1.253314137	20	1.039560978
5	1.189416077	30	1.025899470
6	1.151242546	40	1.019251423
7	1.125868866	50	1.015319195
8	1.107783657	100	1.007578953
9	1.094241683	200	1.003769634
10	1.083722308	500	1.001503132
11	1.075315287	1000	1.000750782

The noncentral t distribution is essential in understanding hypothesis testing in OLS. However, it has a fundamental connection to detection limits that is far more important. This is elucidated in considerable detail in later chapters, especially Chapters 14 and 15, in regard to the very common situation of σ_0, β, and α being unknown. As well, commonly used amplitude signal-to-noise ratios (SNRs) are noncentral t distributed [2]. Indeed, as amplitude SNRs, and for the operative calibration design value of $S_{XX}^{1/2}$, δ is the theoretical SNR of the CMS and t_{slope} is the corresponding experimental SNR that estimates δ. *Ceteris paribus*, both X_D and X_C are inversely proportional to δ, as per Appendix D.

10.6   RESPONSE DOMAIN CURRIE DECISION LEVEL AND DETECTION LIMIT

The experimental response domain Currie decision level, r_C, is defined as

$$r_C \equiv \hat{\alpha} + z_p \eta^{1/2} \sigma_0 \tag{10.3}$$

and the corresponding experimental response domain Currie detection limit, r_D, is defined as

$$r_D \equiv \hat{\alpha} + (\eta^{1/2} z_p + z_q/M^{1/2})\sigma_0 = r_C + z_q \sigma_0/M^{1/2} \tag{10.4}$$

As in Chapter 8, the $\eta^{1/2}$ factor *only* multiplies z_p, *not* the sum of z_p and z_q. This *always* happens in the response domain, whenever α must be estimated by $\hat{\alpha}$.

10.7   NET RESPONSE DOMAIN CURRIE DECISION LEVEL AND DETECTION LIMIT

The theoretical net response domain Currie decision level, Y_C, is defined as

$$Y_C \equiv z_p \eta^{1/2} \sigma_0 \tag{10.5}$$

and the theoretical net response domain Currie detection limit, Y_D, is defined as

$$Y_D \equiv (z_p + z_q)\eta^{1/2}\sigma_0 = \frac{z_p + z_q}{z_p}Y_C \tag{10.6}$$

Both Y_C and Y_D are errorless and greater than zero.

10.8   CONTENT DOMAIN CURRIE DECISION LEVEL AND DETECTION LIMIT

The content domain expressions follow immediately from those in the net response domain. Thus, the experimental content domain Currie decision level, x_C, is defined as

$$x_C \equiv Y_C/b = z_p\eta^{1/2}\sigma_0/b \tag{10.7}$$

and the experimental content domain Currie detection limit, x_D, is defined as

$$x_D \equiv \frac{Y_D}{b} = \frac{(z_p + z_q)\eta^{1/2}\sigma_0}{b} = \frac{z_p + z_q}{z_p}x_C \tag{10.8}$$

As in the previous chapter, both x_C and x_D are scaled reciprocals of Gaussian variates.

10.9   GRAPHICAL ILLUSTRATIONS OF THE DECISION LEVEL AND DETECTION LIMIT EQUATIONS

The only domain in which graphical illustrations of the Currie equations make much sense is the net response domain. This is shown for eqns 10.5 and 10.6 in Fig. 10.2.

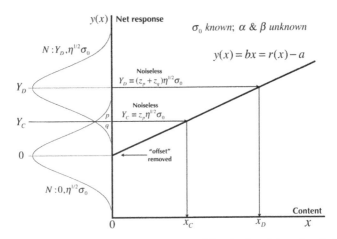

Figure 10.2   Graphical illustration of the Currie decision level and detection limit in the net response domain.

Table 10.3   Currie Decision and Detection Expressions.

Parameter	Known	Unknown	Estimator
σ_0	✓		–
β		✓	b
α		✓	a or $\hat{a}$

	Decision Level	Detection Limit
Response domain	$r_C = \hat{\alpha} + z_p \eta^{1/2} \sigma_0$	$r_D = \hat{\alpha} + (\eta^{1/2} z_p + z_q / M^{1/2}) \sigma_0$
Net response domain	$Y_C = z_p \eta^{1/2} \sigma_0$	$Y_D = (z_p + z_q)\eta^{1/2}\sigma_0$
Content domain	$x_C = z_p \eta^{1/2} \sigma_0 / b$	$x_D = (z_p + z_q)\eta^{1/2}\sigma_0 / b$

Both a and b are usually obtained from OLS with N standards. In this case, with $M = 1$ assumed, $\eta^{1/2} \equiv (1 + N^{-1} + \overline{X}^2/S_{XX})^{1/2}$. This is a special case of $\eta^{1/2} \equiv (M^{-1} + N^{-1} + \overline{X}^2/S_{XX})^{1/2}$.

The net response function shown in Fig. 10.2 is simply one possible such function; every time OLS is performed on a new set of X_i,y_i data, new values of a and b are obtained. But the net response function will always pass through the origin.

10.10   TABULAR SUMMARY OF THE EQUATIONS

A convenient summary of the Currie decision level and detection limit equations is provided in Table 10.3.

10.11   SIMULATION CORROBORATION OF THE EQUATIONS IN TABLE 10.3

The Monte Carlo simulation, with $M = 1$, is shown in Fig. 10.3. Since α and β are assumed unknown, OLS is performed in order to obtain a and b. In the simulation, the $N = 5$ standards had values of 1, 2, 3, 4, and 5. The parameters used in the OLS computation block, which is labeled "Linear calibration curves S," were $\alpha = 1$, $\beta = 3$, and $\sigma_0 = 0.1$. The noise pooling factor, $\eta^{1/2}$, was approximately 1.449138.

The results of running the 10 simulations are shown directly in Fig. 10.3 and are summarized in Table 10.4. The results demonstrate that the Currie decision level and detection limit equations in Table 10.3 are correct.

10.12   CHAPTER HIGHLIGHTS

In this chapter, σ_0 was assumed to be known, but both α and β were assumed unknown, so they were estimated via OLS. The essential concepts and definitions pertaining to OLS were presented in Table 10.1. The noncentral t distribution made its first appearance, since the OLS t test statistic for the slope, t_{slope}, is so

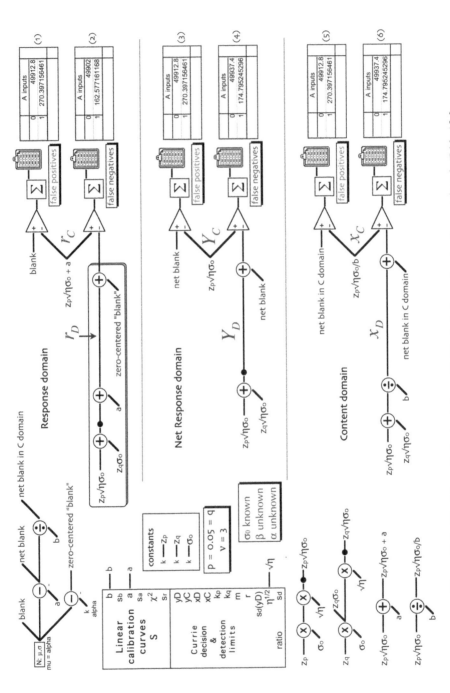

Figure 10.3 The Monte Carlo simulation, with $M = 1$, that corroborates the equations in Table 10.3.

Table 10.4   Summary of Simulation Results for False Positives and False Negatives.

Domain	False Positives ± 1 Standard Deviation	False Negatives ± 1 Standard Deviation
Response	49 912.8 ± 270.4	49 902.0 ± 162.6
Net response	49 912.8 ± 270.4	49 937.4 ± 174.8
Content	49 912.8 ± 270.4	49 937.4 ± 174.8

distributed. Estimation of the noncentrality parameter, δ, was briefly discussed along with the reason why knowing δ, or having an accurate estimate of it, would be highly advantageous. The full set of decision level and detection limit equations was presented, tested via Monte Carlo simulations, and found to be correct.

REFERENCES

1. D.C. Montgomery, E.A. Peck, *Introduction to Linear Regression Analysis*, John Wiley and Sons, New York, 1982.

2. S. Nadarajah, S. Kotz, "Computation of signal-to-noise ratios", *MATCH Commun. Math. Comput. Chem.* **57** (2007) 105–110.

11

IF ONLY THE POPULATION STANDARD DEVIATION IS UNKNOWN

11.1 INTRODUCTION

In Chapters 7–10, it was assumed that σ_0 was known. The remaining two model parameters, α and β, each had the possibility of being known or unknown. Before examining the remaining four combinations, it is worthwhile to consider the important points that have been revealed thus far. These are summarized as follows:

- The Currie expressions in the four net response domains are errorless theoretical equations, regardless of whether α or β is known or unknown.
- For $M = 1$, the noise pooling factor, $\eta^{1/2}$, only appears in the Currie expressions when α is unknown and must be estimated.
- The Currie expressions in the content domain are either errorless theoretical equations, if β is known, or else they are scaled reciprocal Gaussian variates, if $\hat{\beta}$ is Gaussian.
- The Currie expressions in the response domain are more awkward and less symmetric than those in the net response domains, especially in any that contain $\eta^{1/2}$.
- With homoscedastic additive Gaussian white noise (AGWN), as consistently assumed throughout, $Y_D = 2Y_C$, if $q = p$. This is exact for all degrees of freedom.
- Lastly, with homoscedastic AGWN, as consistently assumed throughout, $X_D = 2X_C$ (Chapters 7 and 9) and $x_D = 2x_C$ (Chapters 8 and 10), if $q = p$. These are exact for all degrees of freedom.

Limits of Detection in Chemical Analysis, First Edition. Edward Voigtman.
© 2017 John Wiley & Sons, Inc. Published 2017 by John Wiley & Sons, Inc.
Companion Website: www.wiley.com/go/Voigtman/Limits_of_Detection_in_Chemical_Analysis

11.2   ASSUMING σ_0 IS UNKNOWN, HOW MAY IT BE ESTIMATED?

There are two commonly used methods of estimating σ_0. The first method involves making M_0 replicate measurements of an analytical blank and computing their sample standard deviation, s_0, as per the familiar eqn A.12. In this case, $v = M_0 - 1$. Since the noise is homoscedastic, any convenient standard may be used in place of the analytical blank. The second common method involves using s_r, obtained by performing ordinary least squares (OLS) on N standards. In the present case, this would obviously not be done, because both α and β are known, but it is certainly viable when they are both unknown, as in Chapter 14. For the OLS-based method, $s_0 = s_r$ and $v = N - 2$.

11.3   WHAT HAPPENS IF σ_0 IS ESTIMATED BY s_0?

As shown in eqns A.28 and A.29, s_0 is a negatively biased point estimate of σ_0. This means that, on average, it is less than σ_0 by the bias factor $c_4(v)$. As a convenience, $c_4(v)$ values are given in Table A.5. Accordingly, direct substitution of s_0 in place of σ_0 in any decision level or detection limit expression would immediately result in a biased expression. The same applies to confidence interval expressions, such as eqn A.11. Furthermore, σ_0 *cannot* be replaced by $s_0/c_4(v)$ in decision level, detection limit, or confidence interval expressions because the resulting expressions would still be biased, just less so.

   Figures 11.1–11.4 show simulation models that quantitatively demonstrate the behavior of both unbiased and biased 95% CIs. For each model, a 10-million-step simulation is performed. Thus, each simulation produces 10 million independent 95% CIs, and, for each CI, the simulations keep track of the number of times the true population mean, denoted by μ, is above, within, or below the CI. The expected percentages are 2.5%, 95%, and 2.5%.

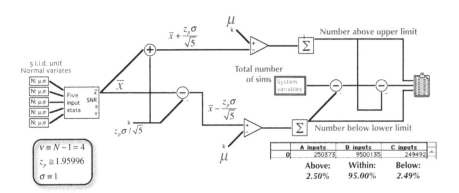

Figure 11.1   Unbiased 95% CI simulations, based on z_p and σ.

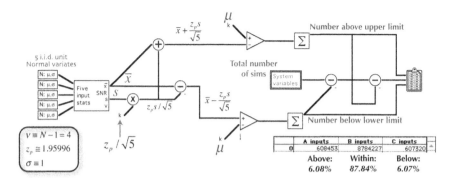

Figure 11.2   Very biased 95% CI simulations, based on z_p and s.

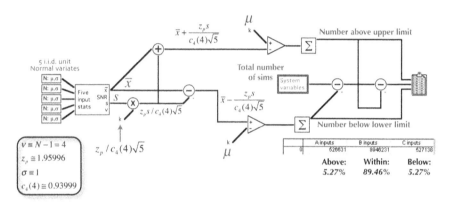

Figure 11.3   Still biased 95% CI simulations, based on z_p and $s/c_4(\nu)$.

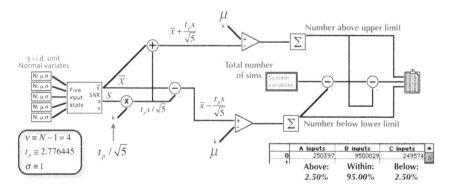

Figure 11.4   Unbiased 95% CI simulations, based on t_p and s.

The first simulation model tests the 95% CI defined by eqn A.11, that is,

$$95\% \text{ CI for } \mu \equiv \bar{x} \pm z_p \sigma / \sqrt{N} \qquad (11.1)$$

for $N \equiv 5$ and $z_p \equiv z_{0.025} \cong 1.959964$. As shown in Fig. 11.1, the obtained percentages are almost perfect. The second simulation model, shown in Fig. 11.2, simply replaces σ by s, with no other changes. Note that 10 million independent s variates are produced; it is *not* the case that just one s variate is used 10 million times. From Fig. 11.2, the results are very biased: only 87.84% of the CIs contain μ. Clearly, the negative bias of s results in the CIs being too narrow.

The third simulation model, shown in Fig. 11.3, naively attempts to eliminate the bias by simply replacing σ by $s/c_4(v)$, where $v \equiv N - 1 = 5 - 1 = 4$ and $c_4(v) \cong 0.939986$. Again note that 10 million independent s variates are produced, each is divided by the constant $c_4(v)$ and then each ratio is tested. The results are still very biased: only 89.46% of the CIs contain μ. This demonstrates that replacing the expectation value of s, $E[s]$, by *any single* value of s, rather than an *average* of *many* s values, does *not* eliminate the negative bias. See Appendix A for details.

Finally, Fig. 11.4 shows what happens when the 95% CI is properly computed via eqn A.14. With s, and the appropriate critical t value, that is, $t_p \equiv t_{0.025} \cong 2.776445$, the results are nicely in accordance with more than a century of incontrovertible statistics [1].

11.4 A USEFUL SUBSTITUTION PRINCIPLE

The simulations in Figs 11.1–11.4 demonstrate a useful substitution principle: if σ is unknown and must be estimated by s, then all product terms of the form $z_{\text{critical}} \times \sigma$ must be replaced by terms of the form $t_{\text{critical}} \times s$, with appropriate degrees of freedom. As a direct consequence, when σ_0 is unknown, the Currie decision level and detection limit equations may be obtained by inspection, from the corresponding equations with σ_0 known. In other words, simply make the following substitutions: $\sigma_0 \mapsto s_0, z_p \mapsto t_p$, and $z_q \mapsto t_q$. This substantially simplifies matters in this chapter, and the next three, and is immediately evident below.

11.5 RESPONSE DOMAIN CURRIE DECISION LEVEL AND DETECTION LIMIT

The response domain Currie decision level, r_C, is defined as

$$r_C \equiv \alpha + t_p s_0 \qquad (11.2)$$

Then the corresponding response domain Currie detection limit, r_D, is defined as

$$r_D \equiv \alpha + (t_p + t_q)s_0 = r_C + t_q s_0 \qquad (11.3)$$

These equations are the same as those in Chapter 7, with $\sigma_0 \mapsto s_0$, $z_p \mapsto t_p$, and $z_q \mapsto t_q$. Note that both r_C and r_D are χ distributed random variates, plus the constant α.

11.6 NET RESPONSE DOMAIN CURRIE DECISION LEVEL AND DETECTION LIMIT

The net response domain Currie decision level, y_C, is defined as

$$y_C \equiv t_p s_0 \tag{11.4}$$

and the net response domain Currie detection limit, y_D, is defined as

$$y_D \equiv (t_p + t_q)s_0 = \frac{t_p + t_q}{t_p} y_C \tag{11.5}$$

assuming $M \equiv 1$, that is, $\eta^{1/2} = 1$, in both cases. These equations are the same as those in Chapter 7, with $\sigma_0 \mapsto s_0$, $z_p \mapsto t_p$, and $z_q \mapsto t_q$. Note that y_C and y_D are χ distributed random variates, as per eqn A.31.

11.7 CONTENT DOMAIN CURRIE DECISION LEVEL AND DETECTION LIMIT

The content domain expressions follow immediately from those in the net response domain. Thus, the content domain Currie decision level, x_C, is defined as

$$x_C \equiv y_C/\beta = t_p s_0/\beta \tag{11.6}$$

and the content domain Currie detection limit, x_D, is defined as

$$x_D \equiv \frac{y_D}{\beta} = \frac{(t_p + t_q)s_0}{\beta} = \frac{t_p + t_q}{t_p} x_C \tag{11.7}$$

These equations are the same as those in Chapter 7, with $\sigma_0 \mapsto s_0$, $z_p \mapsto t_p$, and $z_q \mapsto t_q$. As above, both x_C and x_D are χ distributed random variates, as per eqn A.31.

11.8 MAJOR IMPORTANT DIFFERENCES FROM CHAPTER 7

In Chapter 7, it was assumed that σ_0, β, and α were all known, making it simple to graphically illustrate the Currie decision level and detection limit expressions. These were given in Figs 7.4–7.6. Figure 11.5 shows the net response axis and PDFs from Fig. 7.5.

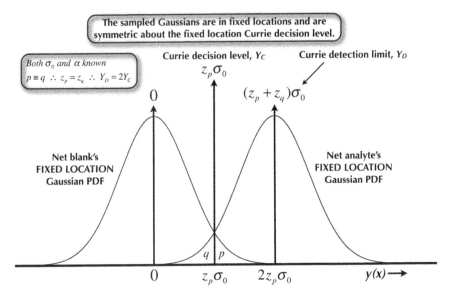

Figure 11.5 The net response axis and PDFs from Fig. 7.5. The PDFs are symmetric about Y_C because $p = q$.

This figure illustrates how sampling is performed, for the purpose of testing for false positives and false negatives, when σ_0 is known. The net blank and net analyte distributions are in *fixed* locations, 0 and Y_D, respectively. If a random sample from the net blank's distribution exceeds Y_C, it is a false positive. Similarly, if a random sample from the net analyte's distribution falls below Y_C, it is a false negative. Hypothesis testing is described in much more detail in Chapter 17, but the above suffices for the moment.

In this chapter, σ_0 is assumed to be unknown and, therefore, estimated by the negatively biased, χ distributed sample standard deviation, s_0. As a consequence, the errorless constants Y_C and Y_D are unavailable and, instead, are replaced by the χ distributed random variates y_C and y_D, respectively. The expressions for y_C and y_D are given in eqns 11.4 and 11.5.

In Fig. 11.6, a simple simulation is shown that generates 10 million independent net blanks, y_C variates and y_D variates. These are individually binned into respective histograms, as shown in Fig. 11.7. The simulation parameters are as shown at the left side of the figure. Note that $y_D = (t_p + t_q)y_C/t_p$, so that $y_D = 2y_C$ if $p = q$, as assumed in the simulation.

The histograms in Fig. 11.7 clearly show that the net blank is Gaussian distributed, as expected, and the y_C and y_D histograms are χ distributed, also as expected. Curve fits are not shown, for reasons of clarity. Testing for false positives and false negatives is analogous to the procedure discussed following Fig. 11.5, but with a somewhat subtle twist, as discussed below.

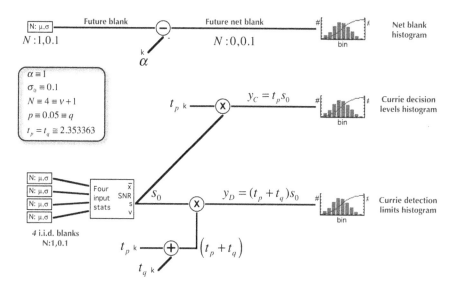

Figure 11.6 Simulation showing the generation of histograms of net blanks, y_C and y_D variates.

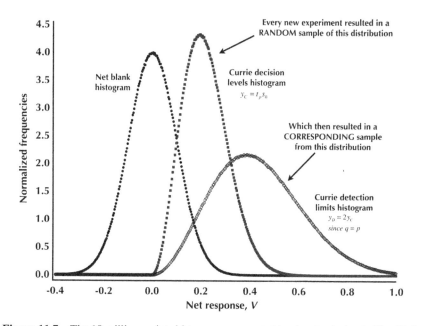

Figure 11.7 The 10 million variate histograms generated by the simulation in Fig. 11.6.

11.9   TESTING FOR FALSE POSITIVES AND FALSE NEGATIVES

Testing for false positives involves comparing a random sample from the net blank's distribution with a random sample from the χ distribution of Currie decision levels. Every new experiment involves new samples from the two distributions, that is, the s_0 value *cannot* be repeatedly used, as though it were a constant. If the net blank variate exceeds its associated y_C variate, it is a false positive.

Testing for false negatives is where the subtlety arises because y_D is χ distributed and "locked" to y_C by eqn 11.5. Therefore, every new experiment results in a new sample from the distribution of y_C variates and this, in turn, immediately results in a *corresponding y_D* variate. Then y_D serves as the *random location parameter* of a Gaussian distribution that has the *same width* as that of the net blank. This is illustrated in Fig. 11.8, with y_C *just happening* to be such that $p = 0.05 = q$.

In Fig. 11.8, the net blank is distributed as $N{:}0, \sigma_0$, and therefore has its location fixed at zero. In contrast, the net analyte is distributed as $N{:}y_D, \sigma_0$, so its *location randomly shifts*, from experiment to experiment, ultimately due to s_0, and therefore y_C, being χ distributed. A false negative occurs when a random sample from $N{:}y_D, \sigma_0$ falls below the *corresponding* random y_C variate. Again note that $y_D = (t_p + t_q)y_C/t_p$, so y_D and y_C are *not* independent χ variates; once s_0 is determined, y_C follows via eqn 11.4, and then y_D via eqn 11.5. Therefore, the net analyte distribution may also be expressed as $N{:}(t_p + t_q)y_C/t_p, \sigma_0$.

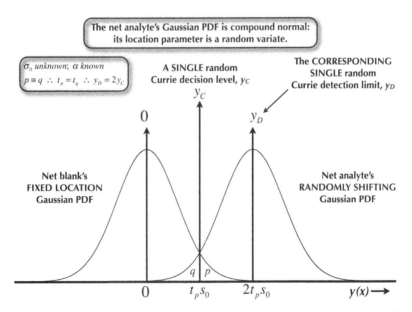

Figure 11.8   Illustration of the two Gaussian distributions that are sampled. The PDFs are symmetric about y_C because $p = q$. In this figure, $p = 0.05 = q$.

11.10 CORRECTION OF A SLIGHTLY MISLEADING FIGURE

In Fig. 11.8, the y_C value depicted happens to be the *unique* one, out of the infinity of possibilities implicit in the y_C distribution that would result in exactly 5% probabilities of both false positives and false negatives. However, y_C is a χ variate, so there is negligible probability that it would be "just right." What matters is the long run: if an *infinite* number of y_C and y_D variates were tested for their performance in regard to false positives and false negatives, then, *on average*, their performance would be as specified *a priori*.

To illustrate this point, Fig. 11.9 shows four possible y_C values and the Gaussian distribution pairs that are symmetrically arranged, in mirror image fashion, about them. In Fig. 11.9a, the y_C value is the unique one shown in Fig. 11.8: the two small tails are each 5%. The remaining three subfigures show tail areas at roughly 10% (Fig. 11.9b), far below 5% (Fig. 11.9c), and far above 5% (Fig. 11.9d). Yet it is the *properly weighted average* of an *infinite number* of such pairs that matters, *not* any one particular y_C value and its pair of *always mirrored* Gaussians.

11.11 AN INFORMATIVE SCREENCAST

A screencast illustrating the above, named "*Currie detection.mov*," is available at the *LightStone*® website: see Appendix B for the URL. It should be viewed in order to fully understand how the Currie detection limit behaves. Then an important point arises; in practice, perhaps only *one* experiment is performed, yielding a single s_0 value. This leads to $y_C = t_p s_0$ and, if $q = p$, then $y_D = 2y_C$. Therefore, if s_0 happens to

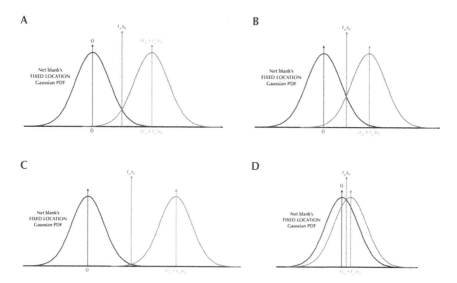

Figure 11.9 Four of the infinity of possible y_C variates and their symmetric Gaussians (due to $p = q$).

be very small, then both y_C and y_D will similarly be very small. Hence, the detection limit would be very low, near the lower limit of its possible range, even though p and q would be well above 0.05. Indeed, a y_C value near zero results in p and q values approaching 50%. Conversely, a high value of s_0 results in a high detection limit. Both situations are possible, though improbable, and this is where knowledge of the PDFs of s_0, y_C, and y_D is advantageous: they make it possible to compute confidence intervals for the theoretical decision levels and detection limits.

11.12   CENTRAL CONFIDENCE INTERVALS FOR σ AND s

Note that Table A.4 provides 95% central confidence intervals for σ, if s is available, and for s, if σ is known. As an example, suppose 7 *i.i.d.* replicate measurements were made from $N{:}\alpha, \sigma_0$, and s_0 was computed in the conventional way, using eqn A.12. Then $v = 6$ and the 95% central CI for σ_0 is from $0.6444s_0$ to $2.2021s_0$.

From Table A.4, it is seen that narrowing the 95% CI for σ_0 requires a great deal of data; even for $v = 100$, it is from $0.8785s_0$ to $1.1607s_0$. Similarly, from Table A.5, the relative precision of s_0 is 7.0% at $v = 100$ and 5.0% at $v = 200$. However, the $c_4(v)$ bias factor, discussed in detail in Appendix A, has a value of 0.99005 for $v = 25$. Hence, the sample standard deviation of 26 *i.i.d.* normal variates should, *on average*, be 99% of σ_0. This can be useful, but it must be emphasized that $c_4(v)$ is *not* a cure for low degrees of freedom and, as shown in Chapter 17, it is very easy to use inappropriately [2–4].

Since σ_0 is unknown, it may be estimated using eqn A.30:

$$\hat{\sigma}_0 = \frac{\bar{s}_0}{c_4(v)} \cong \frac{E[s_0]}{c_4(v)} = \sigma_0 \tag{11.8}$$

where $\bar{s}_0$ is the mean of N' *i.i.d.* s_0 variates, v is the degrees of freedom relevant to the determination of the individual s_0 variates, and $N' \gg 1$. Note that v and N' are unrelated and eqn 11.8 *cannot* be used with $N' = 1$, that is, $E[s_0] \neq s_0$.

11.13   CENTRAL CONFIDENCE INTERVALS FOR Y_C AND Y_D

Given that CIs for σ_0 are already available in Tables A.3 and A.4, it is straightforward to obtain CIs for Y_C and Y_D, taking care to correct for the fact that critical t values must be replaced by critical z values. For example, suppose $N = 7$ *i.i.d.* replicate measurements were made from $N{:}\alpha, \sigma_0$, where σ_0 is unknown and therefore estimated by s_0. Then $v = 6$ and assume also $p \equiv 0.05 \equiv q$, so that $t_p = t_q \cong 1.943180$, $z_p = z_q \cong 1.644854$, and $y_D = 2t_p s_0$. Since the 95% CI for σ_0 is from $0.6444s_0$ to $2.2021s_0$, the 95% CI for $t_p \sigma_0$ is from $0.6444t_p s_0$ to $2.2021t_p s_0$ and the 95% CI for $2t_p \sigma_0$ is from $1.2888t_p s_0$ to $4.4042t_p s_0$. As in Chapter 7, $Y_C = z_p \sigma_0$ and $Y_D = 2z_p \sigma_0$, so the last step is to scale the limits by z_p/t_p. Hence, the 95% CI for Y_C is from $0.6444z_p s_0$ to $2.2021z_p s_0$ and the 95% CI for Y_D is from $1.2888z_p s_0$ to $4.4042z_p s_0$.

Note that these 95% CIs do not violate the stated substitution principle, despite containing $z_p s_0$ products, because the z_p/t_p factor was nothing more than a simple scale factor applied to the 95% CIs for $t_p \sigma_0$ and $2t_p \sigma_0$. The 95% CI for Y_D was tested using a simulation model (not shown) similar to those in Figs 11.1–11.4. For 10 million 95% CIs generated, Y_D was contained within 94.99% of the CIs, was below 2.51% and above 2.50%. Hence, the 95% CI for Y_D worked properly, but was rather wide: $2.120s_0$ to $7.244s_0$. In Chapter 15, a technique is demonstrated that, when applicable, dramatically narrows CIs.

11.14 CENTRAL CONFIDENCE INTERVALS FOR X_C AND X_D

Confidence intervals for X_C and X_D depend also upon whether β is known or estimated. Since β is assumed known in this chapter, CIs for X_C and X_D are directly determined from those for Y_C and Y_D by dividing the limits by β. Thus, for the example above, the 95% CI for X_D is from $1.2888 z_p s_0/\beta$ to $4.4042 z_p s_0/\beta$. As above, the CI is rather wide. When both σ_0 and β are unknown, see Chapter 14.

11.15 TABULAR SUMMARY OF THE EQUATIONS

A convenient summary of the Currie decision level and detection limit equations is provided in Table 11.1.

11.16 SIMULATION CORROBORATION OF THE EQUATIONS IN TABLE 11.1

The Monte Carlo simulation, with $M = 1$, is shown in Fig. 11.10. Since α and β are assumed known, s_0 is computed from 7 *i.i.d.* replicates at a fixed standard value, that is, at $x = X_1$.

Table 11.1 Currie Decision and Detection Expressions

Parameter	Known	Unknown	Estimator
σ_0		✓	s_0
β	✓		–
α	✓		–

	Decision Level	Detection Limit
Response domain	$r_C = \alpha + t_p s_0$	$r_D = \alpha + (t_p + t_q)s_0$
Net response domain	$y_C = t_p s_0$	$y_D = (t_p + t_q)s_0$
Content domain	$x_C = t_p s_0/\beta$	$x_D = (t_p + t_q)s_0/\beta$

1. Make N measurements at any desired X_1 value (that is, homoscedasticity assumed). Then s_0 is their sample standard deviation.

2. If $M > 1$, multiply each s_0 by $\eta^{1/2}$, where $\eta^{1/2} = 1/M^{1/2}$.

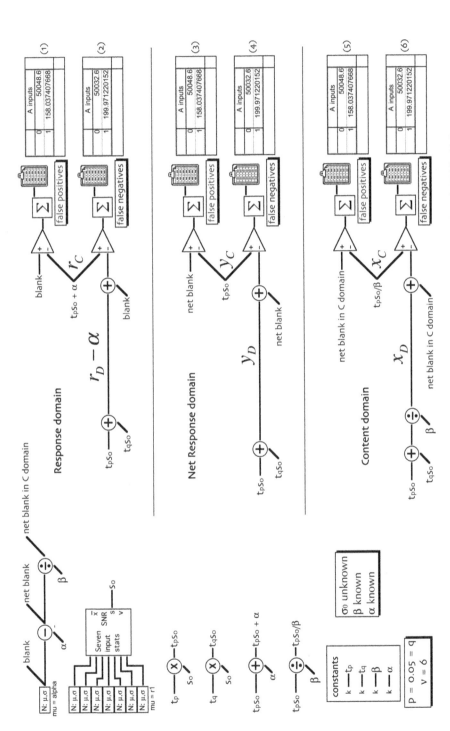

Figure 11.10 The Monte Carlo simulation, with $M = 1$, that corroborates the equations in Table 11.1.

Table 11.2 **Summary of Simulation Results for False Positives and False Negatives**

Domain	False Positives ± 1 Standard Deviation	False Negatives ± 1 Standard Deviation
Response	$50\,048.6 \pm 158.0$	$50\,032.6 \pm 200.0$
Net response	$50\,048.6 \pm 158.0$	$50\,032.6 \pm 200.0$
Content	$50\,048.6 \pm 158.0$	$50\,032.6 \pm 200.0$

The results of running the 10 simulations are shown directly in Fig. 11.10. For clarity, these are summarized in Table 11.2. The results demonstrate that the Currie decision level and detection limit equations in Table 11.1 are correct.

11.17 CHAPTER HIGHLIGHTS

In this chapter, σ_0 was assumed to be unknown, thereby necessitating the use of critical t values whenever σ_0 was estimated by a sample standard deviation, s_0. The bias correction factor, $c_4(v)$, was shown to be *unusable* in the construction of unbiased confidence intervals. An important subtlety in testing for false negatives was discussed in detail, as was the correct method for determining confidence intervals for Y_C and Y_D, and, if β is known, for X_C and X_D. The full set of decision level and detection limit equations was presented, tested via Monte Carlo simulations, and found to be correct.

REFERENCES

1. W.S. Gosset, "The probable error of a mean", *Biometrika* **6** (1908) 1–25.

2. L.A. Currie, "Detection: international update, and some emerging di-lemmas involving calibration, the blank, and multiple detection decisions", *Chemom. Intell. Lab. Syst.* **37** (1997) 151–181.

3. Multi-Agency Radiological Laboratory Analytical Protocols (MARLAP) Manual Volume III, Section 20A.3.1, 20-54–20-58. United States agencies in MARLAP are EPA, DoD, DoE, DHS, NRC, FDA, USGS and NIST. This large document is available as a PDF on-line.

4. L.A. Currie, for IUPAC, "Nomenclature in evaluation of analytical methods including detection and quantification capabilities" *Pure Appl. Chem.* **67** (1995) 1699–1723, IUPAC ©1995.

12

IF ONLY THE TRUE SLOPE IS KNOWN

12.1 INTRODUCTION

Assume σ_0 and α are unknown, but β is known. Then $r(x) = \alpha + \beta x$ is estimated by

$$\widehat{r}(x) = \widehat{\alpha} + \beta x \tag{12.1}$$

where $\widehat{\alpha}$ is the estimate of α. As in Chapter 11, σ_0 must be estimated by s_0. This is done by making M_0 replicate measurements of an analytical blank and computing their sample standard deviation, s_0, with $v = M_0 - 1$. Since the noise is homoscedastic, any convenient standard may be used in place of the analytical blank. Then, the Currie decision level and detection limit equations follow from those in Chapter 8 simply by making the following substitutions: $\sigma_0 \mapsto s_0$, $z_p \mapsto t_p$, and $z_q \mapsto t_q$. There are no additional issues to consider.

12.2 RESPONSE DOMAIN CURRIE DECISION LEVEL AND DETECTION LIMIT

The response domain Currie decision level, r_C, is defined as

$$r_C \equiv \widehat{\alpha} + t_p \eta^{1/2} s_0 \tag{12.2}$$

and the corresponding response domain Currie detection limit, r_D, is defined as

$$r_D \equiv \widehat{\alpha} + (\eta^{1/2} t_p + t_q / M^{1/2}) s_0 = r_C + t_q s_0 / M^{1/2} \tag{12.3}$$

Limits of Detection in Chemical Analysis, First Edition. Edward Voigtman.
© 2017 John Wiley & Sons, Inc. Published 2017 by John Wiley & Sons, Inc.
Companion Website: www.wiley.com/go/Voigtman/Limits_of_Detection_in_Chemical_Analysis

Note that the $\eta^{1/2}$ factor *only* multiplies t_p, *not* the sum of t_p and t_q. Both r_C and r_D are the sums of χ distributed random variates, plus Gaussians (due to $\hat{\alpha}$).

12.3 NET RESPONSE DOMAIN CURRIE DECISION LEVEL AND DETECTION LIMIT

The net response domain Currie decision level, y_C, is defined as

$$y_C \equiv t_p \eta^{1/2} s_0 \tag{12.4}$$

and the net response domain Currie detection limit, y_D, is defined as

$$y_D \equiv (t_p + t_q)\eta^{1/2} s_0 = \frac{t_p + t_q}{t_p} y_C \tag{12.5}$$

Both y_C and y_D are χ distributed random variates.

12.4 CONTENT DOMAIN CURRIE DECISION LEVEL AND DETECTION LIMIT

The content domain Currie decision level, x_C, is defined as

$$x_C \equiv y_C/\beta = t_p \eta^{1/2} s_0/\beta \tag{12.6}$$

and the content domain Currie detection limit, x_D, is defined as

$$x_D \equiv \frac{y_D}{\beta} = \frac{(t_p + t_q)\eta^{1/2} s_0}{\beta} = \frac{t_p + t_q}{t_p} x_C \tag{12.7}$$

Both x_C and x_D are χ distributed random variates.

12.5 GRAPHICAL ILLUSTRATIONS OF THE DECISION LEVEL AND DETECTION LIMIT EQUATIONS

As discussed in Chapter 11, graphical illustrations of the Currie equations, when σ_0 is estimated by s_0, may be inadvertently misleading because y_C is a χ distributed variate. Hence, depicting any *specific* value of y_C may be problematic. For this reason, no further *single* illustrations will be presented.

12.6 TABULAR SUMMARY OF THE EQUATIONS

A convenient summary of the Currie decision level and detection limit equations is provided in Table 12.1.

Table 12.1 Currie Decision and Detection Expressions

Parameter	Known	Unknown	Estimator
σ_0		✓	s_0
β	✓		–
α		✓	$\widehat{\alpha}$

	Decision Level	Detection Limit
Response domain	$r_C = \widehat{\alpha} + t_p \eta^{1/2} s_0$	$r_D = \widehat{\alpha} + (\eta^{1/2} t_p + t_q / M^{1/2}) s_0$
Net response domain	$y_C = t_p \eta^{1/2} s_0$	$y_D = (t_p + t_q) \eta^{1/2} s_0$
Content domain	$x_C = t_p \eta^{1/2} s_0 / \beta$	$x_D = (t_p + t_q) \eta^{1/2} s_0 / \beta$

1. No OLS calibration curve since β is known.
2. Make N measurements at any desired X_1 value (that is, homoscedasticity assumed). Then s_0 is their sample standard deviation and $\widehat{\alpha} = \overline{r}(X_1) - \beta X_1$.
3. $\eta^{1/2} \equiv (1 + N^{-1})^{1/2}$ if $M = 1$. This is a special case of $\eta^{1/2} \equiv (M^{-1} + N^{-1})^{1/2}$.

12.7 SIMULATION CORROBORATION OF THE EQUATIONS IN TABLE 12.1

The Monte Carlo simulation, with $M = 1$, is shown in Fig. 12.1. Since β is known, OLS is not performed. Rather, for every simulation step, $N \equiv M_{X_1} = 7$ *i.i.d.* replicate responses at $x = X_1$ are averaged to yield $\overline{r}(X_1)$ values and concomitantly used to compute s_0 variates. Then, the successive estimates of $\widehat{\alpha}$, shown as "a" in Fig. 12.1, are $\widehat{\alpha} \equiv a = \overline{r}(X_1) - \beta X_1$. Thus, $a \sim N{:}\alpha, \sigma_a$, where $\sigma_a \equiv \sigma_0 / M_{X_1}^{1/2}$. The noise pooling factor, $\eta^{1/2}$, is $(8/7)^{1/2}$.

The results of running the 10 simulations are shown directly in Fig. 12.1 and are summarized in Table 12.2. The results demonstrate that the Currie decision level and detection limit equations in Table 12.1 are correct.

12.8 CHAPTER HIGHLIGHTS

In this chapter, only β was assumed to be known, making it easy to obtain the full set of decision level and detection limit equations. These were presented, tested via Monte Carlo simulations, and found to be correct.

Table 12.2 Summary of Simulation Results for False Positives and False Negatives

Domain	False Positives ± 1 Standard Deviation	False Negatives ± 1 Standard Deviation
Response	$50\,019.8 \pm 245.6$	$49\,915.9 \pm 233.1$
Net response	$50\,019.8 \pm 245.6$	$49\,971.4 \pm 160.7$
Content	$50\,019.8 \pm 245.6$	$49\,971.4 \pm 160.7$

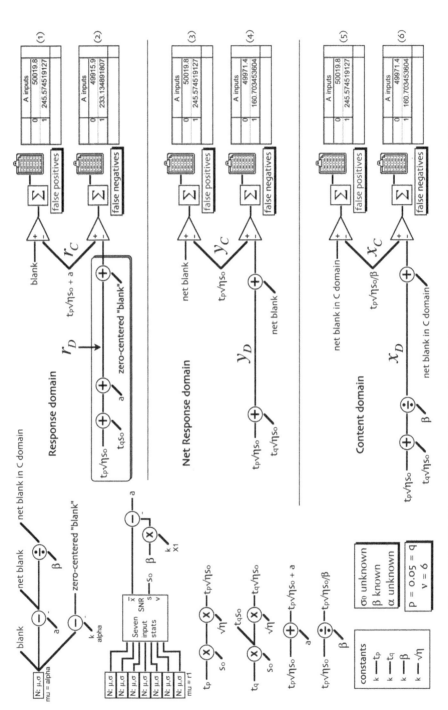

Figure 12.1 The Monte Carlo simulation, with $M = 1$, that corroborates the equations in Table 12.1.

13

IF ONLY THE TRUE INTERCEPT IS KNOWN

13.1 INTRODUCTION

Assume σ_0 and β are unknown, but α is known. Then $r(x) = \alpha + \beta x$ is estimated by

$$\widehat{r}(x) = \alpha + bx \tag{13.1}$$

where b is the estimate of the slope. There are two commonly used methods for obtaining unbiased, normally distributed b estimates. The first is to simply perform ordinary least squares (OLS) on N calibration standards, denoted $X_1, X_2, \ldots, X_N$, forcing the regression line through α. Then $b \sim N{:}\beta, \sigma_b$, with σ_b being the population standard error of the slope:

$$\sigma_b = \sigma_0 / \left(\sum_{i=1}^{N} X_i^2 \right)^{1/2} \tag{13.2}$$

for forced regression through α. The equation for b is eqn 9.10.

Alternatively, given that α is known and two points define a straight line, pick the highest constant x value in the actual linear range, for example, X_H, make M_{X_H} replicate measurements of $r(X_H)$, and compute the slope:

$$b = \frac{\bar{r}(X_H) - \alpha}{X_H} \tag{13.3}$$

where $\bar{r}(X_H)$ is the sample mean of the M_{X_H} replicate measurements at $x \equiv X_H$. The distribution of b is $N{:}\beta, \sigma_0 / X_H M_{X_H}^{1/2}$, showing that X_H should be as large as feasible.

Limits of Detection in Chemical Analysis, First Edition. Edward Voigtman.
© 2017 John Wiley & Sons, Inc. Published 2017 by John Wiley & Sons, Inc.
Companion Website: www.wiley.com/go/Voigtman/Limits_of_Detection_in_Chemical_Analysis

In addition, σ_0 is assumed unknown and therefore must be estimated by s_0. As in Chapter 11, this may be done by making M_0 replicate measurements of an analytical blank and computing their sample standard deviation, s_0, with $v = M_0 - 1$. Since the noise is homoscedastic, any convenient standard may be used in place of the analytical blank. A better alternative, which avoids any possibility of inconsistent degrees of freedom, is given further below.

13.2 RESPONSE DOMAIN CURRIE DECISION LEVEL AND DETECTION LIMIT

The Currie decision level and detection limit equations follow from those in Chapter 9 simply by making the following substitutions: $\sigma_0 \mapsto s_0$, $z_p \mapsto t_p$, and $z_q \mapsto t_q$. Therefore, the response domain Currie decision level, r_C, is defined as

$$r_C \equiv \alpha + t_p s_0 \tag{13.4}$$

and the response domain Currie detection limit, r_D, is defined as

$$r_D \equiv \alpha + (t_p + t_q)s_0 = r_C + t_q s_0 \tag{13.5}$$

Both r_C and r_D are the sums of χ distributed random variates, plus the constant α.

13.3 NET RESPONSE DOMAIN CURRIE DECISION LEVEL AND DETECTION LIMIT

The net response domain Currie decision level, y_C, is defined as

$$y_C \equiv t_p s_0 = r_C - \alpha \tag{13.6}$$

and the net response domain Currie detection limit, y_D, is defined as

$$y_D \equiv (t_p + t_q)s_0 = r_D - \alpha = \frac{t_p + t_q}{t_p}y_C \tag{13.7}$$

Both y_C and y_D are χ distributed random variates.

13.4 CONTENT DOMAIN CURRIE DECISION LEVEL AND DETECTION LIMIT

The content domain Currie decision level, x_C, is defined as

$$x_C \equiv y_C/b = t_p s_0/b \tag{13.8}$$

and the content domain Currie detection limit, x_D, is defined as

$$x_D \equiv \frac{y_D}{b} = \frac{(t_p + t_q)s_0}{b} = \frac{t_p + t_q}{t_p} x_C \tag{13.9}$$

Both x_C and x_D are *modified noncentral t variates* [1, 2]. Their distribution functions are eqns D.5 and D.6, respectively. Much more will be said about them in the following chapter.

13.5 TABULAR SUMMARY OF THE EQUATIONS

A convenient summary of the Currie decision level and detection limit equations is provided in Table 13.1.

13.6 SIMULATION CORROBORATION OF THE EQUATIONS IN TABLE 13.1

The Monte Carlo simulation, with $M = 1$, is shown in Fig. 13.1. Since α is known, b is estimated as in Chapter 9 [3, eqn 2.50]:

$$b = \left(\sum_{i=1}^{N} y_i x_i \right) / \left(\sum_{i=1}^{N} x_i^2 \right) \tag{13.10}$$

with s_0 estimated via [3, eqn 2.52]:

$$s_0 = \left[\frac{1}{N-1} \left(\sum_{i=1}^{N} y_i^2 - b \sum_{i=1}^{N} y_i x_i \right) \right]^{1/2} \tag{13.11}$$

with $y_i = r_i - \alpha$ in both equations.

Table 13.1 Currie Decision and Detection Expressions

Parameter	Known	Unknown	Estimator
σ_0		✓	s_0
β		✓	b (from forced OLS)
α	✓		–

	Decision Level	Detection Limit
Response domain	$r_C = \alpha + t_p s_0$	$r_D = \alpha + (t_p + t_q)s_0$
Net response domain	$y_C = t_p s_0$	$y_D = (t_p + t_q)s_0$
Content domain	$x_C = t_p s_0/b$	$x_D = (t_p + t_q)s_0/b$

1. The value of b is from OLS forced through α.
2. The estimators b and s_0 are determined using eqns 13.10 and 13.11.
3. If $M > 1$, multiply each s_0 by $\eta^{1/2}$, where $\eta^{1/2} = 1/M^{1/2}$.

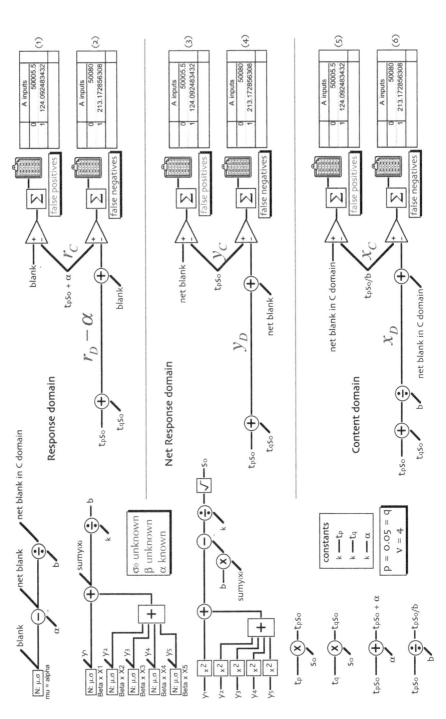

Figure 13.1 The Monte Carlo simulation, with $M = 1$, that corroborates the equations in Table 13.1.

Table 13.2 Summary of Simulation Results for False Positives and False Negatives

Domain	False Positives ± 1 Standard Deviation	False Negatives ± 1 Standard Deviation
Response	50 005.5 ± 124.1	50 080.0 ± 213.2
Net response	50 005.5 ± 124.1	50 080.0 ± 213.2
Content	50 005.5 ± 124.1	50 080.0 ± 213.2

The results of running the 10 simulations are shown directly in Fig. 13.1 and are summarized in Table 13.2. The results in Table 13.2 demonstrate that the Currie decision level and detection limit equations in Table 13.1 are correct.

13.7 CHAPTER HIGHLIGHTS

In this chapter, only α was assumed to be known, making it easy to obtain the full set of decision level and detection limit equations. These were presented, tested via Monte Carlo simulations, and found to be correct.

REFERENCES

1. E. Voigtman, "Limits of detection and decision. Part 1", *Spectrochim. Acta, B* **63** (2008) 115–128.
2. E. Voigtman, "Limits of detection and decision. Part 2", *Spectrochim. Acta, B* **63** (2008) 129–141.
3. D.C. Montgomery, E.A. Peck, *Introduction to Linear Regression Analysis*, John Wiley and Sons, New York, 1982.

14

IF ALL THREE PARAMETERS ARE UNKNOWN

14.1 INTRODUCTION

In this chapter, we examine what happens when all three parameters are unknown. This is the most realistic situation in practice because population parameters are almost never known, aside from computer simulations and the like. The Currie decision level and detection limit equations follow from those in Chapter 10 simply by making the following substitutions: $\sigma_0 \mapsto s_0$, $z_p \mapsto t_p$, and $z_q \mapsto t_q$.

Assume σ_0, β, and α are all unknown. Then $r(x) = \alpha + \beta x$ is estimated by

$$\hat{r}(x) = a + bx = \hat{\alpha} + \hat{\beta}x \qquad (14.1)$$

where a (or $\hat{\alpha}$) is the estimate of α and b (or $\hat{\beta}$) is the estimate of β. By far, the most common method of estimation of a and b is ordinary least squares (OLS) on a set of N calibration standards: $X_1, X_2, \ldots, X_N$. Degenerate standards are allowed, that is, standards may be used more than once. As in Chapter 10, the restriction of only one measurement per standard ensures that σ_r, the population standard error about regression, equals σ_0. Hence, s_r equals s_0, with due attention to degrees of freedom. It is assumed, therefore, that OLS on N calibration standards is used to provide the necessary estimates of σ_0, β, and α, that is, s_0 ($=s_r$), b and a. Table 10.1, provides a convenient summary of relevant OLS definitions, distributions, and relationships.

14.2 RESPONSE DOMAIN CURRIE DECISION LEVEL AND DETECTION LIMIT

The response domain Currie decision level, r_C, is defined as

$$r_C \equiv \hat{\alpha} + t_p \eta^{1/2} s_0 \qquad (14.2)$$

Limits of Detection in Chemical Analysis, First Edition. Edward Voigtman.
© 2017 John Wiley & Sons, Inc. Published 2017 by John Wiley & Sons, Inc.
Companion Website: www.wiley.com/go/Voigtman/Limits_of_Detection_in_Chemical_Analysis

and the response domain Currie detection limit, r_D, is defined as

$$r_D \equiv \hat{\alpha} + (\eta^{1/2}t_p + t_q/M^{1/2})s_0 = r_C + t_q s_0/M^{1/2} \qquad (14.3)$$

Note that the $\eta^{1/2}$ factor *only* multiplies t_p, *not* the sum of t_p and t_q. Both r_C and r_D are the sums of χ distributed random variates, plus Gaussians (due to $\hat{\alpha}$).

14.3   NET RESPONSE DOMAIN CURRIE DECISION LEVEL AND DETECTION LIMIT

The net response domain Currie decision level, y_C, is defined as

$$y_C \equiv t_p \eta^{1/2} s_0 \qquad (14.4)$$

and the net response domain Currie detection limit, y_D, is defined as

$$y_D \equiv (t_p + t_q)\eta^{1/2} s_0 = \frac{t_p + t_q}{t_p} y_C \qquad (14.5)$$

Both y_C and y_D are χ distributed random variates, as per eqn A.31.

14.4   CONTENT DOMAIN CURRIE DECISION LEVEL AND DETECTION LIMIT

The content domain Currie decision level, x_C, is defined as

$$x_C \equiv y_C/b = t_p \eta^{1/2} s_0/b \qquad (14.6)$$

and the content domain Currie detection limit, x_D, is defined as

$$x_D \equiv \frac{y_D}{b} = \frac{(t_p + t_q)\eta^{1/2} s_0}{b} = \frac{t_p + t_q}{t_p} x_C \qquad (14.7)$$

Both x_C and x_D are *modified noncentral t variates* [1, 2], as per eqns D.5 and D.6.

14.5   THE NONCENTRAL t DISTRIBUTION REAPPEARS FOR GOOD

As discussed in Chapter 10, performing OLS on a set of N calibration standards yields values of a, b, s_a, s_b, s_r, r (or related variants), and t_{slope}. The t_{slope} variate is noncentral t distributed, as follows:

$$t_{\text{slope}} \sim t(u|v, \delta) = t(u|N - 2, \beta/\sigma_b) \qquad (14.8)$$

where v is degrees of freedom and δ is the noncentrality parameter. This led immediately to a discussion of the essential role that the noncentral t distribution plays in
hypothesis testing of the slope of the calibration curve.

Now, however, the noncentral t distribution makes its fundamental and most
important appearance in detection limit theory and applications. As shown in
Appendix D, when σ_0 is estimated by the χ distributed variate s_0 and $\beta > 0$ is estimated by the independent Gaussian distributed variate b, then s_0/b is the *reciprocal*
of a noncentral t variate. Since x_C and x_D are simply scaled variants of s_0/b, as per
eqns 14.6 and 14.7, they are *scaled* reciprocals of noncentral t variates. These are
referred to as *modified noncentral t variates*. Explicitly

$$x_C \sim (t_p \eta^{1/2} S_{XX}^{1/2}/x_C^2) t (t_p \eta^{1/2} S_{XX}^{1/2}/x_C | v, \delta) \qquad (14.9)$$

and

$$x_D \sim ((t_p + t_q)\eta^{1/2} S_{XX}^{1/2}/x_D^2) t ((t_p + t_q)\eta^{1/2} S_{XX}^{1/2}/x_D | v, \delta) \qquad (14.10)$$

where $S_{XX}^{1/2}$ is the square root of the S_{XX} summation defined in Table 10.1, $v \equiv N - 2$
and $\delta \equiv \beta/\sigma_b = \beta S_{XX}^{1/2}/\sigma_0$. These are eqns D.5 and D.6, respectively.

14.6 AN INFORMATIVE COMPUTER SIMULATION

Consider a simple CMS with the following parameters (with arbitrary, but consis-
tent, units): $\alpha \equiv -0.05$, $\beta \equiv 3.9$, and $\sigma_0 \equiv 0.3$. If all three parameters are assumed
unknown, then the most common method used to estimate them is construction of a
calibration curve, followed by OLS. Accordingly, assume $N = 4$ standards are to be
used, with errorless values 0.03, 0.1, 0.5, and 1.3. Then $v \equiv N - 2 = 2$, $\overline{X} = 0.4825$,
$S_{XX} = 1.019675$, $S_{XX}^{1/2} \cong 1.009790$, and, from eqn C.13:

$$\eta^{1/2} = \left[\frac{1}{M} + \frac{1}{N} + \frac{\overline{X}^2}{S_{XX}}\right]^{1/2} = \left[\frac{1}{1} + \frac{1}{4} + \frac{(0.4825)^2}{1.019675}\right]^{1/2} \cong 1.215859 \qquad (14.11)$$

since $M \equiv 1$ future blank is assumed. With $p \equiv 0.05 \equiv q$ and $v = 2$, $t_p = t_q \cong 2.919986$. Then in eqn 14.9, $t_p \eta^{1/2} S_{XX}^{1/2} \cong 3.58505$, and in eqn 14.10,
$(t_p + t_q)\eta^{1/2} S_{XX}^{1/2} \cong 7.17010$. The value of δ is 13.12726, computed from
$\delta = \beta S_{XX}^{1/2}/\sigma_0 = 3.9 \times 1.009790/0.3$.

The simulation that generates and tests 10 million x_C and x_D variates is shown in
Fig. 14.1, with the dialog box for the OLS block shown in Fig. 14.2. Ten simulations
are performed, with 1 million simulation steps each. For every simulation step, an
independent 4-point calibration curve is constructed, and all relevant test statistics
are computed and saved or processed, as necessary. The content domain x_C and x_D
variates are computed as per eqns 14.6 and 14.7, respectively, and binned into histograms. They are also tested, "on the fly," for their performance in regard to false

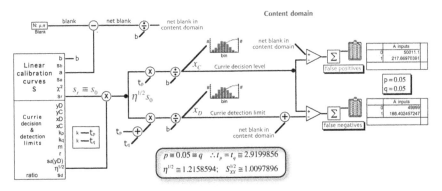

Figure 14.1   Simulation that generates and tests 10 million x_C and x_D variates.

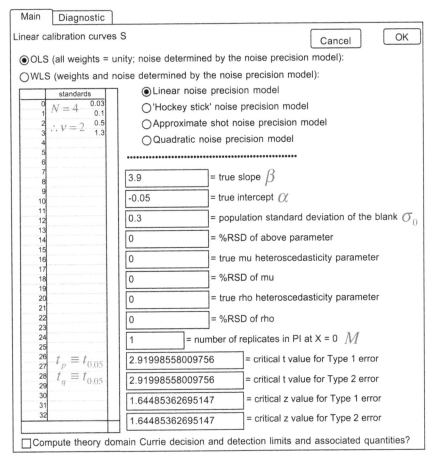

Figure 14.2   Dialog box showing population parameters and standards for the simulation in Fig. 14.1.

positives and false negatives. False positives are those blanks that exceed their associated x_C variate, while false negatives are those compound normal variates that fall below their associated x_C variate. As Fig. 14.1 clearly shows, the number of false positives obtained, expressed as sample mean ± 1 sample standard deviation, was $50\,011.1 \pm 217.7$, while the number of false negatives obtained was $49\,999.0 \pm 166.4$. These are statistically equivalent to the mutual expected values of $50\,000$.

Since $t_p \eta^{1/2} S_{XX}^{1/2} \cong 3.58505$, $(t_p + t_q)\eta^{1/2} S_{XX}^{1/2} \cong 7.17010$, $\delta \cong 13.12726$, and $\nu = 2$, the modified noncentral t distribution expressions for x_C and x_D, eqns 14.9 and 14.10, respectively, are

$$x_C \sim (3.58505/x_C^2)t(3.58505/x_C|2, 13.12726) \tag{14.12}$$

and

$$x_D \sim (7.17010/x_D^2)t(7.17010/x_D|2, 13.12726) \tag{14.13}$$

These expressions are shown in Fig. 14.3, plotted over the respective histograms of 10 million x_C and x_D variates from the simulation in Fig. 14.1. There is *nothing that can be adjusted* and it is evident that no adjustment would be necessary: the results are effectively perfect.

From Fig. 14.3, the distributions are clearly asymmetric for $\nu = 2$. They become narrower and more symmetric as ν increases. The mean values of x_C and x_D are 0.2434 and 0.4869, respectively.

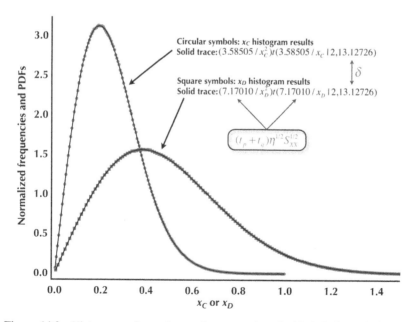

Figure 14.3 Histograms of x_C and x_D variates, overplotted with their theoretical PDFs.

14.7   CONFIDENCE INTERVAL FOR x_D, WITH A MAJOR PROVISO

If δ is unknown and only a single x_D variate is available, it is currently unknown how to construct a valid CI for x_D, and the same applies to construction of valid CIs for X_D. Previously, it was shown that $x_D < [(t_p + t_q)/t_p]\eta^{1/2}S_{XX}^{1/2}$ [1, eqn 30], but this is not useful in construction of a CI: for the simulation in Fig. 14.1, it implies that $x_D < 2.46$, which is quite obvious in Fig. 14.3. Badocco *et al.* provide an approximate expression [3, eqn 5] for the sample standard deviation of x_D, but it appears to be unusable in constructing asymmetric CIs for x_D or X_D.

Working with a quantitative bioassay involving a four-parameter logistic curve response function, Holstein *et al.* [4] reported detection limits and what were asserted to be their approximate 95% CIs. They also stated that "our method can also be applied to any other type of calibration curve for which an explicit equation can be determined." [4, p. 9796]. However, there was no reported validation and verification consisted of using "real data from … our laboratory to illustrate the steps and features of the method." [4, p. 9795]. Repeated attempts to apply their method to the linear CMS in Fig. 14.1 have been unsuccessful. A possible reason is that their method, which is based on the estimation of an asymptotic variance [4, eqn 10], may not be applicable to linear CMSs with small values of v. In any event, the asymmetric x_D histogram in Fig. 14.3 must have an asymmetric 95% CI.

In contrast to the above difficulties, it is easy to find CIs for the x_D histogram in Fig. 14.3: simply compute the histogram's CDF and invert it at the appropriate two bracketing points, for example, a central 95% CI extends from $CDF^{-1}(0.025)$ up to $CDF^{-1}(0.975)$. The same method is applicable to the PDF given by eqn 14.13, resulting in the 95% CI for x_D being from 0.0868 to 1.069, that is, $0.487 + 0.582 - 0.400$. Unfortunately, though, the 95% CI for x_D does not lead to *any* CI for X_D: the inversion method used in Chapter 11, and discussed in Appendix A, does not apply. Furthermore, eqn 14.10 is unusable for constructing CIs for x_D unless δ is known. However, $X_D = (z_p + z_q)\eta^{1/2}S_{XX}^{1/2}/\delta$, as per eqn D.8, so if δ or X_D is known, then the other is known as well. Hence, if δ is unknown, as is almost always the case, a histogram of x_D variates must suffice to compute CIs for x_D.

Accordingly, suppose a histogram of *many* experimental x_D variates was available and δ was unknown. Then solving $X_D = (z_p + z_q)\eta^{1/2}S_{XX}^{1/2}/\delta$ for δ and substituting into eqn 14.10 yields

$$x_D \sim ((t_p + t_q)\eta^{1/2}S_{XX}^{1/2}/x_D^2)t((t_p + t_q)\eta^{1/2}S_{XX}^{1/2}/x_D | v, (z_p + z_q)\eta^{1/2}S_{XX}^{1/2}/X_D) \quad (14.14)$$

Now X_D explicitly appears in the PDF for x_D and every constant quantity in eqn 14.14, other than X_D, is known as soon as the experiment is designed. Therefore, X_D may be estimated by using eqn 14.14, with X_D as the *only* adjustable parameter, to curve fit the experimental histogram of x_D variates. The following chapter demonstrates a much simpler variant of this, that is, direct estimation of δ, resulting in remarkably accurate estimation of X_D in a bootstrapped laser-excited molecular fluorescence experiment.

14.8 CENTRAL CONFIDENCE INTERVALS FOR PREDICTED x VALUES

Suppose $x \equiv X_i$ is unknown and $\bar{r}_i$ is the mean of M i.i.d. measurements of X_i. Then $\bar{r}_i \sim N{:}\alpha + \beta X_i, \sigma_0/M^{1/2}$ and $\bar{y}_i = (\bar{r}_i - a) \sim N{:}\beta X_i, \eta^{1/2}\sigma_0/M^{1/2}$, with $\eta^{1/2}$ as in eqn C.14. The back-calculated estimate of X_i, denoted by x_i or $\hat{X}_i$, is given by $\hat{X}_i \equiv x_i = \bar{y}_i/b$. This is *not* Gaussian distributed [1], but might be roughly approximated as such [5], in which case the approximate 95% CI for X_i is [6]

$$95\% \text{ for } X_i \approx \hat{X}_i \pm t_{0.025}\frac{s_r}{b}(M^{-1} + N^{-1} + (\hat{X}_i - \overline{X})^2/S_{XX})^{1/2} \qquad (14.15)$$

Note that eqn 14.15 is *not* a valid way to estimate content domain CIs for decision levels or detection limits. Even so, it is sometimes misused for precisely this purpose.

14.9 TABULAR SUMMARY OF THE EQUATIONS

A convenient summary of the Currie decision level and detection limit equations is provided in Table 14.1.

14.10 SIMULATION CORROBORATION OF THE EQUATIONS IN TABLE 14.1

The Monte Carlo simulation, with $M = 1$, is shown in Fig. 14.4. As noted above, OLS is performed in order to obtain s_0, b, and a. In the simulation, the $N = 5$ standards had values of 1, 2, 3, 4, and 5. The parameters used in the OLS computation block, which

Table 14.1 Currie Decision and Detection Expressions

Parameter	Known	Unknown	Estimator
σ_0		✓	s_0
β		✓	b
α		✓	a or $\hat{a}$

	Decision Level	Detection Limit
Response domain	$r_C = \hat{a} + t_p\eta^{1/2}s_0$	$r_D = \hat{a} + (\eta^{1/2}t_p + t_q/M^{1/2})s_0$
Net response domain	$y_C = t_p\eta^{1/2}s_0$	$y_D = (t_p + t_q)\eta^{1/2}s_0$
Content domain	$x_C = t_p\eta^{1/2}s_0/b$	$x_D = (t_p + t_q)\eta^{1/2}s_0/b$

1. Both a and b are usually obtained from OLS with N standards. In this case, with $M = 1$ assumed, $\eta^{1/2} \equiv (1 + N^{-1} + \overline{X}^2/S_{XX})^{1/2}$. This is a special case of $\eta^{1/2} \equiv (M^{-1} + N^{-1} + \overline{X}^2/S_{XX})^{1/2}$.
2. Make N measurements at any desired X_i value (that is, homoscedasticity assumed). Then s_0 is their sample standard deviation. Or use $s_0 = s_r$, the sample standard error about regression in OLS.

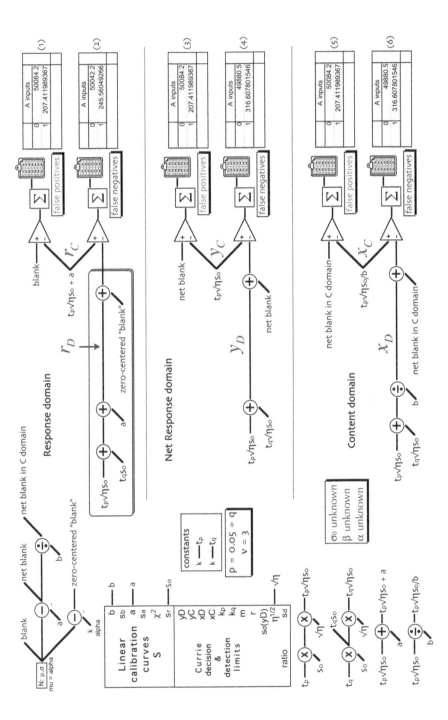

Figure 14.4 The Monte Carlo simulation, with $M = 1$, that corroborates the equations in Table 14.1.

Table 14.2   Summary of Simulation Results for False Positives and False Negatives

Domain	False Positives ±1 Standard Deviation	False Negatives ±1 Standard Deviation
Response	50 084.2 ± 207.4	50 042.2 ± 245.6
Net response	50 084.2 ± 207.4	49 880.5 ± 316.6
Content	50 084.2 ± 207.4	49 880.5 ± 316.6

is labeled "Linear calibration curves S," were $\alpha = 1$, $\beta = 3$, and $\sigma_0 = 0.1$. The noise pooling factor, $\eta^{1/2}$, was approximately 1.449138.

The results of running the 10 simulations are shown directly in Fig. 14.4 and are summarized in Table 14.2. The results demonstrate that the Currie decision level and detection limit equations in Table 14.1 are correct.

14.11   AN EXAMPLE: DIN 32645

Brüggemann *et al.* [7] compared the detection capabilities of German standard DIN 32645 [8], ISO 11843 [9], and Commission Decision 2002/657/EC [10]. For this purpose, they used the calibration data set given in Table 14.3 and the noise was assumed to be homoscedastic.

Since σ_0, β, and α are unknown, OLS was performed on the data, resulting in $s_0 = s_r = 192.29$, $b = 9661.9$, and $\hat{\alpha} = a = 2480.9$. These values, with $\nu = 8$, $M \equiv 1$, $\eta^{1/2} = 1.211060$, $t_{0.05} = 1.859548$, and $t_{0.01} = 2.896459$, were used to evaluate the *six* expressions under Table 14.1, with the results shown in Table 14.4.

From the results presented in Table 14.4, the following conclusions are evident:

Table 14.3   Calibration Data Set (DIN 32645)

X	y
0.05	3060
0.10	3522
0.15	3707
0.20	4280
0.25	5058
0.30	5510
0.35	5703
0.40	6205
0.45	7156
0.50	7178

Table 14.4   Decision Levels and Detection Limits for the Data in Table 14.3

Variate	Table 14.1 Expressions $p = 0.05 = q; p = 0.01 = q$	DIN 32645 [7, Table 1] $p = 0.05 = q; p = 0.01 = q$	ISO 11843 [7, Table 1] $p = 0.05 = q; p = 0.01 = q$
r_C	2913.9; 3155.4	2913.9; 3155.4	2913.9; 3155.4
r_D	3271.5; 3712.4	Not given	Not given
y_C	433.1; 674.5	Not given	Not given
y_D	866.1; 1349.1	Not given	Not given
x_C	0.0448; 0.0698	0.0448; 0.0698	0.0448; 0.0698
x_D	0.0896; 0.1396	0.0896; 0.1396	0.0896; 0.1396

The ISO 11843 x_D results are calculated by "approximating" $\delta_{p,p,v}$ by $2t_p$ in both cases.
As shown in Chapter 17, actual $\delta_{p,p,v}$ values are irrelevant and biased and the use of $2t_p$ results in the expressions becoming rigorously correct, as per Table 14.1, and unbiased for all degrees of freedom.
If $\delta_{0.05,0.05,8} = 3.61713$ were used in place of $2t_{0.05}$, x_D would have been biased low: 0.871.
If $\delta_{0.01,0.01,8} = 5.71003$ were used in place of $2t_{0.01}$, x_D would have been biased low: 0.1376.
No results are shown for Commission Decision 2002/657/EC.

1. For r_C, x_C, and x_D, DIN 32645 perfectly agrees with the correct expressions in Table 14.1.

2. The same applies to ISO 11843 *if* biased δ_{critical} values are replaced with critical t values.

3. ISO 11843 is irrelevant due to its use of δ_{critical} values: see Chapter 17.

In contrast to DIN 32645 and ISO 11843, Commission Decision 2002/657/EC is not an analytical detection standard; it is a collection of recommendations "focused on the evaluation of screening- and confirmation procedures judging the capability for the detection of contaminations refer to a permitted concentration level." [7, p. 104]. Moreover, it assumes that σ_0 is known, which is rarely the case, and it does not have fixed p and q values, since these are "dependent on the specific application of the analytical method." [7, p. 103]. Accordingly, Commission Decision 2002/657/EC is not relevant in the present context.

14.12   CHAPTER HIGHLIGHTS

This chapter dealt with the scenario that is by far the most commonly encountered one: all three population parameters are unknown and must be estimated. It was shown that, in the content domain, decision levels and detection limits were scaled reciprocals of noncentral t variates. The theoretical PDFs for x_C and x_D were in excellent agreement with x_C and x_D histograms obtained by computer simulation. The full set of decision level and detection limit equations was presented, tested via Monte Carlo simulations, and found to be correct. A detailed example was used to illustrate the equations and compare two detection limit standard protocols. The result was that DIN 32645 was in perfect agreement with the r_C, x_C, and x_D expressions in Table 14.1. In contrast, ISO 11843 was biased: see Chapter 17 for full details.

REFERENCES

1. E. Voigtman, "Limits of detection and decision. Part 1", *Spectrochim. Acta, B* **63** (2008) 115–128.

2. E. Voigtman, "Limits of detection and decision. Part 2", *Spectrochim. Acta, B* **63** (2008) 129–141.

3. D. Badocco, I. Lavagnini, A. Mondin, P. Pastore, "Estimation of the uncertainty of the quantification limit", *Spectrochim. Acta, B* **96** (2014) 8–11.

4. C.A. Holstein, M. Griffin, J. Hong, P.D. Sampson, "Statistical method for determining and comparing limits of detection of bioassays", *Anal. Chem.* **87** (2015) 9795–9801.

5. L.A. Currie, "Detection: International update, and some emerging di-lemmas involving calibration, the blank, and multiple detection decisions", *Chemom. Intell. Lab. Syst.* **37** (1997) 151–181.

6. I. Lavagnini, F. Magno, "A statistical overview on univariate calibration, inverse regression, and detection limits: application to gas chromatography/mass spectrometry technique", *Mass Spectrom. Rev.* **26** (2007) 1–18.

7. L. Brüggemann, P. Morgenstern, R. Wennrich, "Comparison of international standards concerning the capability of detection for analytical methods", *Accred. Qual. Assur.* **15** (2010) 99–104.

8. DIN 32645:2008-11, *Chemische Analytik-Nachweis-, Erfassungsund Bestimmungsgrenze*, Normenausschuss Materialprüfung im DIN, 2008.

9. L.A. Currie, for IUPAC, "Nomenclature in evaluation of analytical methods including detection and quantification capabilities" *Pure Appl. Chem.* **67** (1995) 1699–1723, IUPAC ©1995.

10. Commission Decision 2002/657/EC, implementing Council Directive 96/23/EC concerning the performance of analytical methods and the interpretation of results.

15

BOOTSTRAPPED DETECTION LIMITS IN A REAL CMS*

*This chapter is reprinted, with kind permission from Elsevier, from Voigtman, E. and Abraham, K.T. (2011) Statistical behavior of ten million experimental detection limits. *Spectrochim. Acta, B* **66**, 105–113. Aside from correction of a typographical error (*mea culpa*), and now consistent usage of "*r*" for responses and "*y*" for net responses, only trivial changes are made for purposes of notational and stylistic consistency with previous chapters. Subsequent material is discussed in an added postscript, followed by chapter highlights.

Abstract

Using a lab-constructed laser-excited fluorimeter, together with bootstrapping methodology, the authors have generated many millions of experimental linear calibration curves for the detection of rhodamine 6G tetrafluoroborate in ethanol solutions. The detection limits computed from them are in excellent agreement with both previously published theory and with comprehensive Monte Carlo computer simulations. Currie decision levels and Currie detection limits, each in the theoretical, chemical content domain, were found to be simply scaled reciprocals of the noncentrality parameter of the noncentral t distribution that characterizes univariate linear calibration curves that have homoscedastic additive Gaussian white noise. Accurate and precise estimates of the theoretical content domain Currie detection limit for the experimental system, with 5% (each) probabilities of false positives and false negatives, are presented.

15.1   INTRODUCTION

In 2008, Voigtman published a series of papers [1–4] presenting the detailed theory of limits of detection of the simplest chemical measurement systems. The theoretical developments were based on the assumption of a univariate linear system in which the measurement noise was additive Gaussian white noise (AGWN). Most of the work presented dealt with homoscedastic AGWN, but four common heteroscedastic noise precision models were also investigated. The detection limits, and associated decision levels, were as per Currie's program [5], thereby accommodating *a priori* specification and control of both false positives and false negatives. So-called "traditional" detection limits, wherein false negatives are entirely ignored, were shown to be just a special case of the Currie scheme in that the Currie decision level served double duty as both the decision level itself and the detection limit. It was also shown that all detection limits of the general form ks_{blank}/b, where k is a "coverage factor," s_{blank} is a sample standard deviation of the blank, and b is a regression-determined slope of an experimental linear calibration curve, are distributed as a modified noncentral t distribution and that the k value is actually a product that depends on both critical t or z values and on precisely how blank subtraction is performed.

The theory was tested extensively with Monte Carlo computer simulations. Literally tens of billions of calibration curves were generated and used to challenge and test the derived theory and, in every case, the computer simulations were in agreement with theory. However, despite the outstanding agreement between the derived theory and the computer simulation results, there were no experimental results. This was due to what, at the time, appeared to be an insurmountable obstacle, that is, the need for an experimental system that met the fundamental assumptions upon which theory was based and that could be used to generate millions of independent experimental detection limits in a reasonable time period.

The solution to the dilemma was a powerful resampling statistics technique known as bootstrapping [6]. Ordinarily, calibration curves are prepared by making measurements on a series of errorless standards. Even if every standard were to be measured only once, as is the case in all the work reported below, they would be measured sequentially. Then, ordinary least squares (that is, linear regression) would be applied to the acquired data and the desired statistical estimates would be computed. The drawback of this customary method is its general slowness: generating a million or more calibration curves is usually infeasible.

The bootstrapping solution is to measure each standard many times, using a suitable analog-to-digital converter (ADC) and computer, saving the resulting experimental data files on the computer. Then, a given calibration curve, and its associated detection limit and other desired statistics, is computed by randomly selecting (with replacement) a single measurement result from each of the desired standards files. The process is repeated many times; in the present work, 10 million calibration curves, and their associated statistics, were computed in 10 sets of one million each. One major advantage of the bootstrapping procedure is that all temporal correlations in the files for the standards are destroyed, thereby causing the measurement noise to be white. This is important because preamplifier noises are never truly white due to low frequency noise and, possibly, interferences.

15.2 THEORETICAL

15.2.1 Background

The fundamental assumptions, notations, and theory of Currie scheme decision levels and detection limits are as previously described in considerable detail [1, pp. 115–118]. Briefly, the univariate linear chemical measurement system is as described by eqn 15.1:

$$r = \alpha + \beta X + \text{noise} \tag{15.1}$$

where X is the analyte content (concentration, quantity, or amount), α is the true intercept, β is the true slope, r is the measured response, and the noise is AGWN having σ_0 population standard deviation. Thus, any specific measurement r_i is distributed as $r_i \sim N{:}\alpha + \beta X, \sigma_0$. The average of M such independent measurements, denoted by $\bar{r}$, is distributed as $\bar{r} \sim N{:}\alpha + \beta X, \sigma_0/M^{1/2}$. For the important special case of M replicate measurements of an analytical blank, $X = 0$, so that $\bar{r}_{\text{blank}} \sim N{:}\alpha, \sigma_0/M^{1/2}$.

15.2.2 Blank Subtraction Possibilities

If α were known, then the net response for the mean of M *future* replicate measurements of a true blank would be computed as $\bar{r}_{\text{future blank}} - \alpha$ and would be distributed as $\bar{r}_{\text{future blank}} - \alpha \sim N{:}0, \sigma_0/M^{1/2}$. The canonical and most reasonable choice for M is simply unity, that is, one *future* measurement of a true blank, in which case $r_{\text{future blank}} - \alpha \sim N{:}0, \sigma_0$. Throughout the present work, it is assumed that $M = 1$. However, for generality, the relevant equations will retain the general M.

If α is unknown, as is almost certainly the case, then blank subtraction is performed by subtracting an unbiased experimental estimate, denoted by $\hat{\alpha}$, from either $\bar{r}_{\text{blank}}$ or r_{blank}, depending on the value of M. Since $\hat{\alpha}$ is an estimate, the net response is noisier than the corresponding uncorrected response. Furthermore, there are several common ways to obtain an experimental value for $\hat{\alpha}$, so it becomes convenient to define a factor that groups together the effects of M and the error factor associated with $\hat{\alpha}$. This composite "blank subtraction" factor is denoted by $\eta^{1/2}$ and it has the general form

$$\eta^{1/2} \equiv [M^{-1} + \text{error factor due to } \hat{\alpha}]^{1/2} \tag{15.2}$$

One common method for obtaining a value for $\hat{\alpha}$ is by preparing an experimental calibration curve from N standards, performing linear regression, and then taking $\hat{\alpha}$ as the intercept of the calibration curve. In this case, $\eta^{1/2}$ is given by

$$\eta^{1/2} = (M^{-1} + N^{-1} + \bar{X}^2/S_{XX})^{1/2} \tag{15.3}$$

where $\overline{X}$ is the mean of the N standards (for $N > 2$) and S_{XX} is the sum of the squared differences: $S_{XX} \equiv \sum_{i=1}^{N} (X_i - \overline{X})^2$, with X_i being the ith standard. Note that

$$S_{XX} = N\sigma_X^2 = (N - 1)s_X^2 \qquad (15.4)$$

where σ_X^2 is the population variance of the errorless standards and s_X^2 is the corresponding sample variance of the errorless standards. This blank subtraction method is used in the present work.

Another common method for obtaining a value for $\hat{\alpha}$ is as the average of M_0 independent replicate measurements of a true blank. In this case, $\eta^{1/2}$ is given by

$$\eta^{1/2} = [M^{-1} + M_0^{-1}]^{1/2} \qquad (15.5)$$

For a more complicated blank subtraction possibility, involving pooling, see Ref. [4, p. 163].

15.2.3 Currie Decision Levels and Detection Limits

In Currie detection limit theory, M future independent replicate measurements are to be made on a true blank. The resulting average is blank subtracted with α or $\hat{\alpha}$, to yield the net response, which is then compared to the Currie decision level in the net response domain. A false positive occurs if the net response exceeds the decision level; otherwise, the net response is correctly classified as a true blank. Alternatively, and with equal validity, the comparison may be performed in the chemical content domain by simply dividing both the net response, and the decision level, by the slope of the calibration curve.

The present work focuses on the chemical content domain rather than on the net response domain, since the former is generally of greater interest to practicing analysts. In this case, the Currie decision levels (X_C and x_C) and Currie detection limits (X_D and x_D) are

$$X_C = z_p \eta^{1/2} \sigma_0 / \beta \quad \text{and} \quad x_C = t_p \eta^{1/2} s_0 / b \qquad (15.6)$$

$$X_D = (z_p + z_q) \eta^{1/2} \sigma_0 / \beta \quad \text{and} \quad x_D = (t_p + t_q) \eta^{1/2} s_0 / b \qquad (15.7)$$

where the equations on the left are the theoretical domain expressions (uppercase X) and those on the right are the experimental domain expressions (lowercase x). In the above equations, p = false positives probability, q = false negatives probability, z_p and z_q are the critical z values for p and q, respectively, t_p and t_q are the critical t values for p and q, respectively, b is the linear regression slope of the calibration curve, s_0 is the sample standard deviation of the noise, and σ_0, β, and $\eta^{1/2}$ are as defined previously. Note that these four expressions are exact, for all degrees of freedom.

Throughout the present work, and without loss of generality, it is assumed that $p = q = 0.05$ exactly. It is also assumed that blank subtraction involves the use of α only for the theoretical domain expressions on the left side of eqns 15.6 and 15.7.

However, for the experimental domain expressions on the right side of eqns 15.6 and 15.7, it is assumed that blank subtraction involves use of the experimental linear regression intercept $\hat{a}$. Hence the two theoretical expressions simplify, since $\eta^{1/2} = 1$, but the experimental expressions have $\eta^{1/2} > 1$, as given by eqn 15.3 with $M = 1$. Thus,

$$X_C = z_p \sigma_0/\beta \quad \text{and} \quad x_C = t_p \eta^{1/2} s_0/b \tag{15.8}$$

$$X_D = (z_p + z_q)\sigma_0/\beta \quad \text{and} \quad x_D = (t_p + t_q)\eta^{1/2} s_0/b \tag{15.9}$$

Note that if σ_0, β, and α were known, which is virtually never the case, then the two theoretical expressions on the left would be the complete solution to the detection limit problem. Indeed, finding a way to accurately estimate X_C and X_D from the experimental data alone is *the* outstanding detection limit problem.

15.3 EXPERIMENTAL

15.3.1 Experimental Apparatus

For a fair experimental test of the previously published theoretical results, it was necessary to construct a chemical measurement system that was linear over a reasonably wide range of analyte concentrations and that met the requirement of being limited by homoscedastic AGWN. Laser-excited molecular fluorescence of commonly used laser dyes is well known to meet the criterion of linearity easily, and moderate power laser diodes are readily available to serve as the excitation source, so this was the system chosen for testing. A block diagram of the experimental system is shown in Fig. 15.1, with component specifications given in Table 15.1.

Almost every component in the experimental system was lab constructed, inexpensive, or readily available in many laboratories, with nothing exotic or state of the art. This was deliberate; for present purposes, the more ordinary and obvious the experimental apparatus, the better. Indeed, even without the large amount of noise deliberately added for purposes of testing the detection limit theory, the fluorimeter in Fig. 15.1 was much inferior in detection power to *any* commercial fluorimeter.

15.3.2 Experiment Protocol

A total of 12 laser dye solutions were carefully prepared, using digital pipettes and class A volumetric flasks. Starting with the most dilute standard, 3 mL of solution was placed in a Suprasil cuvette, which was then placed in the sample compartment. The output of the summing amplifier was digitized at the rate of 1000 samples per second for a period of 12 s, thereby resulting in data files of 12 001 data per file. The resulting 12 data files were then used in all subsequent bootstrapping experiments. The standards, their concentrations and statistics (that is, sample means and sample standard deviations for each 12 001 point data file) are given in Table 15.2 and Fig. 15.2 shows the calibration curve that results from using all 12 standards in the calibration design. Note that the error bars are ± 1 standard deviation, as given in Table 15.2. Clearly, the

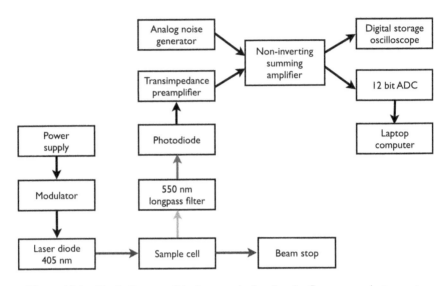

Figure 15.1   Block diagram of the laser-excited molecular fluorescence instrument.

Table 15.1   Components, Reagents, and Instruments Used

Item	Description
Laser diode	Blu-ray, 405 nm, <100 mW, eBay
Laser driver	FlexMod2, Nautilus Integration
Cuvette	Suprasil, 1.00 cm, Janus Laboratories, Inc.
Fluorophore	Rhodamine 6G tetrafluoroborate, Exciton Chemical Co., Inc.
Solvent	Ethanol, 200 proof, Pharmco-Aaper, Pharmco Products Inc.
Emission filter	20 CGA-550, 550 nm long pass, Oriel Corporation of America
Photodiode	UDT-10DP/SB, 0.5 A/W, United Detector Technology
Transimpedance preamplifier	Lab constructed, 1.10 MΩ transimpedance, 1.0 s output time constant
Noise generator	Lab constructed, ~289 mV rms
Summing amplifier	Lab constructed, signal channel gain = +1.00, noise channel gain = +7.10
ADC hardware	LabPro, 12 bit, ±10 V full scale, 1000 samples per second, Vernier Software & Technology
ADC software	Logger Pro, v. 2.2, Vernier Software & Technology
Laptop for data acquisition	Macintosh Powerbook G3 w/Mac OS 8.6, Apple Inc.
Desktop computer	Macintosh iMac 24 (white) w/Mac OS 10.5.8, Apple Inc.
Spreadsheet	Excel 2004 for Mac, v. 11.6, Microsoft Corp.
General computation	LightStone, E. Voigtman's freeware add-on for Extend
Software platform	Extend, v. 6.0.8, Imagine That, Inc.
Graphing	IGOR Pro, v. 6.20, Wavemetrics, Inc.
Oscilloscope	HM 205-2, 20 MHz digital storage, Hameg Instruments GmbH

Table 15.2   Fluorescence Standards (Rhodamine 6G Tetrafluoroborate in Ethanol)

Standard	Concentration (mg/100 mL)	Mean (V)	Standard Deviation (V)
1	0.03392	0.1091	0.2901
2	0.04240	0.1416	0.2913
3	0.0848	0.2779	0.2885
4	0.1272	0.4272	0.2897
5	0.2120	0.7374	0.2899
6	0.2968	1.041	0.2918
7	0.3604	1.261	0.2881
8	0.4240	1.481	0.2896
8a	0.6360	2.330	0.2908
9	0.8480	3.202	0.2876
9a	1.060	4.002	0.2859
10	1.272	4.847	0.2879

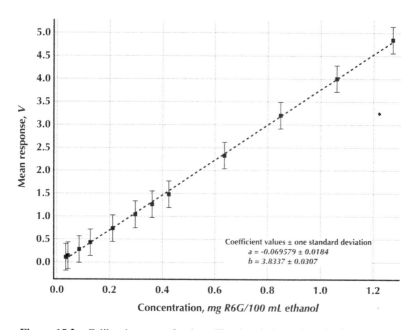

Figure 15.2   Calibration curve for the calibration design using all 12 standards.

standard deviations well approximate the requirement of homoscedasticity ("same scattering"), that is, the standard deviations are independent of the standard's concentrations. The mean of the 12 standard deviations is 0.2893 V and the standard deviation of the 12 standard deviations is 0.0017 V.

15.3.3   Testing the Noise: Is It AGWN?

With regard to the remaining noise properties, a simple analog noise generator, shown in Fig. 15.3 (upper half), was designed to provide Gaussian noise with approximately white power spectral density (PSD). The lower half of Fig. 15.3 shows the transimpedance preamplifier and noninverting summing amplifier. Noise performance was evaluated by first subtracting the mean values (that is, fluorescence signals) from each of the acquired experimental data files for the standards. The resulting 12 files were then concatenated, starting with that from the lowest standard (1) and working up to the highest standard (10), to yield a file of 144 012 points. The merged noise file was processed by computing and averaging 70 sequential 2048 point unilateral, mean-subtracted PSDs, without zero-filling or apodization. The result, shown in Fig. 15.4, is not perfectly white, and power supply interferences are evident at 60, 180, and 300 Hz, but the resampling process involved in bootstrapping automatically whitens the noise, as noted above. This was confirmed by direct test: a total of 700 unilateral, mean-subtracted PSDs, of 16 384 points each, were averaged, resulting in a white PSD having mean value 1.673×10^{-4} V^2/Hz and rms value of 0.2893 V.

The merged noise file was also binned into a histogram to obtain the empirical normalized probability density function (PDF), shown in the inset in Fig. 15.4, together with an overplotted Gaussian. The Gaussian curve fit standard deviation was 0.2897 V, in excellent agreement with the directly determined 144 012 point standard deviation of 0.2893 V. The magnitude of the added noise was designed to be more than

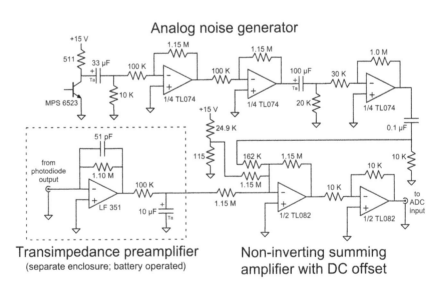

Figure 15.3   The analog noise generator, transimpedance preamplifier, and noninverting summing amplifier circuitry.

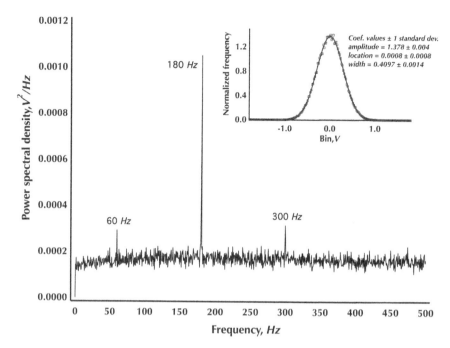

Figure 15.4 The power spectral density and probability density function (inset) of the noise on the digitized fluorescence response files.

sufficient to dominate the relatively small intrinsic noise (approximately 2.6 mV) of the transimpedance preamplifier. Hence, the requirement of AGWN is satisfied, since additivity is obviously satisfied by construction.

The time constant of the fluorimeter was 1 s, so that the fluorimeter's output was approximately the true fluorescence response, while the sampling frequency was 1 kHz. Without the large added amount of approximately white noise, an additional gain stage would have been required before the ADC and the digitized samples would have been strongly correlated. However, proper testing of the theory required that the noise be both Gaussian and uncorrelated, and hence independent. Therefore, the small intrinsic noise was deliberately submersed in a large excess of added noise with approximately white PSD, in order to have maximum control over and knowledge of, the operative noise. As noted above, the bootstrapping process entirely eliminated correlations in the resulting sum. As a consequence, the digitized samples were both Gaussian distributed and uncorrelated, as required to fairly test the theory.

15.3.4 Bootstrapping Protocol in the Experiments

Computation of the bootstrapped experimental calibration curves, and associated statistics, takes place in the computer program shown in Fig. 15.5. Up to 12

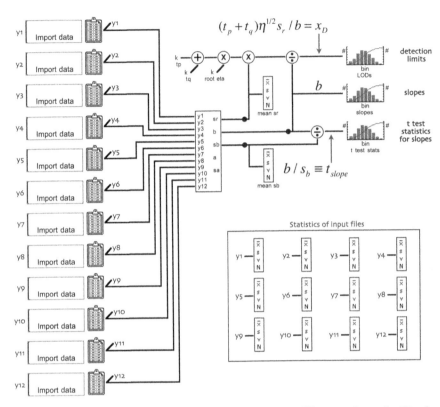

Figure 15.5 The bootstrapping program used to compute 10 million experimental calibration curves, x_D variates, and t_{slope} variates.

standards files may be imported, allowing generation of experimental calibration curves with as few as 3 standards and as many as 12 (an arbitrarily chosen upper limit). Additional inputs required are the number of standards to use (N), the concentration values of the standards (that is, the X_i values), and the values of $\eta^{1/2}$, t_p, and t_q. A simple Excel spreadsheet template, an example of which is shown in Fig. 15.6, was developed in order to streamline computation of necessary quantities such as $\eta^{1/2}$ and $S_{XX}^{1/2}$.

In the present work, 10 simulations of one million steps each were performed for each desired calibration curve design. This generated 10 million experimental calibration curves, each with its own t_{slope} test statistic and x_D detection limit, computed via eqn 15.9, since each standard was measured only once per calibration curve and therefore $s_0 = s_r$ (the sample standard error about regression). These were separately binned into histograms and the histograms exported as ASCII text files for subsequent analysis and plotting. The required simulation times were less than 10 min on a first-generation iMac 24.

Standard #	Concentration x	Mean of 12001 data Mean y	Standard deviation s
1	0.03392	0.1090611142753353	0.2901372242211251
4	0.1272	0.4271995076584413	0.2896582681766666
8a	0.6360	2.3296467646691020	0.290772599434172
10	1.272	4.8471448996151180	0.2878832320609865

Quantity	Value	Units
linear regression intercept = a =	-0.0541386855405068	V
linear regression slope = b =	3.8323572614831400	V/(mg/100mL)
number of standards = N =	4	none
degrees of freedom = v =	2	none
false positives probability = p =	0.0500000000000000	default = 5%
critical t value for false positives = t_p =	2.9199858580099756	none
critical z value for false positives = z_p =	1.6448362695147	none
false negatives probability = q =	0.0500000000000000	default = 5%
critical t value for false negatives = t_q =	2.9199858580099756	none
critical z value for false negatives = z_q =	1.6448362695147	none
population mean of standards = μ_x =	0.5172800000000000	mg/100mL
population standard deviation of standards = σ_x =	0.4923149431 0045	mg/100mL
sum of squared deviations = S_{xx} =	0.9694960128000000	square of above
"scale factor" = square root of S_{xx} =	0.98462988620090	mg/100mL
number of future blank replicates = M =	1	default = 1
factor due to blank subtraction = η =	1.52599762646492	none
square root of factor due to blank subtraction = $\eta^{1/2}$ =	1.2353127646463287	none
sum of critical t values = $t_p + t_q$ =	5.8399711601951	none
"coverage factor" = $(t_p + t_q)\eta^{1/2}$ = k =	7.2141909192769	none
(curve fit of t_{slope} histogram) estimate of δ =	12.838	none
population standard deviation for simulations = σ_0 =	0.2939284514140469	V
estimate of theoretical Currie Decision Level =	0.1262	mg/100mL
estimate of theoretical Currie Limit of Detection =	0.2523	mg/100mL

mg/100 mL	Volts	Volts

average s = 0.28961260310 7988

If the net response (i.e., blank-corrected response) is computed by subtraction of the OLS intercept "a" rather than α, then the $\eta^{1/2}$ factors MUST be used in the Currie Decision Level and Currie Limit of Detection. These are below:

estimate of theoretical Currie Decision Level = 0.1558

estimate of theoretical Currie Limit of Detection = 0.3117

Figure 15.6 The MS Excel template for computation of decision and detection limits and necessary calibration design values.

15.3.5 Estimation of the Experimental Noncentrality Parameter

For any given calibration curve design, the files of binned t_{slope} variates were then fitted with the noncentral t distribution with $N-2$ degrees of freedom. This provided an estimate, denoted by $\hat{\delta}$, of the noncentrality parameter, δ, of the noncentral t distribution. The reason δ is important is that the theoretical PDF for x_D is distributed as a modified noncentral t variate [2, eqn 26]:

$$p_{x_D}(x_D) = ((t_p + t_q)\eta^{1/2}S_{XX}^{1/2}/x_D^2)t((t_p + t_q)\eta^{1/2}S_{XX}^{1/2}/x_D|N - 2, \delta) \qquad (15.10)$$

where $\delta \equiv \beta/\sigma_b$ and σ_b is the population standard error of the slope. Thus, the normalized histogram of experimental x_D variates must be compared with eqn 15.10 and an estimate of δ is essential. This comparison was done using the computer program shown in Fig. 15.7, with $k \equiv (t_p + t_q)\eta^{1/2}$ and "scale factor" $= S_{XX}^{1/2} = N^{1/2}\sigma_X = (N - 1)^{1/2}s_X$, using eqn 15.4.

15.3.6 Computer Simulation Protocol

The program in Fig. 15.7 also allowed for overplotting of x_D variates generated by computer simulation. Computer-simulated calibration curves, plus x_C and x_D variates, were generated as before [1–4]. As each simulated calibration curve was generated, a separate true blank was also generated along with an x_C variate and an x_D variate. This allowed for immediate testing of the specified probabilities of false positives and false negatives. As was true for the experimental x_D variates, 10 million computer-simulated x_D variates, in 10 sets of one million each, were generated for each calibration curve design and the required simulation times were less than 10 min. The computer program for generation of the simulated data is shown in Fig. 15.8.

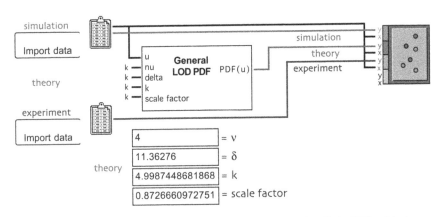

Figure 15.7 The program that overplots the theoretical detection limit PDFs with the normalized 10 million event histograms of experimental and simulation detection limits.

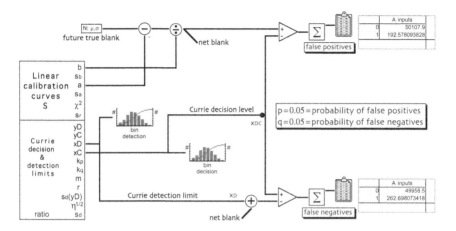

Figure 15.8 The Monte Carlo computer simulation program, with automatic testing of probabilities of false positives and false negatives.

15.4 RESULTS AND DISCUSSION

15.4.1 Results for Four Standards

Figure 15.9 shows an overplot of histogram data and theory for the calibration design using standards 1, 4, 8a, and 10. For this calibration design, the number of standards and their specific concentrations (see Table 15.2) result in $t_p = t_q = 2.919986$ (since $N = 4$ and $p = q = 0.05$), $\eta^{1/2} = 1.235313$, $S_{XX}^{1/2} = 0.984630$, and $k \equiv (t_p + t_q)\eta^{1/2} = 7.214191$, so that, using eqn 15.10, x_D is distributed as

$$p_{x_D}(x_D) = (7.10331/x_D^2)t(7.10331/x_D|2, 12.838) \tag{15.11}$$

where $\hat{\delta} = 12.838$ was used as the estimate of δ in eqn 15.10. A plot of this expression is shown in Fig. 15.9 as a solid curve. As noted above, the value for $\hat{\delta}$ was obtained by curve fitting the experimental histogram of 10 million binned t_{slope} variates for this calibration design.

The 10 million binned experimental x_D are overplotted in Fig. 15.9 in solid square markers and the 10 million binned simulation x_D are also overplotted in solid circular markers. For the computer simulations, the necessary inputs are the specific values of the standards, t_p, t_q, β, and σ_0. An unbiased estimate of β, denoted by b, was obtained by performing linear regression on the 12 001 point means versus concentrations (given in Table 15.2) for standards 1, 4, 8a, and 10. The resulting slope, b, is the estimate of β. This gave $\beta \approx b = 3.83236 \, V/(mg/100\,mL)$. Then σ_0 was estimated by noting that

$$\delta \equiv \beta/\sigma_b \approx b/\sigma_b = bS_{XX}^{1/2}/\sigma_r = bS_{XX}^{1/2}/\sigma_0 \tag{15.12}$$

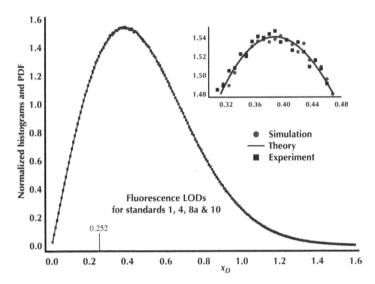

Figure 15.9   The overplotted theoretical (solid line) detection limit PDF for x_D (eqn 15.10), together with the normalized 10 million event histograms of experimental (filled squares) and simulation (filled circles) detection limits for calibration with 4 standards.

where σ_r is the population standard error about regression and only one measurement is made per standard, so that $\sigma_0 = \sigma_r$. Hence, $\hat{\sigma}_0 \approx b S_{XX}^{1/2}/\hat{\delta}$. For this calibration design, $\hat{\sigma}_0 \approx 0.293928$ V and the agreement among experiment, theory, and computer simulation is essentially perfect.

Very importantly, $\hat{\sigma}_0$ is higher than the average of the four standard deviations for the standards 1, 4, 8a, and 10 (that is, 0.28961 V) and also higher than the average standard deviation for all 12 standards (that is, 0.28927 V). Neither of these latter values is an accurate or acceptable substitute for $\hat{\sigma}_0$ in the computer simulations.

15.4.2   Results for 3–12 Standards

Figure 15.10 shows the results for calibration designs having 3–12 standards, with the standards used shown in Table 15.3. Remarkably, this figure shows 100 million experimental limits of detection (LODs) plus 100 million computer simulation LODs; for each of the 10 calibration designs, the slightly fuzzy "curves" are triple plots of theoretical PDFs, experimental histograms, and computer simulation histograms, with methodology as described in Section 15.4.1.

Clearly, the results are in outstanding agreement in all cases. The 3-point calibration design is very broad and monotonically decreasing as x_D increases. All the other calibration designs have modes well above zero and it is clear that as the number of standards increases, the PDFs become narrower, taller, and more symmetric. Presumably, if the number of standards were to increase without limit, the x_D PDF would

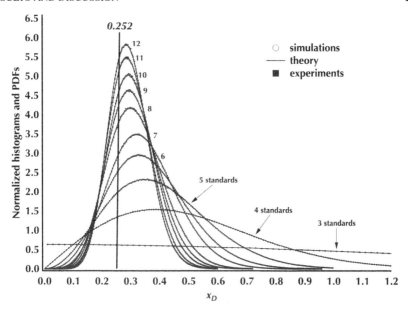

Figure 15.10 The overplotted detection limits for theory (solid lines), experiment (filled squares) and simulation (open circles), for calibration designs with 3–12 standards.

Table 15.3 Calibration Designs, Required Factors and Estimates of δ

Number of Standards	Standards Used	R^2	k $k = (t_p + t_q)\eta^{1/2}$	$S_{XX}^{1/2}$	$\hat{\delta}$
3	1, 6, 10	0.9997	16.312	0.922490	11.971
4	1, 4, 8a, 10	0.9997	7.2142	0.984630	12.838
5	1, 4, 7, 8a, 10	0.9994	5.6455	0.994578	12.991
6	1, 3, 5, 7, 8a, 9a	0.9994	4.9987	0.872666	11.363
7	2, 4, 6, 8, 8a, 9, 10	0.9993	4.7402	1.06206	13.912
8	1, 2, 3, 4, 6, 8, 9, 10	0.9994	4.3167	1.18632	15.509
9	1, 2, 3, 4, 6, 7, 8, 9, 10	0.9992	4.1816	1.18667	15.515
10	1, 2, 3, 5, 6, 7, 8a, 9, 9a, 10	0.9996	4.1268	1.33775	17.572
11	1, 2, 3, 4, 5, 6, 7, 8a, 9, 9a, 10	0.9996	4.0131	1.38047	18.147
12	1, 2, 3, 4, 5, 6, 7, 8, 8a, 9, 9a, 10	0.9994	3.9534	1.38074	18.103

become a delta function at X_D, the theoretical content domain Currie detection limit. This hypothesis has not yet been tested.

15.4.3 Toward Accurate Estimates of X_D

Accurate estimation of X_D is the foremost goal of detection limit investigations. The reason is simple: X_D is an errorless real number, greater than zero, that is entirely independent of calibration design and, therefore, ideally suited for its purpose of

Table 15.4   Estimates of X_D and Simulation Probabilities of False Positives and Negatives

Number of Standards	$\widehat{X}_D$ mg/100 mL	Percent Area Below $\widehat{X}_D$	False Positives Percentage (Left); False Negatives Percentage (Right)
3	0.2535	15.9	5.002 ± 0.015; 5.007 ± 0.016
4	0.2523	18.8	4.989 ± 0.014; 5.012 ± 0.026
5	0.2519	20.5	4.989 ± 0.022; 4.996 ± 0.022
6	0.2527	21.7	4.996 ± 0.018; 4.991 ± 0.026
7	0.2511	21.5	5.014 ± 0.022; 4.998 ± 0.020
8	0.2516	25.8	5.009 ± 0.023; 5.002 ± 0.023
9	0.2516	26.5	5.005 ± 0.028; 5.011 ± 0.026
10	0.2504	26.2	4.984 ± 0.021; 5.009 ± 0.012
11	0.2503	27.7	5.012 ± 0.017; 4.994 ± 0.022
12	0.2509	27.7	4.999 ± 0.019; 5.001 ± 0.021

serving as an invariant figure of merit for comparison purposes. It does depend upon the analyst's specified probabilities of false positives and false negatives, but these could certainly be standardized by IUPAC or others of that ilk. In the present work, these probabilities were arbitrarily defined as 5% each. Note in Table 15.4, however, that the second column contains estimates of X_D, for the 10 calibration designs shown in Fig. 15.10, and these average 0.2516 mg/100 mL, with a standard deviation of only 0.0010 mg/100 mL. Accordingly, the 99% confidence interval for X_D is $0.251_6 \pm 0.001_0$ mg R6G/100 mL ethanol. The value of 0.252 is plotted as a vertical line in Fig. 15.10 and as small vertical line in Fig. 15.9. The widths of these lines are actually wider than the 99% confidence interval.

15.4.4   How the X_D Estimates Were Obtained

These remarkably consistent estimates were obtained by first noting that the theoretical PDFs in Fig. 15.10 are in excellent agreement with their corresponding experimental data histograms. Hence, $\widehat{\delta}$, which was obtained from the bootstrapped experimental t_{slope} data, must be a relatively accurate estimate of δ. Likewise, the agreement between computer-simulated data and both theory and actual experimental data was also a direct result of the accuracy of $\widehat{\delta}$. But, by definition,

$$X_D \equiv (z_p + z_q)\sigma_0/\beta \tag{15.13}$$

assuming, as we have done above, that $M = 1$ and α is used for blank subtraction. Then, if only one measurement is made per standard, $\sigma_0 = \sigma_r$, and it is always true that $\sigma_r = \sigma_b S_{XX}^{1/2}$. Therefore,

$$X_D = (z_p + z_q)\sigma_0/\beta = (z_p + z_q)\sigma_r/\beta = (z_p + z_q)\sigma_b S_{XX}^{1/2}/\beta = (z_p + z_q)S_{XX}^{1/2}/\delta \tag{15.14}$$

The estimate of X_D, denoted by $\widehat{X}_D$, is obtained merely by substituting $\widehat{\delta}$ for δ in eqn 15.14, because everything else is constant and known as soon as the calibration design is set by the analyst, before any measurements have even been made. Therefore,

$$X_D \cong \widehat{X}_D = (z_p + z_q)S_{XX}^{1/2}/\widehat{\delta} = (z_p + z_q)N^{1/2}\sigma_X/\widehat{\delta} = (z_p + z_q)(N-1)^{1/2}s_X/\widehat{\delta}$$
(15.15)

using eqn 15.4. The values of $\widehat{X}_D$ for the 10 calibration designs shown in Fig. 15.10 are given in Table 15.4. Note that X_C is similar

$$X_C \cong \widehat{X}_C = z_p S_{XX}^{1/2}/\widehat{\delta} = z_p N^{1/2}\sigma_X/\widehat{\delta} = z_p(N-1)^{1/2}s_X/\widehat{\delta}$$
(15.16)

so that if false negatives are of no concern, that is, if one prefers the "traditional" detection limits, then these are, by definition, simply the Currie decision levels and eqns 15.8 and 15.16 are used.

15.4.5 Ramifications

Precisely as expected, the detection limit obtained in the present work is very poor due to the large amount of deliberately added noise and the underlying unexceptional fluorimeter itself. However, even if the noise were reduced by several orders of magnitude, the fundamental curves in Fig. 15.10 would not change except for scaling of the axes. In other words, experimental detection limits, x_D, are random samples from intrinsically broad distributions unless large numbers of standards are used in construction of the calibration curves. Table 15.4 shows the approximate percent of the area, under the relevant PDFs, that lies below $\widehat{X}_D$. For the present system, these are roughly the probabilities that any given experimental detection limits might fall below the true theoretical detection limit X_D. Clearly, it is considerably more probable that a given experimental detection limit will be above X_D, even with 12 standards.

Note that Table 15.3 also shows the "coverage factor" k for each of the calibration designs in the present work. These depend on $\eta^{1/2}$ and, therefore, upon how blank subtraction is performed. Table 15.4 shows that using the properly defined k values in Table 15.3, which are trivial to compute, yields excellent agreement between the expected 5% probabilities of false positives and false negatives and the Monte Carlo computer simulation results discussed above.

15.5 CONCLUSION

Thanks to bootstrapping, the moderately sized experimental data files were used to produce high-quality detection limit histograms, of 10 million events each, comparable to those produced by computer simulation. All of the histograms were found to be in excellent agreement with previously published theory. From eqns 15.15 and 15.16, it is then clear that detection limits and decision levels are simply scaled reciprocals of the noncentrality parameter of the modified noncentral t distribution that characterizes univariate linear calibration curves that have homoscedastic AGWN. Hence, any technique that accurately estimates δ automatically estimates X_C and X_D accurately, and thereby solves the general detection limit estimation problem for the system described above. What is remarkable is that there is only one parameter that needs to be estimated, and, with hindsight, this should have been obvious: detection limits and decision levels depend only on the quotient of σ_0 and β, not on the individual values. Furthermore, even if blank subtraction is effected by subtracting an estimate of α, rather than α itself, this just causes $\eta^{1/2}$ to be other than unity: $\eta^{1/2}$ is still easily computed from the calibration design parameters, eqns 15.6 and 15.7 are controlling and the rest of the analysis proceeds as before, *mutatis mutandis*.

There can be little doubt, then, that the previously published theory [1–4] provides an accurate description of the fundamental statistical behavior of univariate linear chemical measurement systems that are characterized by homoscedastic AGWN. These systems are ubiquitous in modern analysis and it has been a vexing problem that Currie's program [5] could not be brought to fruition until now. Many labored to find the key to accurately estimating X_D, but there were too many false leads, statistical errors, notational confusions, and completely untested assumptions, the latter an irony since computers have been available since before 1968. With hindsight, it is obvious that the noncentral t distribution was lurking at the heart of the whole matter and it should have been unmasked many years sooner.

The bootstrapping technique was crucial to the success of the above work. When properly used, it is a powerful tool for investigating the distribution of random variates, which, in the present case, were the x_D variates. Bootstrapping experiments are currently in progress to provide an experimental test of the heteroscedastic theory [3, 4] and to clear up the limit of quantitation chaos so clearly reviewed by Mermet [7].

ACKNOWLEDGMENTS

This research is dedicated to the memory of Prof. Mitchell Evan "Mitch" Johnson.

REFERENCES

1. E. Voigtman, "Limits of detection and decision. Part 1", *Spectrochim. Acta, B* **63** (2008) 115–128.

2. E. Voigtman, "Limits of detection and decision. Part 2", *Spectrochim. Acta, B* **63** (2008) 129–141.

3. E. Voigtman, "Limits of detection and decision. Part 3", *Spectrochim. Acta, B* **63** (2008) 142–153.

4. E. Voigtman, "Limits of detection and decision. Part 4", *Spectrochim. Acta, B* **63** (2008) 154–165.

5. L.A. Currie, "Limits for qualitative and quantitative determination – application to radiochemistry", *Anal. Chem.* **40** (1968) 586–593.

6. B. Efron, R.J. Tibshirani, *An Introduction to the Bootstrap*, 1ˢᵗ Ed., *Chapman and Hall/CRC*, Boca Raton, FL, © 1994.

7. J.-M. Mermet, "Limit of quantitation in atomic spectrometry: an unambiguous concept?", *Spectrochim. Acta, B* **63** (2008) 166–182.

15.6   POSTSCRIPT

After the primary reference for this chapter was published, the coauthors prepared four videos and seven screencasts documenting exactly how everything was done. The files, listed in Table 15.5, are freely available online and they should be viewed in numerical order.

The bootstrapped heteroscedastic experiments, mentioned in Section 15.5, were performed and published in 2011 [6 and 7 in Chapter 7] and Chapter 19 discusses a possible resolution of the limit of quantitation "chaos."

15.7   CHAPTER HIGHLIGHTS

This chapter demonstrated how to estimate X_D, with low uncertainty, for an experimental laser-excited molecular fluorescence experiment. Using the bootstrapping technique, it was possible to obtain tens of millions of experimental x_D variates and, likewise, tens of millions of simulation x_D variates. Their histograms were in excellent agreement with the theoretical predictions of Chapter 14 and this fact was used

Table 15.5   Videos and Screencasts (see Appendix B for the Website URL)

Number	Filename
1	01 Experiment movie.mpg
2	02 Experiment movie.mpg
3	03 Experiment movie.mpg
4	04 Experiment movie.mpg
5	05 Calibration 1.mov
6	06 Calibration 2.mov
7	07 Bootstrapping experiment.mov
8	08 Estimation of delta.mov
9	09 Calibration 3.mov
10	10 Simulation (Monte Carlo).mov
11	11 STE comparisons.mov

in order to determine that the 99% CI for X_D was $0.251_6 \pm 0.001_0$ mg R6G/100 mL ethanol.

An important point is that any technique that accurately estimates δ automatically estimates X_C and X_D accurately, and thereby solves the general detection limit estimation problem for the CMS defined in Chapter 7. Furthermore, δ was simple to estimate since bootstrapping made it easy to obtain the highly accurate t_{slope} histograms necessary for curve fitting with the noncentral t distribution. The following chapter illustrates how the curve fitting was performed on the t_{slope} histogram for seven standards and this is also shown in video 8 in Table 15.5.

Without bootstrapping, the above experiment would have resulted in 10 individual x_D values, that is, one random sample from each of the 10 distributions in Fig. 15.10. However, the x_D distributions in Fig. 15.10 are relatively wide, so these x_D values, though perfectly legitimate, would have been devoid of deep significance. In particular, they would *not* have led to an accurate estimate of X_D. In contrast, bootstrapping was the essential key to estimating X_D with low uncertainty: even with 3 standards, $\widehat{X}_D$ was less than 0.8% higher than the average for all 10 experiments. Yet there is an important caveat: bootstrapping is not a panacea, particularly with very small data sets, and the extensive literature on bootstrapping should be consulted for guidance if the use of bootstrapping is contemplated in a future experiment.

16

FOUR RELEVANT CONSIDERATIONS

16.1 INTRODUCTION

This chapter deals with four relevant considerations arising from the limit of detection (LOD) results presented in Chapters 7–15. First, how important is it, in practice, to satisfy the various assumptions made in the previous nine chapters? Desimoni and Brunetti suggest that as many as 13 "quite severe theoretical assumptions" [1, p. 2] may be involved in estimating σ_0 and β for a model chemical measurement system (CMS) like that defined in Chapter 7. They state that their listed assumptions "are hardly satisfied in real work, but they are usually accepted as valid *a priori*, since allowing an easier and simplified estimation of the LOD" [1, p. 2]. This is consistent with their previous remarks on the subject [2, 3]. Accordingly, it is worthwhile to assess if any of their listed assumptions is truly problematic in terms of the applicability of our model CMS.

Second, how should δ be estimated? This is important because, for our model CMS and its ancillary assumptions (as per Table 7.1), it has been shown that *any* technique that accurately estimates δ automatically estimates X_C and X_D accurately. Consequently, this solves the general detection limit estimation problem for our model CMS.

Third, Chapters 7–14 presented a total of 24 expressions for the Currie decision level, and another 24 expressions for the Currie detection limit. It was worthwhile to exhibit them explicitly and subject them to Monte Carlo testing, but it also became apparent that they exhibited considerable redundancy. Accordingly, it would be beneficial to focus attention on a subset of the expressions that is most relevant in practice.

Lastly, what is the legitimacy of lowering detection limits by artful exploitation of the detection limit expressions, measurement protocols, or processing of the outputs of a CMS? What is reasonable, justifiable, and ethical in this regard?

Limits of Detection in Chemical Analysis, First Edition. Edward Voigtman.
© 2017 John Wiley & Sons, Inc. Published 2017 by John Wiley & Sons, Inc.
Companion Website: www.wiley.com/go/Voigtman/Limits_of_Detection_in_Chemical_Analysis

16.2   THEORETICAL ASSUMPTIONS

The assumptions listed by Desimoni and Brunetti [1] are summarized in Table 16.1. To begin, #3 is always problematic, as discussed in Chapter 2. Aside from constantly being on guard for it, there is no general way to guarantee freedom from systematic error. Thanks to the central limit theorem, Gaussian distributed noise is "normal," so #1 and #8 are quite commonly satisfied and normality testing is easily performed with standard statistics software. As for #2, it only applies if α and σ_0 are actually required in specific cases, as in Chapter 7.

It is sometimes impossible to obtain a true blank, that is, one with absolutely no analyte content. As well, sometimes blanks may be slightly contaminated with analyte, and this may be unknown to the analyst, at least initially. But it is also commonly the case that blanks may be obtained that have negligible analyte content. So #4 depends on the specifics of the CMS and analyte. This also applies to #9: matrix matching scenarios range from trivial to impossible. For #7, it is usually the case that calibration standards are prepared carefully enough so that their uncertainties are negligible.

If the CMS is supposed to have homoscedastic noise, as per #'s 5, 8, and 10, this may be easily checked via an F test [4, eqn 4]. If the noise is revealed to be heteroscedastic, then see Chapter 18. To check linearity, as per #11, Olivieri provides useful expressions [5, eqns 3–5]. If the noise is additive Gaussian white noise (AGWN), the ordinary least squares (OLS) slope and intercept are known to be the lowest variance, linear unbiased estimators [6, p. 16], so #12 is satisfied. As for #13, if #11 is true and α is *known*, then there is no point in t testing the OLS intercept, a, versus α: even "failing" the t test would only mean that a was relatively improbable. Finally, #6 simply means using an adequate amount of data, and avoiding, if possible, having low degrees of freedom.

Table 16.1   **Theoretical Assumptions Pertaining to the Model CMS**

Number	Description
1	*Noise* $\sim N{:}\alpha, \sigma_0$
2	Both α and σ_0 are known
3	No systematic errors
4	Analytical blank has negligible analyte content
5	Noise is homoscedastic, or almost so, at low analyte content
6	"A consistent number of independent measurements is necessary" [1, p. 2]
7	Calibration standards have negligible error
8	$r(x) \sim N{:}\alpha + \beta x, \sigma_0$
9	Calibration standards must be matrix matched
10	Noise is homoscedastic over the range of the calibration standards
11	$r(x) = \alpha + \beta x$
12	The OLS slope and intercept must be "good estimates" of β and α, respectively
13	The OLS intercept, a, must be statistically equivalent to α

In summary, all of the assumptions in Table 16.1, with the possible exception of #3, are relatively mild, so it is not surprising that the model CMS defined in Chapter 7 is widely useful. In particular, note that the laser-excited molecular fluorescence experiment in Chapter 15 was deliberately designed to be in full compliance with the CMS model defined in Chapter 7, and agreement among experiment, theory, and computer simulation was excellent.

16.3   BEST ESTIMATION OF δ

In Chapter 15, the method used to estimate δ in the 10 different calibration designs was mentioned, but not explicitly shown; for space reasons, it was not present in the original publication [7]. However, the screencast named "*08 Estimation of delta.mov*" shows exactly how it was accomplished, for the calibration design having 7 standards, as per Table 15.3. It should be noted that there was a trivial scaling error (*mea culpa*) in the running χ^2 value in the screencast, resulting in that particular χ^2 value being about 20% too high, but the error did not affect the estimated δ value of 13.912. The histogram of 10^7 t_{slope} variates, obtained through the use of the bootstrapping technique, is shown in Fig. 16.1.

It is quite clear that the noncentral t distribution, with $\nu \equiv 5$ and $\hat{\delta} \cong 13.912$, provides an excellent fit, overall, to the histogram of t_{slope} variates. This is also supported

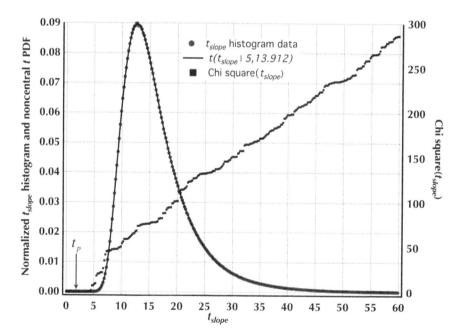

Figure 16.1   Histogram of 10^7 t_{slope} variates, overplotted with the noncentral t distribution that minimized the (correct) running χ^2 value. With $\nu = 5$ and $p = 0.05$, $t_p \cong 2.015$.

by the χ^2 value of 286.373, with 290 fitted histogram bins, that is, 2.1–59.9, with the bin width of 0.2. The only region where the fit could be better is in the initial rise, that is, roughly from 4 to 7.5. From Fig. 16.1, virtually every t_{slope} variate is above t_p, so there is negligible risk in rejecting $H_\varnothing : \beta = 0$, and virtually 100% confidence in not rejecting $H_a : \beta > 0$.

An alternative is to estimate $\bar{t}_{slope}$ as the sample mean of N' *i.i.d.* t_{slope} variates, with $N' \gg 1$. For the $N' = 10^7$ t_{slope} variates above, $\bar{t}_{slope}$ was 16.54786. From eqn A.21, with $v \equiv 5, \bar{t}_{slope} \cong E[t_{slope}] = 1.189416077 \times \delta$. This yields a second estimate of δ, that is, $\hat{\delta} = 13.913$. But an important caution is in order: estimating $E[t_{slope}]$ with a *single* t_{slope} value is an *extremely poor idea*, even worse than estimating $E[s_0]$ by a single value of s_0. The reason is the same in both situations; their biases are significantly large and are *not* eliminated.

It is generally preferable to estimate δ by fitting a histogram of independent t_{slope} variates with the noncentral t distribution. This avoids potential problems, for small v, caused by the long upper tail of the noncentral t distribution. Fortunately, curve fitting is particularly easy since there are only two parameters, one of which is known in advance, that is, $v \equiv N - 2$. That leaves δ as the *only fitting parameter*.

Once an estimate of δ is available, the theoretical probability density function (PDF) of x_D may be computed via eqn 14.10, because everything else needed is available as soon as the experiment has been designed, that is, once the standards and measurement protocol are specified, but *prior* to the experiment even being performed. In other words, $t_p, t_q, \eta^{1/2}, v$, and $S_{XX}^{1/2}$ can be known *before* the experiments are performed. Consequently, there is *no adjustability*: the theoretical x_D PDF either fits the histogram of experimental x_D variates or it does not. The same applies to fitting the histogram of simulation x_D variates.

This fitting procedure is illustrated in Fig. 16.2, where the simulation data histogram (solid circles) is plotted first, then the theoretical PDF (line trace) on top of it, and, lastly, the experimental data histogram (solid squares) on top of both. Each histogram has 10^7 binned variates and the fits are excellent. The results in Fig. 16.2 are also shown as part of Fig. 15.10, with the value 0.252 being the best estimate of X_D. Obviously, it is straightforward to obtain confidence intervals, as illustrated in Fig. 24.3.

16.4   POSSIBLE REDUCTION IN THE NUMBER OF EXPRESSIONS?

As noted above, there are 24 Currie decision level expressions in Chapters 7–14, and the same number of Currie detection limit expressions. However, they display considerable redundancy, and some useful patterns are evident:

- If σ_0 is known, then the errorless constants Y_C and Y_D are known.
- If σ_0 is unknown, then the χ distributed variates $\bar{y}_C$ and y_D are easily obtainable from s_0 or s_r.
- If $p = q$, then $Y_D = 2Y_C$ and $y_D = 2y_C$, for all degrees of freedom.

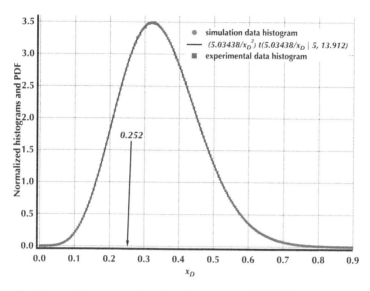

Figure 16.2 Content domain detection limit histograms, with 10^7 binned variates each, and the theoretical PDF computed from eqn 14.10.

- The net response domain expressions are simpler and more symmetric than those in the response domain, particularly when α is unknown, as is usually the case.

- The content domain expressions are simply obtained from those in the net response domain, by dividing by β, if known, or $\hat{\beta}$, otherwise.

- The two most useful combinations of parameter knowledge are the two extremes, that is, when all three parameters are known (as in Chapter 7) and when all three are unknown (as in Chapter 14).

The response domain is where all measurements are made, so there is no denying its fundamental importance. However, the intercept is usually of no particular importance and what really matters is the response *relative to* the intercept, that is, the net response. The net response is simply the response *referenced to* α, if known, or $\hat{\alpha}$, otherwise.

The above considerations are the underlying rationale for de-emphasizing presentation of the response domain expressions and also focusing on the two parameter knowledge extremes. The scenario where σ_0, β, and α are all known is called the "theoretical" case, while the scenario where σ_0, β, and α are all unknown is the "experimental" case. This leads to the so-called "quadrant" diagrams, such as that shown in Fig. 16.3.

From Fig. 16.3, the symmetry and simplicity of the expressions is immediately evident; the true population parameters, σ_0, β, and α, are on the left, while their respective point estimate test statistics, s_0, b, and a, are on the right. Indeed, in passing

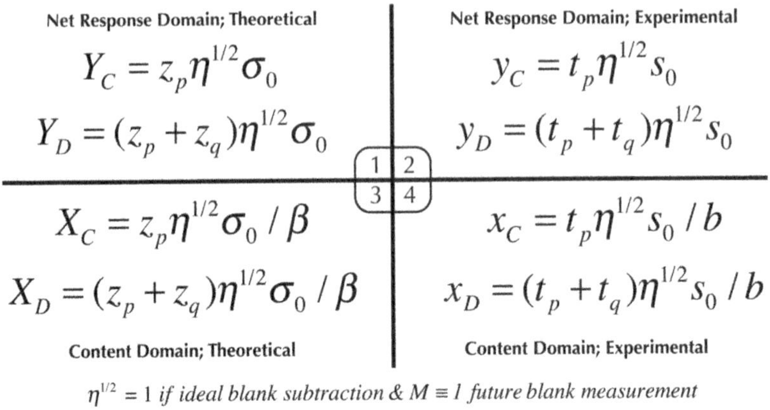

Net Response Domain; Theoretical

$$Y_C = z_p \eta^{1/2} \sigma_0$$

$$Y_D = (z_p + z_q) \eta^{1/2} \sigma_0$$

Net Response Domain; Experimental

$$y_C = t_p \eta^{1/2} s_0$$

$$y_D = (t_p + t_q) \eta^{1/2} s_0$$

1	2
3	4

$$X_C = z_p \eta^{1/2} \sigma_0 / \beta$$

$$X_D = (z_p + z_q) \eta^{1/2} \sigma_0 / \beta$$

Content Domain; Theoretical

$$x_C = t_p \eta^{1/2} s_0 / b$$

$$x_D = (t_p + t_q) \eta^{1/2} s_0 / b$$

Content Domain; Experimental

$\eta^{1/2} = 1$ *if ideal blank subtraction & M ≡ 1 future blank measurement*

Figure 16.3   The quadrant diagram for Currie decision levels and detection limits. Voigtman, 2008 [8]. Reproduced with permission of Elsevier.

from the theoretical expressions, in quadrants 1 and 3, to the experimental expressions, in quadrants 2 and 4, all that happens in the expressions is $\sigma_0 \mapsto s_0$, $z_p \mapsto t_p$, $z_q \mapsto t_q$, and so on. Thus, errorless composite constants, for example, Y_D and X_D, are replaced with the random variates y_D and x_D, respectively. Of course, all of the other expressions are still available, but the expressions in Fig. 16.3 suffice for most purposes, which is why they were emphasized from the beginning [8] and consistently thereafter [9–11].

16.5   LOWERING DETECTION LIMITS

The large majority of CMSs are deliberately designed to yield the highest signal-to-noise ratio (SNR) that is feasible, both in terms of state-of-the-art engineering technology and product price point in the market. Enhancement of an SNR is generally achieved either by reducing the dominant noise's power spectral density (PSD) or by appropriately modifying the measurement bandwidth, and much is known about the various ways that this can be accomplished, for example, signal averaging, filtering, integration, curve fitting, modulation, oversampling, phase-sensitive detection, working at low temperatures, and many others. However, it is ultimately counterproductive to try to incorporate such specifics into the detection limit definition, even for white noise alone, because it would require knowing both the PSD and the noise bandwidth (see Chapter 21) and it is commonly the case that neither is accurately known.

Computations of noise variances can be rather complicated [12, 13], even in simple cases, and it is often infeasible to know what signal processing is taking place in any given CMS; instrument manufacturers do not necessarily wish to divulge proprietary information or trade secrets. This means that, to some extent, a given CMS must be treated as the figurative "black box"; it is the outputs of the CMS, or user-defined functions of them, that constitute the noisy measurements. This raises an important issue that is probably irresolvable: How much processing of the outputs of a CMS is justifiable, in the sense of not *artificially* lowering the detection limit?

As a simple example, consider the averaging that arises in connection with the analyst-defined value of M in the blank subtraction factor, $\eta^{1/2}$:

$$\eta^{1/2} = (M^{-1} + \text{error term(s) due to } \hat{\alpha})^{1/2} \tag{16.1}$$

The canonical and best choice for elementary measurements is $M \equiv 1$, because any other choice inessentially lowers $\eta^{1/2}$ and thereby also speciously lowers decision levels and detection limits, as per Fig. 16.3.

A second example involves replacement of the blank's sample standard deviation, s_0, with its sample standard error:

$$\text{LOD}_e \equiv \frac{(t_p + t_q)\eta^{1/2}s_0/M_0^{1/2}}{b} = \frac{x_D}{M_0^{1/2}} \tag{16.2}$$

where $s_0/M_0^{1/2}$ is the sample standard error. So long as the noise on the blank remains white, that is, low-frequency noises and drift are negligible, then LOD_e decreases as M_0 increases, while x_D would be only slightly refined. At some point, increasing M_0 becomes ineffectual or even counterproductive, due to inevitable low-frequency noises. Knowing when to stop averaging is one reason why the Allan variance concept [14] was invented, but, in practice, it may be difficult to know when to stop. In any event, s_0 should not be replaced by $s_0/M_0^{1/2}$ unless there is a *compelling reason* for doing so. As in the previous example, it is easy to find publications where this advice is violated.

In the two examples discussed, the tacit effect was to lower the noise's PSD by lowering the effective value of σ_0. This is also true when measurements are averaged, even if they are partially correlated. For example, consider a CMS as in Chapter 7 and suppose 4 response measurements are made for each of 6 calibration standards. If each response is considered to be from an independent standard, and OLS is performed, then $\nu = 22$ and s_r estimates s_0. But if each set of four replicates per standard is averaged, and OLS is performed on the resulting six average responses, then $\nu = 4$ and s_r estimates $s_0/4^{1/2}$. Both methods of processing the OLS data are valid in this hypothetical scenario, yet the latter method clearly involves estimation via a standard error and there is nothing illegitimate about it. Consequently, what is of paramount importance is honest consistency; whatever valid method is employed, no attempt should be made to favor the method that gives the lowest detection limits.

In addition to the above, fundamental issues may arise simply as a consequence of exactly how a CMS is defined. To illustrate, consider Fig. 16.4. Nominally, the CMS is purely analog, with the indicated noisy response function. Then the output of the CMS may be digitized using an analog-to-digital converter (ADC), producing a series

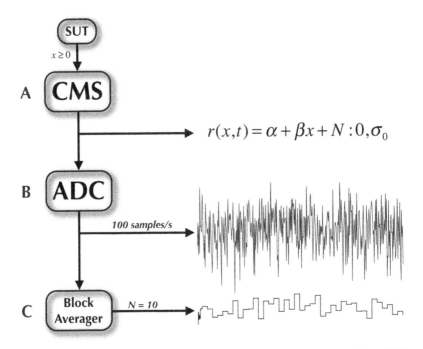

$$r(x,t) = \alpha + \beta x + N : 0, \sigma_0$$

Figure 16.4   Illustration of the inherent arbitrariness encountered in defining a CMS.

of samples, that is, elemental measurements. Optionally, for purposes of noise reduction, successive blocks of N samples may be block averaged, producing a series of compound measurements. In Fig. 16.4, elemental measurements are shown as being produced at the rate of 100 samples per second, while compound measurements, for $N = 10$, are produced 10 times more slowly.

However, Fig. 16.4 is somewhat biased in that there are two alternative ways to define the CMS. In the first of these, the analog CMS is followed by the ADC, resulting in a digital CMS. In the second alternative, all three functionalities are in sequence, resulting in a digital CMS with internal signal processing. Both digital alternatives are very commonly encountered. Indeed, as noted in Chapter 2, an inexpensive digital voltmeter typically averages over 0.4 s to eliminate possible 50 or 60 Hz line frequency interferences.

This inherent arbitrariness in defining the CMS causes difficulties in deciding how much processing of the output of the CMS is justifiable. For example, Laude *et al.* [15] note that "real-time oversampling" significantly reduces white noise, down to levels comparable to $1/f$ and other "environmental" noises. Although the technique is not new [16], and is well known as block averaging, it is clearly an effective digital filter means of reducing white noise; this is shown in Fig. 16.4, where the $N = 10$ block averaged trace was obtained from the noisier trace above it.

Hence, if block averaging is *internally* used in the CMS, it is certainly legitimate, whether or not the analyst knows that such filtering is taking place [17, 18]. In this case, N may well be entirely unknown to the analyst. However, although the block averaging technique is easily used with data files that have already been collected, that is, in postmortem fashion rather than in real time, it is then the analyst's decision as to the size of the averaged blocks. Thus, how many points are to be averaged in each block?

In a best case scenario, with white noise only, N is given by

$$N = \frac{\text{maximum digitization rate}}{2 \times f_{\text{Nyquist}}} \quad \text{(rounded to an integer)} \quad (16.3)$$

Then the standard deviation of the noise is reduced by a factor of $1/N^{1/2}$. However, this reduction factor does *not* apply if there is significant low-frequency noise; it is replaced by $1/N_{\text{effective}}^{1/2}$, where $N_{\text{effective}}$ may be much less than N. For example, in a computer simulation with $N = 100$, 1000 samples/s and $1/f$ noise in the 1 mHz–10 Hz range, it is found that $N_{\text{effective}} \approx 2.8$. Therefore, if the noise has both white and low-frequency components, block averaging will be much more effective in reducing the white noise, but it would be difficult to know, without detailed knowledge of the noise's nonwhite PSD, what N should be used.

Obvious issues are that arbitrary N values might be used or $N_{\text{effective}}$ might erroneously be assumed to equal N. In any event, if block averaging is employed on the output of a CMS, there should be a supportable rationale for doing so and *the same applies to all other post-CMS signal processing operations*. For example, suppose the output of a *real* CMS is postprocessed by integration for a fixed period of time, for example, 1 or 3 s. This would be done for the deliberate purpose of reducing the noise, thereby lowering the detection limit. Then obvious questions are as follows:

1. What was the real CMS's output noise PSD, before the integrator?
2. What was the selected integration time period?
3. Why, specifically, was the integration time period chosen?

All too frequently, only the second question is answered in publications, making it difficult to know how much noise reduction was accomplished. Furthermore, unless the real CMS's PSD was *white* prior to integration, it is not feasible to back-calculate what the detection limit would have been *without* the integration. Then the situation is akin to that discussed after eqn 16.2.

Consequently, it is strongly recommended that the decision level and detection limit equations in Chapters 7–14 should *not* be modified because it would effectively destroy their utility as transferable figures of merit. Little, if anything, can be done about signal processing and noise reduction techniques that already exist inside the CMS: they are simply intrinsic to the CMS. But any post-CMS signal processing operations should, at the very least, be justified and documented fully.

What ultimately matters is honest data analysis and presentation integrity. Publications should contain as much information as possible, concerning experiment conditions, protocols, instrument settings and the like, and figures of merit should be computed and presented properly. This is admittedly imperfect, but it greatly facilitates replication of experiments and, in principle, if not quite in practice, provides what is needed to judge the quality of the presented results.

16.6 CHAPTER HIGHLIGHTS

This chapter dealt with considerations that arose from the material presented in Chapters 7–15. First, is the model system defined in Chapter 7 unrealistically idealized or inapplicable to real CMSs? Second, how may δ be accurately estimated? Third, is it feasible to reduce the number of equations presented in Chapters 7–14? Finally, how much processing of the outputs of a CMS is permissible, in the sense of not artificially lowering the detection limit?

REFERENCES

1. E. Desimoni, B. Brunetti, "About estimating the limit of detection by the signal to noise approach", *Pharm. Anal. Acta* **6** (2015) 355, 4 pp., but no page numbers, *per se*, doi: 10.4172/2153-2435.1000355.
2. E. Desimoni, B. Brunetti, "Considering uncertainty of measurement when assessing compliance or non-compliance with reference values given in compositional specifications and statutory limits: a proposal", *Accred. Qual. Assur.* **11** (2006) 363–366.
3. E. Desimoni, B. Brunetti, "About estimating the limit of detection of heteroscedastic analytical systems", *Anal. Chim. Acta* **655** (2009) 30–37.
4. R.J.N. Bettencourt da Salva, "Spreadsheet for designing valid least-squares calibrations: a tutorial", *Talanta* **148** (2016) 177–190.
5. A.C. Olivieri, "Practical guidelines for reporting results in single- and multi-component analytical calibration: a tutorial", *Anal. Chim. Acta* **868** (2015) 10–22.
6. D.C. Montgomery, E.A. Peck, *Introduction to Linear Regression Analysis*, John Wiley and Sons, New York, 1982.
7. E. Voigtman, K.T. Abraham, "Statistical behavior of ten million experimental detection limits", *Spectrochim. Acta, B* **66** (2011) 105-113.
8. E. Voigtman, "Limits of detection and decision. Part 2", *Spectrochim. Acta, B* **63** (2008) 129-141.
9. E. Voigtman, "Limits of detection and decision. Part 4", *Spectrochim. Acta, B* **63** (2008) 154–165.
10. E. Voigtman, K.T. Abraham, "True detection limits in an experimental linearly heteroscedastic system. Part 2", *Spectrochim. Acta, B* **66** (2011) 828–833.
11. J. Carlson, A. Wysoczanski, E. Voigtman, "Limits of quantitation – yet another suggestion", *Spectrochim. Acta, B* **96** (2014) 69–73.
12. E. Voigtman, J.D. Winefordner, "Time variant filters for analytical measurements – electronic measurement systems", *Prog. Anal. Spectrosc.* **9** (1986) 7–143.
13. E. Voigtman, J.D. Winefordner, "Low-pass filters for signal averaging", *Rev. Sci. Instrum.* **57** (1986) 957–966.
14. D.W. Allan, "Statistics of atomic frequency standards", *Proc. IEEE* **54** (1966) 221–230.
15. N.D. Laude, C.W. Atcherley, M.L. Heien, "Rethinking data collection and signal processing. 1. Real-time oversampling filter for chemical measurements", *Anal. Chem.* **84** (2012) 8422–8426.
16. Logger *Pro*™ User's Manual, Version 2.1, ©1997–2000, Tufts University and Vernier Software & Technology, p. 22.

17. A. Felinger, A. Kilár, B. Boros, "The myth of data acquisition rate", *Anal. Chim. Acta* **854** (2015) 178–182.

18. M.F. Wahab, P.K. Dasgupta, A.F. Kadjo, D.W. Armstrong, "Sampling frequency, response times and embedded signal filtration in fast, high efficiency liquid chromatography: a tutorial", *Anal. Chim. Acta* **907** (2016) 31–44.

17

NEYMAN–PEARSON HYPOTHESIS TESTING

17.1 INTRODUCTION

Currie's detection limit schema [1] is based on classical frequentist statistics and Neyman–Pearson hypothesis testing principles, which require consideration of both false positives and false negatives. Before exhibiting the actual hypotheses that will be tested, it is useful to discuss precisely how such hypothesis testing is implemented in Monte Carlo computer simulations.

17.2 SIMULATION MODEL FOR NEYMAN–PEARSON HYPOTHESIS TESTING

Figure 17.1 shows a typical *LightStone* simulation model that is used to perform hypothesis testing in the net response domain, with arbitrary simulation units of volts. It is closely related to models in the previous chapters, and to other published models [2–5]. The model's parameters are as follows: $\sigma_0 \equiv 0.1\,\text{V}$, $\beta \equiv 1\,\text{V/content unit}$, $\alpha \equiv 1\,\text{V}$, $X_1 \equiv 1, X_2 \equiv 2, \ldots, X_5 \equiv 5$, $N \equiv 5$, $v \equiv N - 2 = 3$, $p \equiv 0.05 \equiv q$, $z_p = z_q \cong 1.644854$, $t_p = t_q \cong 2.353363$, and $\eta^{1/2} \cong 1.449138$. The sample standard error about regression, s_r, equals s_0, the sample standard deviation of the blank, since only one measurement is made per standard. Thus, s_r estimates σ_0.

Ten simulations, of 1 million steps each, resulted in 10 million of each of the following variates: net blank, net analyte at Y_D, net analyte at Y_D^{CHE}, y_D, y_C, net analyte at y_D, net analyte at $\delta_{p,q,v}\eta^{1/2}s_r$, and net analyte at $\delta_{p,q,v}\eta^{1/2}s_r/c_4(3)$. Not all of these variates are pertinent to every tested hypothesis, but they were all concomitantly generated in the set of 10 simulations and binned into separate histograms. Examination of the obtained percentages of false positives and false negatives is deferred until after the necessary hypotheses have been stated and hypothesis testing has been discussed. The total simulation run time was about 5 min.

Limits of Detection in Chemical Analysis, First Edition. Edward Voigtman.
© 2017 John Wiley & Sons, Inc. Published 2017 by John Wiley & Sons, Inc.
Companion Website: www.wiley.com/go/Voigtman/Limits_of_Detection_in_Chemical_Analysis

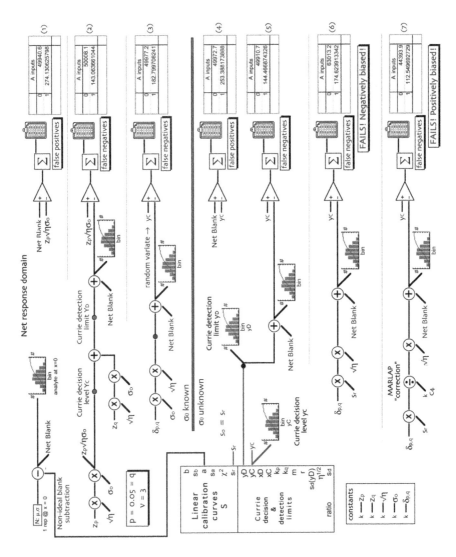

Figure 17.1 Simulation model used to generate all of the histograms discussed in the text.

17.3 HYPOTHESES AND HYPOTHESIS TESTING

Ultimately, statistical hypothesis testing is the basis for specifying decision and detection probabilities, just as Currie noted [1]. Furthermore, every competent researcher knows that meticulous care must be taken in the formulation of hypotheses since an erroneous hypothesis is most likely useless and a waste of research effort, time, and funding. Statements of the hypotheses relevant to detection limit schema are presented for the net response domain only. The relevant equations, with $s_0 = s_r$ in Fig. 17.1, are presented in Fig. 17.2.

17.3.1 Hypotheses Pertaining to False Positives

17.3.1.1 Hypothesis 1   Suppose that σ_0 is known and p is fixed. Then, the theoretical Currie decision level, Y_C, is the smallest positive *constant* such that a random sample from N:0, $\sigma_0\eta^{1/2}$ has no more than p probability of exceeding Y_C. Note that Y_C, which is *optimal*, is given by the equation on the left side of Fig. 17.2. This hypothesis test is illustrated in Fig. 17.3.

The inset expression is simply the concise notational equivalent of the image: the probability of false positives, p, is defined as the probability, Pr, that a random sample from the net blank distribution exceeds the fixed value Y_C, when the true value of y is zero.

17.3.1.2 Hypothesis 2   Suppose that σ_0 is unknown and that p and v are fixed. In this case, σ_0 must be estimated by s_0 ($=s_r$) and z_p must be replaced by t_p, the critical t value for p, with v degrees of freedom. The experimental Currie decision level, y_C, is the χ *distributed random variate* such that there is no more than p probability that a *random sample* from N:0, $\sigma_0\eta^{1/2}$ will exceed a *random sample* from the distribution of y_C. This hypothesis test is illustrated in Fig. 17.4.

For reasons of clarity, the inset expression is somewhat shortened, since y_C is a *random sample* from the following χ distribution:

Always optimal if σ_0 is known	**Always irrelevant and nonoptimal**	**Always optimal if σ_0 is unknown**
$Y_C = z_p\eta^{1/2}\sigma_0$	$y_C = t_p\eta^{1/2}s_0$	$y_C = t_p\eta^{1/2}s_0$
$Y_D = (z_p + z_q)\eta^{1/2}\sigma_0$	$Y_D^{CHE} = \delta_{p,q,v}\eta^{1/2}\sigma_0$	$y_D = (t_p + t_q)\eta^{1/2}s_0$
Theoretical domain	Absolutely requires both domains	Experimental domain

$\eta^{1/2} = 1$ if ideal blank subtraction and $M \equiv 1$ future blank measurement

Figure 17.2   Exact, net response domain decision level ("C" subscript) and detection limit ("D" subscript) equations. Divide by the relevant slope (β or b) to get the content domain equations, if desired.

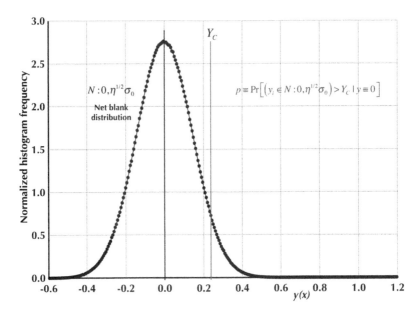

Figure 17.3 Testing for false positives when σ_0 is known. Note: $Y_C = 0.238362$ V.

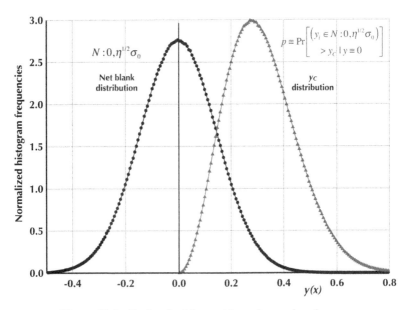

Figure 17.4 Testing for false positives when σ_0 is unknown.

$$p(y_C) = \frac{v^{v/2}y_C^{v-1}e^{-vy_C^2/2(\eta^{1/2}t_p\sigma_0)^2}}{(\eta^{1/2}t_p\sigma_0)^v\,2^{(v-2)/2}\Gamma(v/2)}U(y_C) \qquad (17.1)$$

where $y_C \equiv \eta^{1/2}t_p s_0$, as in Fig. 17.2, and eqn 17.1 is simply eqn A.31, with $u \equiv ks_0$ and $k \equiv \eta^{1/2}t_p$.

17.3.2 Hypotheses Pertaining to False Negatives

17.3.2.1 Hypothesis 3 Suppose that σ_0 is known and that p and q are fixed. The theoretical Currie detection limit, Y_D, is the smallest positive *constant* value of Y such that there is no more than q probability that a random sample from $N{:}Y, \sigma_0\eta^{1/2}$ will fall below Y_C. Similar to Y_C, Y_D is the *optimal* value and is given on the left side of Fig. 17.2. This hypothesis test is illustrated in Fig. 17.5.

17.3.2.2 Hypothesis 4 Suppose that σ_0 is unknown and that p, q, and v are fixed. The experimental Currie detection limit on the right side of Fig. 17.2, y_D, is the χ *distributed random variate*, equal to $(t_p + t_q)y_C/t_p$ for all v, such that there is no more than q probability that a *random sample* from $N{:}y_D, \sigma_0\eta^{1/2}$ will fall below a *random sample* from the χ distribution of y_C. The χ distribution of y_D is given by eqn A.31, with $u = ks_0$ and $k = (t_p + t_q)\eta^{1/2}$. If $q = p$, then $y_D = 2y_C$, for all degrees of freedom. This hypothesis test is the most complicated to illustrate because y_D is the χ *distributed location parameter* of a Gaussian distribution, that is, $N{:}y_D, \sigma_0\eta^{1/2}$ is a

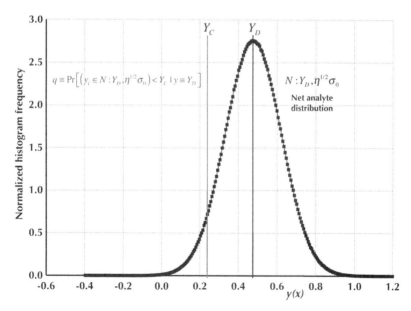

Figure 17.5 Testing for false negatives when σ_0 is known. Note: $Y_C = 0.238362$ V and $Y_D = 0.476724$ V.

compound normal distribution [6]. For any *one* value of y_D, this distribution has the same "width" as that of the net blank, but it *randomly moves around*, always centered on y_D, from experiment to experiment. The y_C, y_D pair is *optimal* when σ_0 is unknown.

Figures 17.6–17.10 illustrate the above situation. Consider Fig. 17.6. This figure shows the net blank, y_C and y_D histograms produced by the model in Fig. 17.1. Each histogram is overplotted by its respective theoretical PDF, with excellent results. Every time a new calibration curve is obtained, a new sample of the s_r variate is obtained. Since $s_0 = s_r$, new samples of y_C and y_D are thereby obtained, as per the equations in Fig. 17.2. The histograms in Fig. 17.6 each contain 10 million binned variates.

For illustrative purposes, three possible values of y_C are shown in Fig. 17.6: $y_C^{(1)} \equiv 0.2\,\text{V}$, $y_C^{(2)} \equiv 0.4\,\text{V}$, and $y_C^{(3)} \equiv 0.6\,\text{V}$. Since $p \equiv q$, $y_D = 2y_C$, so the corresponding y_D values are $y_D^{(1)} \equiv 0.4\,\text{V}$, $y_D^{(2)} \equiv 0.8\,\text{V}$, and $y_D^{(3)} \equiv 1.2\,\text{V}$. These are also shown in Fig. 17.6. Now suppose that $y_C = y_C^{(1)} = 0.2\,\text{V}$. Then the net analyte distribution would be centered on $y_D^{(1)} \equiv 0.4\,\text{V}$, as shown in Fig. 17.7.

The pair of Gaussian distributions are equidistant from $y_C^{(1)} \equiv 0.2\,\text{V}$, that is, they are symmetric about $y_C^{(1)}$. Therefore, the probability that a random sample from the net analyte distribution would fall below $y_C^{(1)}$ is exactly the same as the probability that a random sample from the net blank distribution would exceed $y_C^{(1)}$. Hence $p = q$, but, discussed as follows, they are almost certainly not equal to the *a priori* specified value of 0.05.

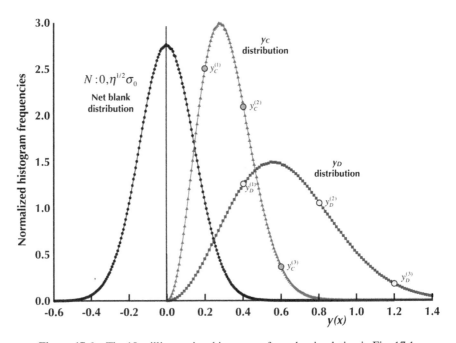

Figure 17.6   The 10 million variate histograms from the simulation in Fig. 17.1.

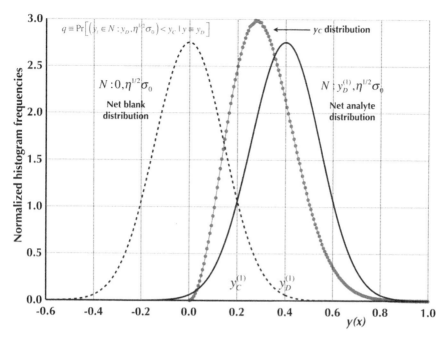

Figure 17.7   Testing for false negatives if $y_C = 0.2$ V in Fig. 17.6 and $y_D = 2y_C = 0.4$ V.

The following two figures show the case where $y_C = y_C^{(2)} = 0.4$ V (Fig. 17.8) and $y_C = y_C^{(3)} = 0.6$ V (Fig. 17.9). For reasons of clarity, the net blank distribution centered on zero is not shown in Figs 17.8 and 17.9, but it is obvious that the symmetry relationship noted in Fig. 17.7 still applies. Then $p = q$ in Figs 17.8 and 17.9 as well, but it is highly improbable that $p = 0.05 = q$ in either figure. Rather, it is the y_C-*weighted average*, over an *infinite ensemble* of possible y_C variates, that results in $p = 0.05 = q$. In other words, it works perfectly in the long run, but almost surely does *not* for any one y_C sample from the y_C distribution. It is recommended to again view the screencast named "*Currie detection.mov*," referenced in Chapter 11, to ensure that this is fully understood.

Since the net analyte distribution is a compound normal distribution, with the χ distributed variate y_D as its location "parameter," it is *not* a normal distribution. This is clearly demonstrated by Model 4 in Appendix B. The relatively broad y_C distribution shown in Fig. 17.6 also results in a *mean* net analyte distribution that is relatively broad. This is shown in Fig. 17.10.

The *mean* net analyte distribution is actually the histogram of net analyte variates at y_D, directly from the simulation in Fig. 17.1. It is clearly asymmetric and *much broader* than the net blank histogram. Also note that the mean net analyte distribution has some lower tail area below zero. In the present case, this is due to occasional $N \equiv 5$ point calibration curves having very small s_r values. With more calibration curve standards, it is less likely to occur. In any event, it is *not* problematic.

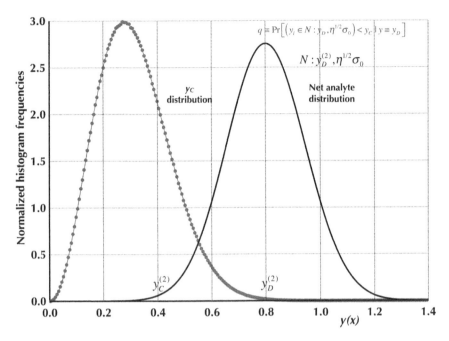

Figure 17.8 Testing for false negatives if $y_C = 0.4$ V in Fig. 17.6 and $y_D = 2y_C = 0.8$ V.

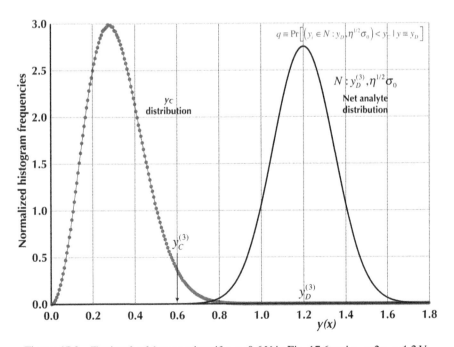

Figure 17.9 Testing for false negatives if $y_C = 0.6$ V in Fig. 17.6 and $y_D = 2y_C = 1.2$ V.

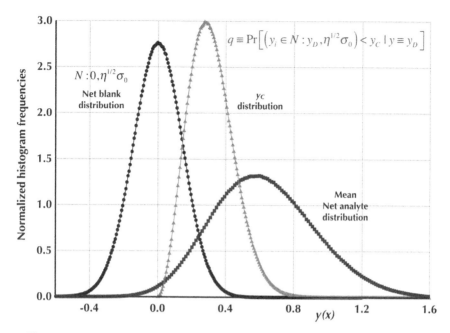

$$q \equiv \mathrm{Pr}\left[\left(y_i \in N : y_D, \eta^{1/2}\sigma_0\right) < y_C \mid y \equiv y_D\right]$$

$N : 0, \eta^{1/2}\sigma_0$

Net blank
distribution

y_C
distribution

Mean
Net analyte
distribution

Figure 17.10   The net blank, y_C and *mean* net analyte 10 million variate histograms.

17.4   THE CLAYTON, HINES, AND ELKINS METHOD (1987–2008)

In 1987, Clayton *et al.* [7] published a detection limit method that has since become the fundamental basis of several detection limit protocols, for example, ISO 11843-2 [8], promulgated by various governmental and nongovernmental standards organizations. For brevity, henceforth the method is referred to as the $\delta_{\mathrm{critical}}$ method. The relevant equations for the $\delta_{\mathrm{critical}}$ method, in the nomenclature used herein, are given in the center of Fig. 17.2. The $\eta^{1/2}$ factor, denoted by w_0 in Ref. [7], is present in every expression in Fig. 17.2 and therefore has no relative effect. Importantly, for *all possible values* of p, q, and v, $(z_p + z_q) < \delta_{p,q,v} < (t_{p,v} + t_{q,v})$.

The $\delta_{\mathrm{critical}}$ method derives from the fact that the noncentral t distribution is the basis of the "statistical power" of the t test, as discussed briefly in Appendix A. However, in order to avoid bias, it *absolutely requires* use of *both* the *population* standard deviation of the (homoscedastic noise), σ_0, *and* the *sample* standard deviation of the noise, s_0. This is shown in the middle of Fig. 17.2. Yet, if σ_0 is *known*, the expressions at the left of Fig. 17.2, the theoretical net response domain equations, are *optimal and unbiased for all degrees of freedom*. Conversely, if σ_0 is *unknown*, then s_0 must be used. In this case, the expressions at the right of Fig. 17.2, the experimental net response domain equations, are *optimal and unbiased for all degrees of freedom*.

Furthermore, if σ_0 is unknown, there is *no possible way* to salvage the $\delta_{\mathrm{critical}}$ method's detection limit expression, Y_D^{CHE}, without causing bias; replacement of σ_0 by s_0 results in negative bias in the detection limit, that is, too many false negatives

[3, 4]. Both Currie [9, 10] and MARLAP [11, p. 20-54–20-55] explicitly acknowl-
edged this negative bias, with MARLAP [11, eqn 20.66] then suggesting replacing
σ_0 by $s_0/c_4(v)$, where the $c_4(v)$ bias correction factor is eqn A.29, repeated as
follows:

$$c_4(v) \equiv \frac{E[s_0]}{\sigma_0} = \left(\frac{2}{v}\right)^{1/2} \frac{\Gamma[(v+1)/2]}{\Gamma[v/2]} \tag{17.2}$$

For convenience, Table A.5 presents selected values of $c_4(v)$. If σ_0 is replaced
by $s_0/c_4(v)$, the resulting detection limits have positive bias, that is, too few false
negatives [3]. This is shown in Fig. 17.11, where the results are from previous work
[3, Table 3], which had $\eta^{1/2} = 1$.

Consequently, the δ_{critical} method is *entirely irrelevant*: it would *never* be used in
preference to the optimal equations, if σ_0 is known, and it *cannot* be used without
causing bias, if σ_0 is unknown.

Verification that the δ_{critical} method is biased, when σ_0 is replaced by *either* s_0 or
$s_0/c_4(v)$, is simple: Fig. 17.12 shows a *Mathematica*® (*v.* 9) program that
compares probabilities of false negatives, assuming blank subtraction via the
sample mean of 4 *i.i.d.* blank replicates, for y_D (Fig. 17.2, lower right), and
the Y_D^{CHE} expression in Fig. 17.2, with σ_0 replaced by s_0 or $s_0/c_4(v)$. In the
program, y_D is *detectionlimitCurrie*, *detectionlimitClayton* $= \delta_{p,q,v} \eta^{1/2} s_0$, and

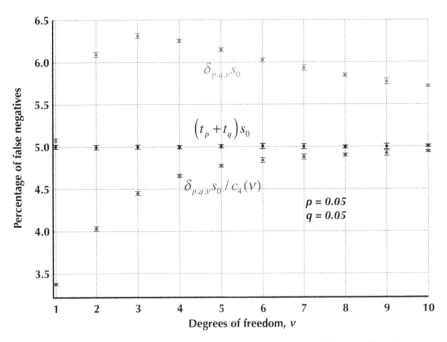

Figure 17.11 Previous results demonstrating the bias in two possible "fixes" of the δ_{critical}
method.

```
alpha = 1.0; sigma0 = 0.1; m0 = 4; nu = m0 - 1; p = 0.05; q = 0.05; m = 1;
numtrials = 1000000; tp = InverseCDF[StudentTDistribution[nu], 1 - p];
tq = InverseCDF[StudentTDistribution[nu], 1 - q];
tcritsum = tp + tq; deltapqnu = 4.45636085942; etaroot = Sqrt[(1.0/m) + (1.0/m0)];
c4nu = (Sqrt[2.0/nu]) Gamma[(nu + 1)/2.0]/Gamma[nu/2.0];
fnCurrie := 0; fnClayton := 0; fnMARLAP := 0;
Do[{s := StandardDeviation[{RandomVariate[NormalDistribution[alpha, sigma0]],
    RandomVariate[NormalDistribution[alpha, sigma0]],
    RandomVariate[NormalDistribution[alpha, sigma0]],
    RandomVariate[NormalDistribution[alpha, sigma0]]}]; s0 = s;
  ers0 := etaroot s0;
  netresponse := RandomVariate[NormalDistribution[0.0, etaroot sigma0]];
  decisionlevel = tp ers0; detectionlimitCurrie = tcritsum ers0;
  detectionlimitClayton = deltapqnu ers0;              [change this line]
  detectionlimitMARLAP = deltapqnu ers0/c4nu;
  shiftednetblankCurrie = detectionlimitCurrie + netresponse;
  shiftednetblankClayton = detectionlimitClayton + netresponse;
  shiftednetblankMARLAP = detectionlimitMARLAP + netresponse;
  If[shiftednetblankCurrie < decisionlevel, fnCurrie++, fnCurrie];
  If[shiftednetblankClayton < decisionlevel, fnClayton++, fnClayton];
  If[shiftednetblankMARLAP < decisionlevel, fnMARLAP++, fnMARLAP];},
 {numtrials}]
qsimCurrie = 1.0 fnCurrie/numtrials
qsimClayton = 1.0 fnClayton/numtrials
qsimMARLAP = 1.0 fnMARLAP/numtrials
```

Figure 17.12   *Mathematica* program demonstrating the bias in the "fixes" of the δ_{critical} method. (N.B. Use care in typing: some spaces are very important!)

$detectionlimitMARLAP = \delta_{p,q,v}\eta^{1/2}s_0/c_4(v)$, for $p = 0.05$, $q = 0.05$, and $v = 3$. All three detection limits use the y_C decision level expression in Fig. 17.2.

The program may be run on the fully functional 30-day free trial version of *Mathematica*. For 1 million trials, with $q = 0.05$, the results were $qsimCurrie = 0.050114$, $qsimClayton = 0.063049$, and $qsimMARLAP = 0.044373$, in excellent agreement with the $v = 3$ results shown in Fig. 17.11. To verify that the δ_{critical} method works *correctly* when *both* σ_0 *and* s_0 are used, replace "detectionlimitClayton = deltapqnu ers0" by "detectionlimitClayton = deltapqnu etaroot sigma0" in the *Mathematica* program in Fig. 17.12; the arrow indicates the line. This is then as per the equations in the middle of Fig. 17.2.

17.5   NO VALID EXTENSION FOR HETEROSCEDASTIC SYSTEMS

The δ_{critical} method *has no valid extension to heteroscedastic systems*; see Chapter 18 for full details. But the experimental system used by Clayton *et al.* [7] was actually heteroscedastic, so they performed a variance stabilizing "shifted square-root" transformation [7, eqn 19] in order to make the system homoscedastic. As a consequence, the noise was no longer Gaussian distributed, so their method was technically inapplicable. Furthermore, σ_0 was unknown, so they simply replaced it with s_0. As a consequence, their experimental results, which always had at least 28 degrees of freedom, did not validate their proposed method.

In summary, then, the $\delta_{critical}$ method is *entirely irrelevant*, regardless of whether the noise is homoscedastic or heteroscedastic. The same applies to ISO 11843-2 [8], IUPAC 1995 [9] and all other standards that are based on the use of critical $\delta_{p,q,v}$ values or pointless approximations thereof.

17.6 HYPOTHESIS TESTING FOR THE $\delta_{critical}$ METHOD

Although the $\delta_{critical}$ method is unsound metrologically, it is useful to examine its hypothesis testing formulations: had this been properly done decades ago, the method never would have become a pernicious obstruction.

17.6.1 Hypothesis Pertaining to False Positives

17.6.1.1 Hypothesis 5   This is exactly the same as hypothesis 2.

17.6.2 Hypothesis Pertaining to False Negatives

17.6.2.1 Hypothesis 6   Assume σ_0 is known, with p, q, and v fixed. Then the *suboptimal* theoretical Clayton, Hines, and Elkins detection limit, Y_D^{CHE}, is the smallest positive *constant* value of Y such that there is no more than q probability that a *random sample* from $N{:}Y, \sigma_0\eta^{1/2}$ will fall below a *random sample* from the χ distribution of y_C. For all values of p, q, and v, $Y_D < Y_D^{CHE}$. This hypothesis test is illustrated in Fig. 17.13.

From the simulation in Fig. 17.1, $Y_D \cong 0.476724\,\mathrm{V} < Y_D^{CHE} \cong 0.645788\,\mathrm{V}$. Hence, there is *never* any reason to use Y_D^{CHE} except for cross-checking computer simulations, where it works very well.

17.7 MONTE CARLO TESTS OF THE HYPOTHESES

What ultimately matters is simply this: Are the *obtained* probabilities of false positives and false negatives statistically equivalent to the *a priori* specified probabilities? The results from the simulation shown in Fig. 17.1 are summarized in Table 17.1. Note that both p and q have *a priori* values of exactly 5%.

The results in Table 17.1 verify that all of the equations in Fig. 17.2 work exactly as expected; properly used, they are free from bias. However, the $\delta_{critical}$ method is biased if σ_0 is replaced by either s_r or $s_r/c_4(3)$. Furthermore, the biased results are in excellent agreement with the $v = 3$ results in Fig. 17.11 and with those obtained from the *Mathematica* program in Fig. 17.12.

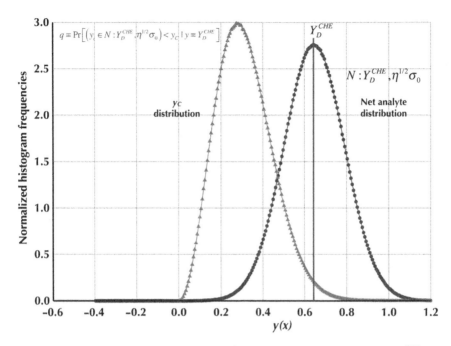

Figure 17.13 Testing for false negatives using the δ_{critical} method's detection limit, Y_D^{CHE}. The value of Y_D^{CHE} was 0.645788 V.

Table 17.1 Summary of Monte Carlo Simulation Results from Fig. 17.1

Data Table in Fig. 17.1	Tested Hypotheses	Errors $\pm$ 1 Standard Deviation
(1)	1	$4.994 \pm 0.027\%$ False positives
(2)	3 (with 1)	$5.001 \pm 0.014\%$ False negatives
(3)	6 (with 5), Both σ_0 and s_r used	$4.994 \pm 0.018\%$ False negatives
(4)	2	$4.997 \pm 0.025\%$ False positives
(5)	4 (with 2)	$4.991 \pm 0.014\%$ False negatives
(6)	6 (with 5), σ_0 replaced with s_r	$6.301 \pm 0.017\%$ False negatives
(7)	6 (with 5), σ_0 replaced with $s_r/c_4(3)$	$4.439 \pm 0.011\%$ False negatives

17.8 THE OTHER PROPAGATION OF ERROR

As above, suppose $p = q$, the chemical measurement system (CMS) is linear, the noise is homoscedastic additive Gaussian white noise (AGWN), with unknown σ_0, and ordinary least squares (OLS) is used to process the calibration data. Then

$\delta_{p,q,v} = \delta_{p,p,v}$, σ_0 is estimated by s_r, $\delta_{p,p,v}\eta^{1/2}\sigma_0$ is unavailable, both $\delta_{p,p,v}\eta^{1/2}s_r$ and $\delta_{p,p,v}\eta^{1/2}s_r/c_4(v)$ are biased relative to $\delta_{p,p,v}\eta^{1/2}\sigma_0$, and the *only* unbiased option is to use the appropriate y_C and y_D equations. Consequently, from Fig. 17.2, $y_D = 2y_C = 2t_p\eta^{1/2}s_r$ and $\delta_{p,p,v}$ is never used.

Unaware of the above, Currie [9, eqn 14] suggested that $\delta_{p,p,v}$ was approximately equal to $2t_p$, especially when v was large. This is true, given the limiting values of t_p and $\delta_{p,p,v}$: see Appendix A. However, both $\delta_{p,p,v}$ and $\delta_{p,q,v}$ are irrelevant, as are unnecessary approximations of them, and $2t_p$ is *not* an approximation. This mistake, that is, trying to force the use of $\delta_{p,p,v}$, illustrates what can happen when the facts are at variance with a preferred theory that has neither been tested nor validated in any meaningful way. Indeed, simple computer modeling would have demonstrated that y_D was *not* an approximation, for *any* set of p, q, and v values. The problem was the erroneous belief that hypothesis 6 was the *only* statistically valid one pertaining to false negatives. Yet, when σ_0 is unknown, hypothesis 4 is the *only* relevant one for false negatives.

An even worse problem involves estimating $\delta_{p,p,v}$ as shown in eqn 17.3 [9, p. 1711]:

$$\hat{\delta}_{p,p,v} \equiv 2t_p \times f(v) \quad \text{for } v > 4 \tag{17.3}$$

where $f(v)$ is defined as

$$f(v) \equiv 4v/(4v + 1) \tag{17.4}$$

Given that accurate $\delta_{p,q,v}$ values are readily available in tables [7, Table 1] or may be computed via simple numerical integration of the noncentral t distribution, eqn 17.3 is *pointless*. The $f(v)$ factor was asserted by Currie to be a "very simple correction factor for $2t$, $4v/(4v + 1)$, which takes into account the bias in s" [9, p. 1711]. Currie *also* used $f(v)$ as an approximation for $c_4(v)$ [10, eqn 12], despite $c_4(v)$ being easily computed via eqn 17.2 and available in tables [11, Table 20.2]. In actuality, $f(v)$ introduces negative bias and, as above, *this would have been revealed had any computer modeling been performed*. Ironically, if $\delta_{p,p,v}$ is approximated as in eqn 17.3 and if $f(v)$ is *also* used to approximate $c_4(v)$, then

$$Y_D^{CHE} = \delta_{p,p,v}\eta^{1/2}\sigma_0 \approx \hat{\delta}_{p,p,v}\eta^{1/2}\hat{\sigma}_0 = [2t_p \times f(v)]\eta^{1/2}[s_r/f(v)]$$
$$= 2t_p\eta^{1/2}s_r = y_D = 2y_C \tag{17.5}$$

so Y_D^{CHE} is effectively *replaced* by the *unbiased* random variate y_D, thereby achieving the best possible estimate and entirely eliminating the rationale for using Y_D^{CHE} in the first place. See also MARLAP eqn 20.66 [11, p. 20–55].

Lamentably, unforced errors such as the above tend to propagate; following IUPAC [9, p. 1711], Bettencourt da Salva assumed that use of $f(v)$ in his LOD equations "aims at correcting the bias of the estimated standard deviation of measurements population relevant from small number of degrees of freedom. The Student's t increases the LOD estimated from a small number of signals" [12, eqns 1,2,5,6]. However, this statement is incorrect. To illustrate, consider Bettencourt da Salva's eqn (1):

$$\text{LOD} = t_{\nu_{bk}}^{95\%;\text{one}} s_{bk} f(\nu_{bk}) + t_{\nu_{\text{LOD}}}^{95\%;\text{one}} s_{\text{LOD}} f(\nu_{\text{LOD}}) \tag{17.6}$$

where s_{bk} is the sample standard deviation of the blank, with ν_{bk} degrees of freedom, s_{LOD} is the sample standard deviation in the vicinity of y_D, with ν_{LOD} degrees of freedom, and $f(\nu)$ as in eqn 17.4. Assuming the noise is homoscedastic AGWN and $\nu_{bk} = \nu_{\text{LOD}} = \nu_\tau$, eqn 17.6 simplifies to Bettencourt da Salva's eqn (2):

$$\text{LOD} = 2t_{\nu_\tau}^{95\%;\text{one}} s_\tau f(\nu_\tau) \tag{17.7}$$

Then Fig. 17.14 shows the Monte Carlo program used to test eqn 17.7 and related variants.

The model's parameters were as follows: $\sigma_0 \equiv 0.1\,\text{V}$, $\beta \equiv 1\,\text{V/content unit}$, $\alpha \equiv 1\,\text{V}$, $N \equiv 7$ $(X_1 \equiv 1, X_2 \equiv 2, \ldots, X_7 \equiv 7)$, $\nu \equiv N - 2 = 5 = \nu_{bk} = \nu_{\text{LOD}} = \nu_\tau$, $f(\nu_\tau) = f(5) \cong 0.952381$, $p \equiv 0.05 \equiv q$, $z_p = z_q \cong 1.644854$, $t_p = t_q = t_{\nu_\tau}^{95\%;\text{one}} \cong 2.015048$, and $\eta^{1/2} \cong 1.309307$. Ten simulations were performed, with 1 million steps each, and the sample standard error about regression, s_r, was again the estimate of σ_0. Note that $s_\tau \equiv s_r$ and the $\eta^{1/2}$ factor is required because α is estimated by the regression intercept, a.

The top two data tables, (1) and (2), show that the y_C and y_D equations from Fig. 17.2 yield excellent results: $5.001 \pm 0.020\%$ false positives and $4.997 \pm 0.017\%$ false negatives. The middle two data tables, (3) and (4), show that eqn 17.7 is badly negatively biased: $10.129 \pm 0.038\%$ false positives and $10.119 \pm 0.020\%$ false negatives. Most of the bias results from omission of the $\eta^{1/2}$ factor. The bottom two data tables, (5) and (6), show what happens when the right-hand side of eqn 17.7 is multiplied by the correct $\eta^{1/2}$ factor: $5.656 \pm 0.021\%$ false positives and $5.649 \pm 0.020\%$ false negatives. Thus, the unbiased version of eqn 17.7 is

$$\text{LOD} = 2t_{\nu_\tau}^{95\%;\text{one}} \eta^{1/2} s_\tau = y_D = (t_p + t_q)\eta^{1/2} s_r = 2t_p \eta^{1/2} s_r \tag{17.8}$$

since $p = q$ and $s_0 = s_r$. Then, the unbiased version of eqn 17.6 is

$$\text{LOD} = t_{\nu_{bk}}^{95\%;\text{one}} \eta^{1/2} s_{bk} + t_{\nu_{\text{LOD}}}^{95\%;\text{one}} \eta^{1/2} s_{\text{LOD}} \tag{17.9}$$

Clearly, $f(\nu)$ should *never* be used, either to approximate $c_4(\nu)$ or to "correct" $2t_p$. In contrast, $\eta^{1/2}$ should be used, as per Chapters 7–14, though it will be unity if $M \equiv 1$ and the true intercept, α, is used for blank subtraction.

In summary, the proper course of action is to ignore ISO 11843-2 [8], IUPAC 1995 [9], and all similarly flawed standards that depend upon critical $\delta_{p,q,\nu}$ values or misleading approximations of them. Then simply use the results in Chapters 7–14: for homoscedastic AGWN, they are correct and unbiased, for all degrees of freedom. For heteroscedastic AGWN, as implied in eqn 17.9, see the following chapter.

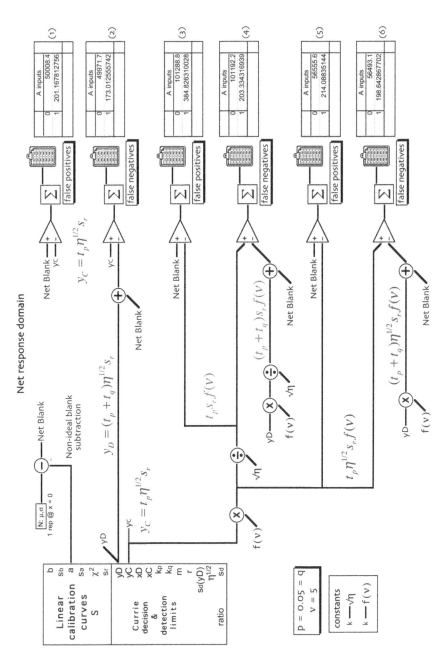

Figure 17.14 Simulation model used to test eqn 17.7 and variants (see text).

17.9 CHAPTER HIGHLIGHTS

In this chapter, Neyman–Pearson hypothesis testing was discussed and all relevant hypotheses were clearly stated and illustrated. It was shown that the $\delta_{critical}$ method is entirely irrelevant, and the same applies to all officially sanctioned detection limit protocols based on critical $\delta_{p,q,v}$ values. Ironically, any analyst who "estimated" $\delta_{p,q,v}$ as $t_p + t_q$ actually avoided the mistake of using $\delta_{p,q,v}$. It was also demonstrated that $f(v)$, defined in eqn 17.4, should *never* be used: it *always* results in needless bias. Finally, computer simulations are absolutely essential; no one who has seriously studied the fundamental aspects of detection limits has had infallible intuition, *most certainly including the author*. Indeed, in regard to detection limit theory and practice, it is fair to say that competently devised and performed computer simulations are the most effective way, *by far*, to avoid fooling oneself.

REFERENCES

1. L.A. Currie, "Limits for qualitative and quantitative determination – application to radiochemistry", *Anal. Chem.* **40** (1968) 586–593.

2. E. Voigtman, "Limits of detection and decision. Part 1", *Spectrochim. Acta, B* **63** (2008) 115–128.

3. E. Voigtman, "Limits of detection and decision. Part 2", *Spectrochim. Acta, B* **63** (2008) 129–141.

4. E. Voigtman, "Limits of detection and decision. Part 3", *Spectrochim. Acta, B* **63** (2008) 142–153.

5. E. Voigtman, "Limits of detection and decision. Part 4", *Spectrochim. Acta, B* **63** (2008) 154–165.

6. N.L. Johnson, S. Kotz, N. Balakrishnan, *Continuous Univariate Distributions,* Volume 1, 2nd Ed., John Wiley & Sons, New York, 1994, p. 163.

7. C.A. Clayton, J.W. Hines, P.D. Elkins, "Detection limits with specified assurance probabilities", *Anal. Chem.* **59** (1987) 2506–2514.

8. ISO 11843-2, "Capability of detection – Part 2: methodology in the linear calibration case" ISO, Genève, 2000.

9. L.A. Currie, for IUPAC, "Nomenclature in evaluation of analytical methods including detection and quantification capabilities" *Pure Appl. Chem.* **67** (1995) 1699–1723, IUPAC ©1995.

10. L.A. Currie, "Detection: international update, and some emerging di-lemmas involving calibration, the blank, and multiple detection decisions", *Chemom. Intell. Lab. Syst.* **37** (1997) 151–181.

11. Multi-Agency Radiological Laboratory Analytical Protocols (MARLAP) Manual Volume III, Section 20A.3.1, 20-54 – 20-58, United States agencies in MARLAP are EPA, DoD, DoE, DHS, NRC, FDA, USGS and NIST. This large document is available as a PDF on-line.

12. R.J.N. Bettencourt da Salva, "Spreadsheet for designing valid least-squares calibrations: a tutorial", *Talanta* **148** (2016) 177–190.

18

HETEROSCEDASTIC NOISES

18.1   INTRODUCTION

In the previous chapters, the noise was assumed to be homoscedastic additive Gaussian white noise (AGWN), with homoscedasticity simply meaning that the population standard deviation of the noise was independent of x, that is, was constant. This is the simplest possible noise precision model (NPM), and it may be expressed in equation form as

$$\sigma(x) \equiv \sigma_0 \tag{18.1}$$

where σ_0 is the constant population standard deviation of the AGWN. Heteroscedasticity means that $\sigma(x)$ varies with x, with $\sigma(0) \equiv \sigma_0$. In principle, $\sigma(x)$ could be any of an infinite number of possible functions, but, in practice, the number of heteroscedastic NPMs that have actually been used in published papers, perhaps representative of reality, is relatively small. The two simplest heteroscedastic NPMs are the most widely used models, so they are the main focus of this chapter.

18.2   THE TWO SIMPLEST HETEROSCEDASTIC NPMs

The linear and "hockey stick" NPMs are expressed in equation form, respectively, as shown in eqns 18.2 and 18.3:

$$\sigma(x) \equiv \sigma_0 + \mu\beta x = \sigma_0 + \mu y(x) \tag{18.2}$$

and

$$\sigma(x) \equiv [\sigma_0^2 + \mu^2\beta^2 x^2]^{1/2} = [\sigma_0^2 + \mu^2 y^2(x)]^{1/2} \tag{18.3}$$

Limits of Detection in Chemical Analysis, First Edition. Edward Voigtman.
© 2017 John Wiley & Sons, Inc. Published 2017 by John Wiley & Sons, Inc.
Companion Website: www.wiley.com/go/Voigtman/Limits_of_Detection_in_Chemical_Analysis

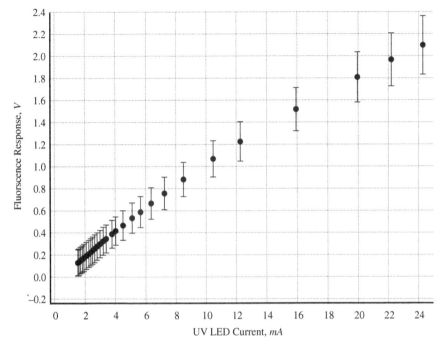

Figure 18.1 Nonlinear response showing mild heteroscedasticity, with roughly linear NPM. Wysoczanski, 2014 [1]. Reproduced with permission of Elsevier.

where μ is the true heteroscedasticity factor, that is, the asymptotic relative standard deviation (RSD). Note that $\mu \geq 0$, and, if $\mu \equiv 0$, eqns 18.2 and 18.3 reduce to eqn 18.1.

In real chemical measurement systems (CMSs), heteroscedasticity may range from mild to severe. A relatively mild degree of heteroscedasticity is apparent in the experimental nonlinear response function [1] shown in Fig. 18.1.

The results shown in Fig. 18.1 are from a real experiment, where both the response function and the NPM were deliberately unknown in a quantitative sense. They were not needed because the detection limit was obtained through the use of receiver operating characteristics (ROCs), as in Chapter 6. However, a question arises: What NPM model should be used, if it is desired to use one? Unfortunately, this is a very difficult question to answer, in general, and it is often made even harder by having little actual data from which to formulate a plausible NPM. This issue is easily illustrated: many real CMSs are tacitly assumed to be homoscedastic, that is, have eqn 18.1 as their NPM, with little or no experimental evidence to support the assumption. Sometimes, this is done simply for convenience or out of ignorance, but, either way, it is problematic.

Any heteroscedastic NPM is necessarily more complicated than eqn 18.1, and, if nothing else, the survey by Thompson [2] demonstrates that a few empirical NPMs

have long found favor, though perhaps inadequate theoretical or experimental validation. More will be said about this, but note that use of an incorrect NPM, or misuse
of an NPM that is known to be correct, may substantially degrade, or even invalidate,
results dependent upon the NPM. The latter situation may arise if weighted regression is used and the weights are inadequately estimated or are simply inappropriate
for the NPM, for example, using reciprocal standard deviations rather than reciprocal
variances.

18.2.1 Linear NPM

For this NPM, the fundamental equations needed to derive the Currie detection limit
have already been given and solved [3, 4], so the following treatment derives from
that prior work. Starting in the net response domain, the expression for Y_C is

$$Y_C \equiv z_p \sigma_d \tag{18.4}$$

where

$$\sigma_d^2 \equiv \sigma_a^2 + \sigma_0^2/M \equiv \eta\sigma_0^2 \tag{18.5}$$

In eqn 18.5, σ_a is the population standard error of the ordinary least squares (OLS)
intercept, assuming OLS is used, and σ_d is the population standard deviation of the
net blank. Regardless of the NPM, eqn 18.4 applies.

For homoscedastic noise, the previous Y_D expression was

$$Y_D = (z_p + z_q)\eta^{1/2}\sigma_0 \equiv (z_p + z_q)\sigma_d = Y_C + z_q\sigma_d = z_p\sigma_d + z_q\sigma_d \tag{18.6}$$

However, for the linear NPM in eqn 18.2, eqn 18.6 is modified in the obvious way:

$$Y_D = Y_C + z_q\sigma_d(Y_D) = z_p\sigma_d + z_q\sigma_d(Y_D) \tag{18.7}$$

where $\sigma_d(Y_D)$ is the population standard deviation of the net analyte at $x = X_D$, that
is, at $y(X_D) = Y_D$. Thus, using eqns 18.2 and 18.5, eqn 18.7 yields

$$\sigma_d^2(Y_D) = \sigma_a^2 + (\sigma_0 + \mu Y_D)^2/M \tag{18.8}$$

If $\mu = 0$, eqn 18.8 reduces to eqn 18.5, that is, the noise is homoscedastic. The following step is to solve eqns 18.4, 18.5, 18.7, and 18.8 for Y_D. The closed-form result
of simple algebra is then [3, eqn 28]

$$Y_D = B^{-1}\{Y_C + A + [(Y_C + A)^2 - Y_C^2 B(1 - z_q^2/z_p^2)]^{1/2}\} \tag{18.9}$$

where A and B are convenience terms defined as

$$A \equiv \sigma_0 z_q^2 \mu/M \tag{18.10}$$

and

$$B \equiv 1 - \mu^2 z_q^2 / M \tag{18.11}$$

If $\mu = 0$, then $A = 0$ and $B = 1$, so that eqn 18.9 simplifies to

$$Y_D = \frac{z_p + z_q}{z_p} Y_C \tag{18.12}$$

as expected. Furthermore, if $p = q$ as well, then $Y_D = 2Y_C$, also as expected. For the content domain, $X_C = Y_C/\beta$ and $X_D = Y_D/\beta$. In the experimental, net response domain, the expressions corresponding to eqns 18.9–18.12 are obtained by the standard substitutions: $\sigma_0 \mapsto s_0$, $z_p \mapsto t_p$, $z_q \mapsto t_q$, $\sigma_a \mapsto s_a$, $\sigma_d \mapsto s_d$, $Y_C \mapsto y_C$, $Y_D \mapsto y_D$, $\sigma_d(Y_D) \mapsto s_d(y_D)$, $A \mapsto \hat{A}$, $B \mapsto \hat{B}$, and $\mu \mapsto m$, where m is the experimental heteroscedasticity, that is, the estimate of μ. The content domain expressions are obtained from those in the net response domain by dividing by β, if known, or by $\hat{\beta}$, otherwise.

18.2.2 Experimental Corroboration of the Linear NPM

The above equations have been tested with both extensive Monte Carlo computer simulations and with real laser-excited molecular fluorescence detection experiments [5, 6]. Several of the most important experimental results [6], based on the combined use of bootstrapping and properly weighted least squares, are presented as follows. It should be noted that the experimental system was deliberately designed to have a known linear NPM, with relatively high, and easily controlled, noise magnitudes. This was done in order to overwhelm the CMS's intrinsic low level noises, thereby avoiding the very common problem of having to settle for whatever NPM the CMS might have. It also facilitated accurate estimation of the necessary weights, which is crucially important.

The first result is shown in Fig. 18.2, where the experimental histogram of 10^7 t_{slope} variates is overplotted with the noncentral t distribution with $\nu \equiv 5$ and the best fitting δ estimate, that is, $\hat{\delta} \cong 22.72719$. The following two figures show the 10^7 variate experimental histograms for y_C and y_D (Fig. 18.3) and x_C and x_D (Fig. 18.4). In each case, the fits are in excellent agreement with the indicated equations shown in the figures.

The significance of the equations and results presented in Figs 18.2–18.4 is easily summarized:

- The Currie decision level and detection limit expressions, for a univariate, linear CMS with linearly heteroscedastic AGWN, were published in 2008 [3].
- Carefully crafted, laser-excited molecular fluorescence experiments then fully corroborated the 2008 theory and simulations, with the results published in 2011 [5, 6].

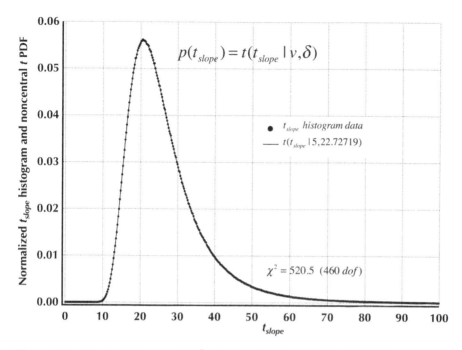

Figure 18.2 The t_{slope} histogram of 10^7 values, in 500 bins, overplotted with the noncentral t distribution, that is, eqn 10 in Voigtman, 2011 [6]. Reproduced with permission of Elsevier.

18.2.3 Hazards with Heteroscedastic NPMs

Before moving on to consider other NPMs, it is worth noting that there are several hazards that must be avoided in dealing with heteroscedastic NPMs. The most serious one, by far, involves finding the "correct" or "best" heteroscedastic NPM. This may be very difficult, in practice, especially if very little noise precision data is available and obtaining more is problematic. Theoretical considerations may be of great help, for example, if it is known, from fundamental principles, that $\sigma(x) \propto x$. Exploratory data analysis methods may also help in choosing among plausible model NPMs, but they do not necessarily find the "best" such NPM and, ultimately, can guarantee nothing. Accordingly, specifying the "correct" NPM is the most difficult aspect of dealing with detection in heteroscedastic noise: a seemingly small modeling mistake may have significant deleterious consequences and unforeseen ramifications.

A lesser problem is that of long-standing folklore, that is, that heteroscedastic NPMs invariably result in recursive equations that do not possess closed-form solutions, and, therefore, must be solved iteratively. However, several of the simplest and most commonly used NPMs have closed-form solutions that are derivable with nothing more than a little algebra. To illustrate, it was noted above that the Y_D expression

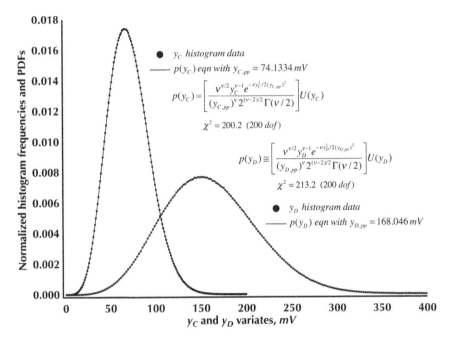

Figure 18.3   The y_C and y_D histograms of 10^7 values each, in 200 bins each, overplotted with eqns 11 and 12 in Voigtman, 2011 [6]. Reproduced with permission of Elsevier.

(eqn 18.9) was obtained in closed form. Every step of the derivation of eqn 18.9 is shown in Fig. 18.5.

18.2.4   Example: A CMS with Linear NPM

Consider a CMS with $\alpha = 0$, $\beta = 1$, $\sigma_0 = 10$, $p = 0.05 = q$, and $M = 1$. Assume α and σ_0 are known. Then $\sigma_a = 0$ and $\eta^{1/2} = 1$, since α would be used for blank subtraction. Also assume a linear NPM: $\sigma(Y) = \sigma_0 + \mu Y$, where μ is the true heteroscedasticity, that is, asymptotic RSD. Since $p = q$, eqn 18.9 simplifies to

$$Y_D = \frac{2(Y_C + A)}{B} \tag{18.13}$$

where A and B are defined in eqns 18.10 and 18.11, respectively. With $z_p = z_q = 1.644854$, eqns 18.4 and 18.5 yield $Y_C = 16.4485363$. Table 18.1 shows Y_D and $\sigma_d(Y_D)$ for various μ values.

The Y_D and $\sigma_d(Y_D)$ results were tested in the obvious way: for a given μ, 10^7 i.i.d. samples from $N{:}Y_D, \sigma_d(Y_D)$, in 10 sets of 10^6 each, were compared with Y_C. The results are shown in Table 18.1, and, in each case, the observed probabilities of false negatives were statistically equivalent to $q = 0.05$.

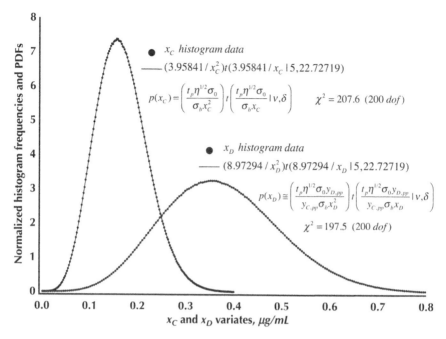

Figure 18.4 The x_C and x_D histograms of 10^7 values each, in 200 bins each, overplotted with eqns 13 and 14, respectively in [6]. The estimated δ values are from the fit in Fig. 18.2. Voigtman, 2011 [6]. Reproduced with permission of Elsevier.

$$Y_C \equiv z_p \sigma_d \qquad \sigma_d^2 = \sigma_a^2 + \sigma_0^2/M \qquad Y_D = Y_C + z_q \sigma_d\left(Y_D\right) \qquad \sigma_d^2\left(Y_D\right) = \sigma_a^2 + \left(\sigma_0 + \mu Y_D\right)^2/M$$

$$\sigma_d^2(Y_D) = \sigma_a^2 + \frac{(\sigma_0 + \mu Y_D)^2}{M} = \sigma_a^2 + \frac{\sigma_0^2}{M} + \frac{2\sigma_0 \mu Y_D}{M} + \frac{(\mu Y_D)^2}{M} \qquad \therefore \sigma_d^2\left(Y_D\right) = \sigma_d^2 + \frac{1}{M}\left[2\sigma_0\mu Y_D + (\mu Y_D)^2\right]$$

$$\therefore Y_D = Y_C + z_q\left\{\sigma_d^2 + \frac{1}{M}\left[2\sigma_0\mu Y_D + (\mu Y_D)^2\right]\right\}^{1/2} \qquad \therefore \left(Y_D - Y_C\right)^2 = z_q^2\left\{\sigma_d^2 + \frac{1}{M}\left[2\sigma_0\mu Y_D + (\mu Y_D)^2\right]\right\}$$

$$\therefore Y_D^2 - 2Y_C Y_D + Y_C^2 = z_q^2\sigma_d^2 + \frac{2z_q^2\sigma_0\mu}{M}Y_D + \frac{z_q^2\mu^2}{M}Y_D^2 \quad \therefore \left(1 - \frac{z_q^2\mu^2}{M}\right)Y_D^2 - \left(2Y_C + \frac{2z_q^2\sigma_0\mu}{M}\right)Y_D + \left(Y_C^2 - z_q^2\sigma_d^2\right) = 0$$

$$\text{But } \left(Y_C^2 - z_q^2\sigma_d^2\right) = Y_C^2 - \frac{z_p^2 z_q^2 \sigma_d^2}{z_p^2} = Y_C^2\left(1 - \frac{z_q^2}{z_p^2}\right) = Y_C^2\left(1 - z_q^2/z_p^2\right)$$

$$\therefore \left(1 - \frac{z_q^2\mu^2}{M}\right)Y_D^2 - \left(2Y_C + \frac{2z_q^2\sigma_0\mu}{M}\right)Y_D + Y_C^2\left(1 - z_q^2/z_p^2\right) = 0 \qquad a\ simple\ quadratic\ equation$$

$$\text{Define 2 terms: } \quad B \equiv 1 - \frac{z_q^2\mu^2}{M} \quad \text{ and } \quad A \equiv \frac{z_q^2\sigma_0\mu}{M} \qquad \therefore BY_D^2 - 2\left(Y_C + A\right)Y_D + Y_C^2\left(1 - z_q^2/z_p^2\right) = 0$$

$$\therefore Y_D = \frac{2(Y_C + A)\pm\left[4(Y_C + A)^2 - 4BY_C^2\left(1 - z_q^2/z_p^2\right)\right]^{1/2}}{2B} \qquad \therefore Y_D = \frac{(Y_C + A)\pm\left[(Y_C + A)^2 - BY_C^2\left(1 - z_q^2/z_p^2\right)\right]^{1/2}}{B}$$

$$\therefore Y_D = B^{-1}\left\{\left(Y_C + A\right) + \left[\left(Y_C + A\right)^2 - BY_C^2\left(1 - z_q^2/z_p^2\right)\right]^{1/2}\right\} \qquad the\ positive\ root\ is\ the\ correct\ one$$

Figure 18.5 Equations 18.4, 18.5, 18.7, and 18.8 are above the line, at the top of the figure.

Table 18.1 Summary of Results for the Example

μ	Y_D	$\sigma_d(Y_D)$	False Negatives ± 1 Standard Deviation (%)
0.5	185.2592	102.6296	4.996 ± 0.0313
0.05	35.84507	11.79230	5.003 ± 0.0275
0.005	33.16987	10.16585	4.996 ± 0.0253
0.0005	32.92415	10.01646	4.995 ± 0.0205
0.00005	32.89978	10.00164	4.991 ± 0.0247
0.000005	32.89734	10.00016	5.017 ± 0.0254
0	32.89707	10	5.003 ± 0.0256

Not knowing of a closed-form solution for a detection limit equation is usually inconsequential [7, 8], but not always. For example, Desimoni and Brunetti compared several published detection limit methods, and their associated x_D expressions, using an experimental system with linear NPM [9]. They were unaware that the correct x_D expression [9, eqn 10] was a special case of previously published equations [3, eqns 33–35], with $x_D \equiv y_D/b$, $s_a \equiv 0$, $M \equiv 1$, and M and M_0 notationally swapped. They then concluded that the ISO 11843-2 method [10] was "most suitable" [9]. However, as seen in the following section, the ISO 11843-2 method can *never* be most suitable because it is based on the δ_{critical} method.

18.3 HAZARDS WITH *ad hoc* PROCEDURES

Neither the CMS with linear NPM example nor the real experimental results presented in Figs 18.2–18.4 involved any use of critical $\delta_{p,q,\nu}$ values. The reason for this is that the δ_{critical} method is inapplicable with heteroscedastic noise. As discussed in the previous chapter, in connection with hypothesis 6, Y_D^{CHE} is the *fixed* location of a net analyte Gaussian distribution for which *i.i.d.* samples are tested against *i.i.d.* samples of the y_C distribution. Since $Y_D^{\text{CHE}} = \delta_{p,q,\nu}\eta^{1/2}\sigma_0$, p, q, and ν *uniquely* specify the operative noncentral t distribution when the noise is homoscedastic. For heteroscedastic NPMs, the net analyte distribution would have to shift to a fixed location *other* than Y_D^{CHE}, yet there is no possibility of a second noncentral t distribution, with different $\delta_{p,q,\nu}$, because p, q, and ν are independent of the NPM. Thus, there are two bad *ad hoc* possibilities:

1. Treat $\delta_{p,q,\nu}$ as an adjustable variable or
2. Use another function in place of the nonexistent second noncentral t distribution.

The first option is used in the MARLAP protocol [11]. The second option violates the very *raison d'être* for using the δ_{critical} method. As well, nothing is known about possible alternative functions or how they would depend upon the relevant NPMs. Despite this, ISO 11843-2 [10] implicitly assumes the second option, with

$Y_D^{CHE} \equiv \delta_{p,q,v} \sigma_d(Y_D^{CHE})$ defined as a heteroscedastic generalization of $Y_D^{CHE} = \delta_{p,q,v} \eta^{1/2} \sigma_0$. Then $\sigma_d(Y_D^{CHE})$ would be estimated using an iteratively reweighted linear regression analysis of empirical NPM data [9, 10]. However, the presumed generalization has no fundamental basis and is biased, as would have been revealed had *any* computer modeling been performed.

Unaware of these issues, Currie promoted the adoption of the $\delta_{critical}$ method [12–14]. Although there is no legitimate extension to heteroscedastic NPMs, Currie posited a conjectural solution in the net response domain: $S_D \approx \delta(\sigma_0 + \sigma_D)/2$ [12]. When he performed a rudimentary Monte Carlo simulation of his conjecture, it failed "by about a factor of two" [12, p. 159]. This is exactly as expected: there is no statistically valid extension of the $\delta_{critical}$ method to heteroscedastic NPMs and eqn 18.9, with $S_D \equiv Y_D$, is the correct solution for a linear NPM.

Currie deliberately did not conjecture an experimental counterpart for S_D: a major reason for his support of the method of critical $\delta_{p,q,v}$ values was that it does not allow for random variate detection limits, even though it absolutely requires the use of random variate decision levels, that is, y_C variates. Indeed, he stated [12, p. 158] "if S_D were defined using s_0 rather than σ_0, it would lead to a varying performance characteristic, subject to the whims of random error – an unacceptable situation." Currie was simply wrong in this: not only is the situation acceptable, it is inevitable; there is no escape from the "whims of random error."

18.4 THE HS ("HOCKEY STICK") NPM

The HS NPM given in eqn 18.3 has found some favor [1, 7, 15], in part because it respects the notion of adding independent variances, that is, the square of eqn 18.3 is simply

$$\sigma^2(x) \equiv \sigma_0^2 + \mu^2 \beta^2 x^2 = \sigma_0^2 + \mu^2 y^2(x) \qquad (18.14)$$

As Fig. 18.6 shows, the "hockey stick" name derives from the standard deviation being roughly constant at small values of x (that is, the "blade"), while the RSD is roughly constant at large values of x (that is, the "stick"). As for a simple RC electronic filter, the "corner analyte content," x_{cac}, is the location of the "knee" in a plot of $\log(\sigma(x))$ versus $\log(x)$ [15, Fig. 2].

Thompson and Ellison [7] used the HS NPM in their paper advocating the use of "uncertainty functions," previously called "characteristic functions" [15], as a possibly superior alternative to more conventional detection limit methodology. In this regard, x_{cac} is a *de facto* type of detection limit that dichotomizes the two error regimes of this CMS. Since "uncertainty functions" are simply renamed NPMs, it is not surprising to find that relevant prior work was published at least as far back as 1981, when Glaser et al. explored the possibility of using a linear NPM in preference to a response function [16]. See also Gautschi et al. [17] and references therein. However, accurate estimation of an NPM usually requires considerably more experimental data than does accurate estimation of a response function. Indeed, as Thompson and Coles state [15]:

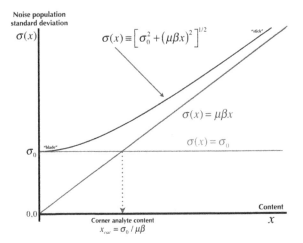

Figure 18.6 Accurate graphical illustration of the HS NPM and its "corner analyte content."

Very large datasets are needed adequately to study the functional form of these relationships, one or more orders of magnitude greater than those resulting from typical method validation studies. This is because individual estimates of standard deviation (at a single concentration) have wide confidence intervals unless a large number of repeat results are available. Moreover, to characterise a potentially curved functional form over a useful concentration range, a large number of closely spaced concentrations need to be studied. Even then, the best that can be achieved is to show that there is no systematic lack of fit of the data to a particular model – other untried functions could conceivably fit the data equally well. In such circumstances, any function selected should have the following properties: it should be parsimonious (i.e. have a small number of empirically determined parameters), theoretically plausible, and applicable in the widest range of analytical circumstances.

The characteristic function has been shown to fulfil these criteria. Moreover, although it requires a large dataset to demonstrate satisfactorily that there is no systematic lack of fit in a specific analytical system, a much smaller dataset (that is, typical of routine validation) can serve to estimate the parameters if the characteristic function can be assumed to be applicable from previous experience.

This is problematic for a variety of reasons. First, the relative precision of the standard deviation, $\sigma_s(v)/\sigma$, is quite poor even with relatively large data sets, that is, from Table A.5, $\sigma_s(v)/\sigma = 5\%$ even at $v = 200$. This underscores the necessity of a large data set if the NPM, and its parameters, need to be determined experimentally. Second, the HS NPM is certainly not the universal NPM for all CMSs. In fact, it is not even unique in having approximately constant standard deviation for small x and approximately constant RSD for large x [18–20]. Third, x_{cac} is incomplete; see Gautschi et al. [17] for a detailed analysis of what else would be required in order to construct a plausible detection limit method based on NPMs. Jiménez-Chacón and Alvarez-Prieto also provide cogent insight into this issue [21]. Lastly, the NPM selection criteria in the above quotation are vague, for example, "theoretical plausibility," "no systematic lack of fit of the data," and "applicable from previous experience." How would such criteria be properly evaluated, and by whom?

In view of the above, and additional issues as well, the "uncertainty functions" method will not receive further consideration. Instead, attention will refocus on how NPMs are correctly used in the Currie schema. In this regard, it is interesting that the HS NPM is one of the few NPMs that have closed-form solutions, contrary to what Thompson and Ellison stated [7, p. 5859].

18.5 CLOSED-FORM SOLUTIONS FOR FOUR HETEROSCEDASTIC NPMs

For sufficiently simple NPMs, it is relatively easy to find closed-form solutions for the relevant theoretical and experimental detection limits in the net response domain. These may then be used, in the obvious way, to obtain the corresponding response and content domain expressions. The following two figures provide the Y_D expressions (Fig. 18.7) and y_D expressions (Fig. 18.8) for four simple NPMs. In each case, Y_C and y_C are as shown at the bottom of their respective figures and the Y_D and y_D expressions simplify considerably if $p = q$, as is usually assumed.

As well, all the Y_D and y_D expressions reduce correctly to the homoscedastic limit. For example, the top three Y_D expressions in Fig. 18.7 reduce to

$$Y_D = Y_C \frac{z_p + z_q}{z_q} \tag{18.15}$$

if $\mu = 0$.

Noise precision model	Theoretical, net response domain limits of detection
Linear $\varepsilon(Y) \sim N : 0, \sigma_0 + \mu Y$	$Y_D = B^{-1} \left\{ Y_C + A + \left[(Y_C + A)^2 - Y_C^2 B \left(1 - z_q^2/z_p^2 \right) \right]^{1/2} \right\}$ $A \equiv \sigma_0 z_q^2 \mu/M$ ¢ $B \equiv 1 - z_q^2 \mu^2/M$
"Hockey stick" $\varepsilon(Y) \sim N : 0, (\sigma_0^2 + \mu^2 Y^2)^{1/2}$	$Y_D = Y_C B^{-1} \left\{ 1 + \left[1 - B \left(1 - z_q^2/z_p^2 \right) \right]^{1/2} \right\}$ $B \equiv 1 - z_q^2 \mu^2/M$
Shot noise *Gaussian approximation* $\varepsilon(Y) \sim N : 0, (\sigma_0^2 + \mu Y)^{1/2}$	$Y_D = Y_C + C + \left[(Y_C + C)^2 - Y_C^2 \left(1 - z_q^2/z_p^2 \right) \right]^{1/2}$ $C \equiv z_q^2 \mu^2/2M$
Root quadratic $\varepsilon(Y) \sim N : 0, (\sigma_0^2 + \lambda Y + \mu^2 Y^2)^{1/2}$	$Y_D = B^{-1} \left\{ Y_C + A + \left[(Y_C + A)^2 - Y_C^2 B \left(1 - z_q^2/z_p^2 \right) \right]^{1/2} \right\}$ $A \equiv z_q^2 \lambda/2M$ and $B \equiv 1 - z_q^2 \mu^2/M$

In every expression, $Y_C = z_p \eta^{1/2} \sigma_0 \equiv z_p \sigma_d$ ¢ $X = Y/\beta$, so that $X_C = Y_C/\beta$ and $X_D = Y_D/\beta$

Figure 18.7 Summary of closed-form Y_D expressions for four heteroscedastic NPMs.

Noise precision model	Experimental, net response domain limits of detection
Linear $\varepsilon(Y) \sim N : 0, \sigma_0 + \mu Y$	$y_D = \widehat{B}^{-1} \left\{ y_C + \widehat{A} + \left[\left(y_C + \widehat{A} \right)^2 - y_C^2 \widehat{B} \left(1 - t_q^2/t_p^2 \right) \right]^{1/2} \right\}$ $\widehat{A} \equiv s_0 t_q^2 m/M \quad \text{and} \quad \widehat{B} \equiv 1 - t_q^2 m^2/M$
"Hockey stick" $\varepsilon(Y) \sim N : 0, \left(\sigma_0^2 + \mu^2 Y^2 \right)^{1/2}$	$y_D = y_C \widehat{B}^{-1} \left\{ 1 + \left[1 - \widehat{B} \left(1 - t_q^2/t_p^2 \right) \right]^{1/2} \right\}$ $\widehat{B} \equiv 1 - t_q^2 m^2/M$
Shot noise *Gaussian approximation* $\varepsilon(Y) \sim N : 0, \left(\sigma_0^2 + \mu Y \right)^{1/2}$	$y_D = y_C + \widehat{C} + \left[(y_C + \widehat{C})^2 - y_C^2 (1 - t_q^2/t_p^2) \right]^{1/2}$ $\widehat{C} \equiv t_q^2 m^2/2M$
Root quadratic $\varepsilon(Y) \sim N : 0, \left(\sigma_0^2 + \lambda Y + \mu^2 Y^2 \right)^{1/2}$	$y_D = \widehat{B}^{-1} \left\{ y_C + \widehat{A} + \left[\left(y_C + \widehat{A} \right)^2 - y_C^2 \widehat{B} \left(1 - t_q^2/t_p^2 \right) \right]^{1/2} \right\}$ $\widehat{A} \equiv t_q^2 t/2M \quad \text{ç} \quad \widehat{B} \equiv 1 - t_q^2 m^2/M$

In every expression, $y_C = t_p \eta^{1/2} s_0 \equiv t_p s_d$ and $x = y/b$, so that $x_C = y_C/b$ and $x_D = y_D/b$

Figure 18.8   Summary of closed-form y_D expressions for four heteroscedastic NPMs.

The bottom Y_D expression does as well, if $\mu = 0$ and $\lambda = 0$. Note that the Y_D expression for the linear NPM is exactly the same as that for the "Root Quadratic" NPM: only the convenience term, A, differs. Hence, if $\lambda \equiv 2\mu\sigma_0$, the square is completed and the latter reduces to the former. Precisely analogous considerations apply to Fig. 18.8. Also note that y_D, for the linear NPM in Fig. 18.8, is the experimental "S_D" that Currie considered anathema.

18.6   SHOT NOISE (GAUSSIAN APPROXIMATION) NPM

Shot noise has a discrete distribution, that is, the Poisson distribution or a more complicated variant thereof. For a Poisson distribution, the mean, λ, equals the variance. Thus, if λ is large, the RSD will be low: it is $\lambda^{-1/2}$. If λ is not so low as to require consideration of the Poisson distribution's discrete nature, then the Gaussian distribution may serve as a useful continuous approximation, with $\mu \equiv \sigma^2 \cong \lambda$. If λ is sufficiently large, the envelope of the discrete distribution is essentially a finely sampled representation of a continuous function. Thanks to the central limit theorem, that continuous function is approximately Gaussian.

In this NPM, the total variance is the sum of σ_0^2 plus μY, since the shot noise variance is due to the mean value of μY. Note that σ_0^2 is optional: it may be set equal to zero for pure shot noise in the Gaussian approximation.

18.7   ROOT QUADRATIC NPM

The Y_D and y_D expressions for this NPM were independently derived, not realizing (*mea culpa*) that the Y_D version had been published in the MARLAP protocol, that is, eqn 20.22, with $S_D \equiv Y_D$, $c \equiv \sigma_0^2$ and $M \equiv 1$ [11, p. 20–21]. The same document also contains an *ad hoc* iterative procedure for estimating Y_D, via the use of *adjustable* critical $\delta_{p,q,v}$ values, and the iterative procedure is illustrated in example 20.13 [11, pp. 20-56–20-58]. However, the iterative procedure is an invalid muddle of parameters, estimates, and critical values. Not surprisingly, it fails in the illustrative example, yielding a converging series of incorrect Y_D estimates: 35.822, 37.242, 37.354, 37.363, 37.364, and 37.364. The correct solution, for y_D, is illustrated as follows.

18.8   EXAMPLE: MARLAP EXAMPLE 20.13, CORRECTED

Seven independent and normally distributed blank replicates have the following values: 58, 43, 64, 53, 47, 66, and 60 [11, p. 20–38]. Then $s_0 \cong 8.59125$, with $v = 6$. The stated NPM is

$$\sigma^2(Y) = aY^2 + bY + c \qquad (18.16)$$

with $a = (0.05)^2$, $b = 1$, and $c = \sigma_0^2$, with σ_0 unknown. This is quadratic in variance, but root quadratic in standard deviation, so, as per Fig. 18.8, $a = \mu^2$, $b = \lambda$, and $c = \sigma_0^2$. With $\mu = 0.05$, $\lambda = 1$, $M = 1$, and $p = 0.05 = q$, $t_p = t_q \cong 1.943180$, $\eta^{1/2} \cong 1.069045$, and $y_C = t_p \eta^{1/2} s_0 \cong 17.84700$. Then $\hat{A} \cong t_q^2 \lambda / 2M \cong 1.887975$, $\hat{B} \cong 1 - t_q^2 \mu^2 / M \cong 0.990560$, and $y_D = 2(y_C + \hat{A})/\hat{B} \cong 39.846$. Note that $y_D \approx 2.23 y_C$ and that neither Y_C nor Y_D can be obtained because σ_0 is unknown. Similarly, MARLAP eqn 20.22 cannot be used when σ_0 is unknown.

The MARLAP iterative procedure also fails for the linear NPM example discussed earlier, even in the $\mu = 0$ limit. Again, this would have been revealed had *any* computer modeling been performed. Consequently, the MARLAP iterative procedure should be spurned.

18.9   QUADRATIC NPM

Aside from a few simple NPMs for which it is possible to find closed-form expressions for Y_D and y_D, all others are dealt with via recursive iterations until satisfactory convergence is achieved. The Quadratic NPM, shown in Fig. 18.9, has no closed-form solution, so it serves to illustrate how the recursive iteration is performed. Thus, to find its Y_D value, do the following steps:

1. Compute σ_d.
2. Compute Y_C and initially estimate Y_D as $2Y_C$.

Quadratic NPM

$$\varepsilon(Y) \sim N:0, \sigma_0 + \mu Y + \rho Y^2$$

$$\sigma_d^2 = \sigma_a^2 + \sigma_0^2 / M$$

$$Y_D = Y_C + z_q \sigma_d(Y_D) = z_p \sigma_d + z_q \sigma_d(Y_D)$$

$$\sigma_d^2(Y_D) = \sigma_d^2 + M^{-1}\left[2\sigma_0 \mu Y_D + \left(2\sigma_0 \rho + \mu^2\right)Y_D^2 + 2\mu\rho Y_D^3 + \rho^2 Y_D^4\right]$$

$\boxed{Y_D \text{ found by iteration}}$

Figure 18.9   The quadratic NPM and its Y_D expression.

3. Use Y_D to evaluate the polynomial and obtain an initial value of $\sigma_d(Y_D)$.
4. Use $\sigma_d(Y_D)$ to evaluate the Y_D expression.
5. Cyclically repeat steps 3 and 4 until Y_D has satisfactorily converged.

In the experimental, net response domain, the corresponding expressions are obtained by the standard substitutions: $\sigma_0 \mapsto s_0$, $z_p \mapsto t_p$, $z_q \mapsto t_q$, $\sigma_a \mapsto s_a$, $\sigma_d \mapsto s_d$, $Y_C \mapsto y_C$, $Y_D \mapsto y_D$, $\sigma_d(Y_D) \mapsto s_d(y_D)$, $\mu \mapsto m$, and $\rho \mapsto r$.

18.10   A FEW IMPORTANT POINTS

First, the main issue in dealing with heteroscedastic noise is that of providing a justifiable answer to the question "what is the noise's actual NPM?" In general, whatever NPM is adopted should be as simple as possible, consistent with the experimental data and any relevant theoretical considerations. Second, contrary to conventional wisdom, several of the simplest and most commonly encountered NPMs have closed-form solutions, given in Figs 18.7 and 18.8. Third, computer simulations are *absolutely essential tools* in detection limit research, as clearly demonstrated throughout this text. Far too many wrong notions have arisen, and been promulgated, as a consequence of uncritical reliance upon intuition and the incompletely checked accuracy of the published, refereed literature. Finally, although a discussion of weighted regression is beyond the scope of the present text, one salient fact is of paramount importance: *weights must be consistent with the true NPM, whatever it happens to be, and need to be as accurate as possible.* This is a stringent requirement, but Asuero and González [22] provide an exceptionally thorough and readable discussion of the topic.

18.11   CHAPTER HIGHLIGHTS

Heteroscedastic noises and their NPMs were discussed, with emphasis on the two most commonly assumed NPMs: linear and "hockey stick." The matter of finding an appropriate NPM was discussed and several attendant hazards were examined. In particular, it was shown that closed-form solutions exist for both linear and "hockey stick" NPMs, despite myths to the contrary, and a worked example was provided for a CMS with linear NPM.

It was shown that the $\delta_{critical}$ method cannot work with heteroscedastic noises. In connection with this, MARLAP example 20.13 was shown to be incorrect, as would have been revealed by simple computer simulations, and the "uncertainty functions" method was found to be problematic.

REFERENCES

1. A. Wysoczanski, E. Voigtman, "Receiver operating characteristic – curve limits of detection", *Spectrochim. Acta, B* **100** (2014) 70–77.

2. M. Thompson, "Uncertainty functions, a compact way of summarising or specifying the behavior of analytical systems", *Trends Anal. Chem.* **30** (2011) 1168–1175.

3. E. Voigtman, "Limits of detection and decision. Part 3", *Spectrochim. Acta, B* **63** (2008) 142–153.

4. E. Voigtman, "Limits of detection and decision. Part 4", *Spectrochim. Acta, B* **63** (2008) 154–165.

5. E. Voigtman, K.T. Abraham, "True detection limits in an experimental linearly heteroscedastic system. Part 1", *Spectrochim. Acta, B* **66** (2011) 822–827.

6. E. Voigtman, K.T. Abraham, "True detection limits in an experimental linearly heteroscedastic system. Part 2", *Spectrochim. Acta, B* **66** (2011) 828–833.

7. M. Thompson, S.L.R. Ellison, "Towards an uncertainty paradigm of detection capability", *Anal. Methods*, **5** (2013) 5857–5861.

8. J. Jiménez-Chacón, M. Alvarez-Prieto, "An approach to detection capabilities estimation of analytical procedures based on measurement uncertainty", *Accred. Qual. Assur.* **15** (2010) 19–28.

9. E. Desimoni, B. Brunetti, "About estimating the limit of detection of heteroscedastic analytical systems", *Anal. Chim. Acta* **655** (2009) 30–37.

10. ISO 11843-2, *Capability of Detection - Part 2: Methodology in the linear calibration case*, International Standards Organization (ISO), Genève, 2000.

11. Multi-Agency Radiological Laboratory Analytical Protocols (MARLAP) Manual Volume III, Section 20A.3.1, 20-54–20-58, United States agencies in MARLAP are EPA, DoD, DoE, DHS, NRC, FDA, USGS and NIST. This huge document is available as a PDF on-line.

12. L.A. Currie, "Detection: international update, and some emerging di-lemmas involving calibration, the blank, and multiple detection decisions", *Chemom. Intell. Lab. Syst.* **37** (1997) 151–181.

13. L.A. Currie, "Nomenclature in evaluation of analytical methods including detection and quantification capabilities (IUPAC Recommendations 1995)", *Pure Appl. Chem.* **67** (1995) 1699–1723.

14. L. A. Currie, "Detection and quantification limits: origins and historical overview", *Anal. Chim. Acta* **391** (1999) 127–134.

15. M. Thompson, B.J. Coles, "Use of the 'characteristic function' for modelling repeatability precision", *Accred. Qual. Assur.* **16** (2011) 13–19.

16. J.A. Glaser, D.L. Foerst, G.D. McKee, S.A. Quave, W.L. Budde, "Trace analyses for wastewaters", *Environ. Sci. Technol.* **15** (1981) 1426–1435.

17. K. Gautschi, B. Keller, H. Keller, P. Pei, D.J. Vonderschmitt, "A new look at the limits of detection (L_D), quantification (L_Q) and power of definition (PD)", *Eur. J. Clin. Chem. Clin. Biochem.* **31** (1993) 433–440.

18. L. Oppenheimer, T.P. Capizzi, R.M. Weppelman, H. Mehta, "Determining the lowest limit of reliable assay measurement" *Anal. Chem.* **55** (1983) 638–643.

19. D.M. Rocke, S. Lorenzato, "A two-component model for measurement error in analytical chemistry", *Technometrics* **37** (1995) 176–184.

20. M.D. Wilson, D.M. Rocke, B. Durbin, H.D. Kahn, "Detection limits and goodness-of-fit measures for the two-component model of chemical analytical error", *Anal. Chim. Acta* **509** (2004) 197–208.

21. J. Jiménez-Chacón, M. Alvarez-Prieto, "Modelling uncertainty in a concentration range", *Accred. Qual. Assur.* **14** (2009) 15–27.

22. A.G. Asuero, G. González, "Fitting straight lines with replicated observations by linear regression. III. Weighting data", *Crit. Rev. Anal. Chem.* **37** (2007) 143–172.

19

LIMITS OF QUANTITATION*

*This chapter is reprinted, with kind permission from Elsevier, from Carlson, J., Wysoczanski, A., and Voigtman, E. (2014) Limits of quantitation – Yet another suggestion. *Spectrochim. Acta, B* **96**, 69–73. Only trivial changes were made for purposes of notational and stylistic consistency with previous chapters. Subsequent material is discussed in an added postscript, followed by chapter highlights.

Abstract

The work presented herein suggests that the limit of quantitation concept may be rendered substantially less ambiguous and ultimately more useful as a figure of merit by basing it upon the significant figure and relative measurement error ideas due to Coleman, Auses, and Gram, coupled with the correct instantiation of Currie's detection limit methodology. Simple theoretical results are presented for a linear, univariate chemical measurement system with homoscedastic Gaussian noise, and these are tested against both Monte Carlo computer simulations and laser-excited molecular fluorescence experimental results. Good agreement among experiment, theory, and simulation is obtained and an easy extension to linearly heteroscedastic Gaussian noise is also outlined.

19.1 INTRODUCTION

Limits of quantitation have always been overshadowed by decision levels and detection limits. The decision level is simply the level above which it is highly improbable to find a true net blank response, while, following Currie [1], the detection limit is such that analyte present at the detection limit is highly unlikely to go undetected. But the quantitation limit is far less well formed as a concept. Currie's "10σ" definition of it was simple, but has no fundamental justification: he simply referenced Adams *et al.* [2] as a reasonable source of the factor of 10. Likewise, the 1980 publication from the ACS Committee on Environmental Improvement simply defines the factor as 10 [3, p. 2247].

In 1970, Kaiser [4] provided a plausible justification of the factor based on Tschebyscheff's inequality, but that highly conservative bounding inequality is essentially a "last resort": it applies even when nothing at all is known of the distribution of the noise. With exceptions, analysts nowadays usually have some knowledge of the relevant noise distribution, thereby enabling more efficient and effective use of experimental data. Furthermore, the rise of nonparametric methodologies, such as that of the receiver operating characteristic [5], has also mitigated against the coupled use of efficient methods with Tschebyscheff's inequality.

Not surprisingly, numerous other definitions of the limit of quantitation (LOQ) have arisen since 1980 and find continued publication popularity. Mermet's thorough and cogent 2008 survey of LOQ definitions, methodologies, and usages in atomic spectrometry revealed a remarkably complicated situation, with no less than five distinct varieties of LOQ [6]. Although Mermet provided a helpful comparison of the relative advantages and disadvantages of the surveyed LOQ methodologies, he concluded that [6] "any LOQ can be selected, provided that both selection and procedure are clearly justified in relation to the analytical needs defined for method validation. Analytical chemistry is a science of rigor, and there is no reason why LOQ, which is a crucial characteristic of an analytical method, should escape from this rigor." Yet apparently it has escaped: in the absence of an explicitly stated, compelling, *a priori* reason for the use of an LOQ in general, analysts have free rein to construct new LOQ definitions or alter existing ones. This seriously undermines the utility and validity of LOQs as figures of merit, especially given the difficulty in interconverting results obtained with the various LOQ methodologies. Indeed, Mermet *et al.* [7] have since provided a detailed study of yet another possibility for quantifying LOQs: the "accuracy profile" method. At the very least, this indicates that none of the previously surveyed LOQ methodologies [6] are obviously superior.

At the risk of adding yet another variant LOQ definition to the current collection, one possible route to standardization of the LOQ is to employ a prescient idea put forth by Coleman *et al.* in 1997 [8, p. 78]: the LOQ "is the lowest concentration at or above which ... measurements have at least 1.0 significant digit (at high confidence), and, equivalently, have limited relative measurement error, RME $\leq 5\%$." This, then, may constitute the fundamental reason for the formulation and usage of the LOQ concept, regardless of the specifics of any particular LOQ methodology. Clearly, if existing LOQ formalisms were brought into compliance with this requirement, it would facilitate meaningful comparisons of LOQs and promote their use as figures of merit.

Coleman *et al.* provided a detailed and carefully reasoned exposition of their idea, and developed it in the context of RME and fractional significant figures. However, they also used the disadvantageous Hubaux and Vos detection limit methodology [9] to find a relationship between their LOQ definition and the Hubaux and Vos detection limit. Subsequently, Voigtman demonstrated, in a series of publications [10–16], the optimum implementation of Currie's program and it has since become clear (*vide infra*) that it is the relationship between the LOQ and the Currie decision level, not detection limit, that is of primary utility.

19.2 THEORY

We assume the chemical measurement system is univariate and linear in chemical content, X:

$$r(X) = \alpha + \beta X + \text{noise} \tag{19.1}$$

where α is the true intercept, β is the true slope, and $r(X)$ is the noisy response. Systematic error is assumed to be zero. The noise is additive Gaussian white noise (AGWN) and is homoscedastic, that is, $\sigma(\beta X) \equiv \sigma_0$, where σ_0 is the population standard deviation of the noise on the blank. As a consequence of the noise, a single measured response, r_i, at any arbitrary value of X, is simply a random sample from a Gaussian (aka "Normal") distribution centered at $\alpha + \beta X$ and having population standard deviation σ_0. If ideal blank subtraction is performed, to obtain the net response, then this simply means that the errorless value α is subtracted from r_i, so the net response, $r_i - \alpha$, is a random sample of a Gaussian distribution centered at βX, with population standard deviation σ_0. The shorthand notation for this is $(r_i - \alpha) \sim N{:}\beta X, \sigma_0$, where "$\sim$" means "is distributed as."

If α is unavailable, as is typically the case, then blank subtraction may be performed by subtracting $\hat{\alpha}$, an unbiased estimate of α. Then $(r_i - \hat{\alpha}) \sim N{:}\beta X, \sigma_d$, where σ_d is the population standard deviation of the difference. The relationship between σ_d and σ_0 is $\sigma_d = \eta^{1/2}\sigma_0$, where the factor $\eta^{1/2}$ is of the general form:

$$\eta^{1/2} \equiv [M^{-1} + \text{error term(s) due to } \hat{\alpha}]^{1/2} \tag{19.2}$$

with M being the number of future blank measurements. Optimally, M is defined as unity, so that $\eta^{1/2} = 1$ only if ideal blank subtraction is performed and is otherwise greater than unity.

For maximum generality, it will be assumed that nonideal blank subtraction is performed, that is, $\hat{\alpha}$ is used for blank subtraction even if β and σ_0 are known. Therefore, a *single* net response at βX will be a random sample from $N{:}\beta X, \sigma_d$. As is well known, there is a 95% probability, over many repetitions of the process, that the net response will be in the (central) 95% confidence interval (CI) defined by $\beta X \pm z'_p \sigma_d$, where $z'_p \equiv z_{0.025} \cong 1.959964$. Following the lead of Coleman *et al.* [8], the absolute measurement error is *arbitrarily* defined as $z'_p \sigma_d$, that is, the *half-width* of the 95% CI, and the maximum relative measurement error, RME(X), is

$$\text{RME}(X) \equiv \frac{z'_p \sigma_d}{\beta X} \cong \frac{1.96\sigma_d}{\beta X} \tag{19.3}$$

while still remaining within the 95% CI. Note that the RME is simply defined as the *half-width* of the CI, divided by its *center* value. As noted above [8], in order for the LOQ to guarantee *at least* one significant figure, $RME(X)$ must not exceed 0.05. Therefore, the theoretical content domain LOQ, denoted by X_Q, is defined as that value of X such that $RME(X_Q) \equiv 0.05$. Hence

$$X_Q \equiv \frac{z'_p \sigma_d}{0.05\beta} = 20\frac{z'_p \sigma_d}{\beta} = 20\frac{z'_p \eta^{1/2} \sigma_0}{\beta} \cong 39.2\frac{\eta^{1/2}\sigma_0}{\beta} \tag{19.4}$$

since $\sigma_d = \eta^{1/2}\sigma_0$. However, the theoretical content domain Currie decision level is defined as

$$X_C \equiv \frac{z_p \eta^{1/2}\sigma_0}{\beta} \tag{19.5}$$

where z_p is the critical z value for probability p of false positives [10]. Typically, $p \equiv 0.05$, so that $z_p = z_{0.05} \cong 1.6448536$. As a result,

$$X_Q = 20\frac{z'_p}{z_p}X_C \cong \frac{39.2}{z_p}X_C \tag{19.6}$$

In the theoretical net response domain, $Y_C = \beta X_C$ and likewise $Y_Q = \beta X_Q$. Hence,

$$Y_Q = 20\frac{z'_p}{z_p}Y_C \cong \frac{39.2}{z_p}Y_C \tag{19.7}$$

Values of z_p are easily found using *Microsoft Excel*, that is, $z_p = -\text{NORMSINV}(p)$, or via standard tables.

The experimental content domain decision level, x_C, is given by [10, 11]

$$x_C \equiv \frac{t_p \eta^{1/2} s_0}{b} \tag{19.8}$$

where b is an experimentally determined, unbiased estimate of β, s_0 is the sample standard deviation determined with v degrees of freedom (dof), and t_p is the critical t value for probability p of false positives and v dof. Then the corresponding experimental content domain LOQ, denoted by x_Q, is

$$x_Q \equiv \frac{t'_p \eta^{1/2} s_0}{0.05b} = 20\frac{t'_p \eta^{1/2} s_0}{b} = 20\frac{t_{0.025} \eta^{1/2} s_0}{b} \tag{19.9}$$

where t'_p is the critical t value for 95% confidence (analogous to $z_{0.025} \simeq 1.96$) with v dof. Combining eqns 19.8 and 19.9 then yields

$$x_Q = 20\frac{t'_p}{t_p}x_C \tag{19.10}$$

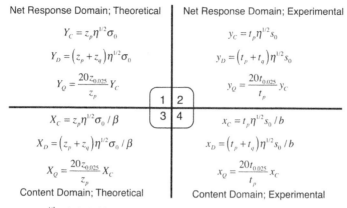

Figure with quadrant labels:

Net Response Domain; Theoretical | Net Response Domain; Experimental

$$Y_C = z_p \eta^{1/2} \sigma_0$$
$$Y_D = (z_p + z_q)\eta^{1/2}\sigma_0$$
$$Y_Q = \frac{20 z_{0.025}}{z_p} Y_C$$

$$y_C = t_p \eta^{1/2} s_0$$
$$y_D = (t_p + t_q)\eta^{1/2}s_0$$
$$y_Q = \frac{20 t_{0.025}}{t_p} y_C$$

Quadrants: 1 | 2 / 3 | 4

$$X_C = z_p \eta^{1/2}\sigma_0 / \beta$$
$$X_D = (z_p + z_q)\eta^{1/2}\sigma_0 / \beta$$
$$X_Q = \frac{20 z_{0.025}}{z_p} X_C$$

$$x_C = t_p \eta^{1/2} s_0 / b$$
$$x_D = (t_p + t_q)\eta^{1/2}s_0 / b$$
$$x_Q = \frac{20 t_{0.025}}{t_p} x_C$$

Content Domain; Theoretical | Content Domain; Experimental

$\eta^{1/2} = 1$ *if ideal blank subtraction & M = 1 future blank measurement*

Figure 19.1 Master summary of expressions in the four detection quadrants.

with v dof for both critical t values. In the experimental net response domain, $y_C = bx_C$ and $y_Q = bx_Q$. Therefore,

$$y_Q = 20\frac{t'_p}{t_p}y_C \tag{19.11}$$

Values of t_p are easily found using *Microsoft Excel*, that is, $t_p = \text{TINV}(2p, v)$, or via standard tables.

Figure 19.1 collects together eqns 19.6, 19.7, 19.10, and 19.11, and also gives the Currie decision level expressions ("*C*" subscripts) and detection limit expressions ("*D*" subscripts) in all four detection quadrants. The critical values z_q and t_q in the detection limit expressions are for probability q of false negatives. These are correct and unbiased for all dof, including $v = 1$. Note that all of the theoretical domain expressions are errorless real numbers that require population parameters (that is, true values, since systematic error is assumed to be negligible), while the expressions in quadrant 2 are χ variates and those in quadrant 4 are modified noncentral t variates [10, 11].

19.3 COMPUTER SIMULATION

The theory presented above is very simple and perfectly suited to testing via computer simulation. Accordingly, the following ideal model parameters were used, ignoring units: $\alpha \equiv -0.05$, $\beta \equiv 3.85$, $\sigma_0 \equiv 0.03$, $p \equiv 0.05$, $z_p \cong 1.644854$, $p' \equiv 0.025$, $z'_p \equiv z_{0.025} \cong 1.959964$, $M \equiv 1$ future blank replicate, and $\hat{\alpha}$ the sample mean of $N = 7$ i.i.d. blank replicates. In this case, $\eta^{1/2} \equiv [M^{-1} + N^{-1}]^{1/2} = [1 + (1/7)]^{1/2} \equiv 1.069045$, so that, from Fig. 19.1, $X_C = z_p\eta^{1/2}\sigma_0/\beta \cong 0.013702$ and $X_Q = 20(z'_p/z_p)X_C \cong 0.3265386$.

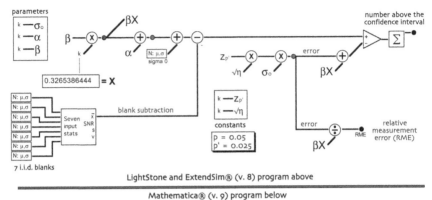

Figure 19.2 Monte Carlo programs for X_Q tests: *LightStone and ExtendSim*® (upper) and *Mathematica*® (lower).

Figure 19.2 shows two different simulation programs designed to test the model and its theoretically predicted X_Q. The upper half of Fig. 19.2 shows a Monte Carlo program implemented using the author's free *LightStone* component libraries, which are an add-on for the commercial *ExtendSim*® simulation software. This combination has been used extensively in prior work [10–16] and a free demo copy of *ExtendSim* is available from *Imagine That Inc.* (www.extendsim.com). The lower half of Fig. 19.2 shows a Monte Carlo *Mathematica*® (v. 9) program with *exactly* the same functionality. It may be run on the fully functional 30-day free trial version of *Mathematica* available from *Wolfram Research* (www.wolfram.com/).

For each of 1 million steps (*LightStone/ExtendSim*) or trials (*Mathematica*), a new blank subtracted variate is generated at the user-specified value of X. The total number of events that fall *above* the 95% CI centered on βX is recorded. Since the 95% CI is a *central* CI, the expectation is that 25 000, that is, the upper-tail 2.5% of 1 million total events, will fall above the 95% CI. For the *LightStone/ExtendSim* program in Fig. 19.2, with $X \equiv X_Q \cong 0.3265386$, the mean number of events above the 95% CI was $24\,948.6 \pm 159.2$, for 10 sets of 1 million steps each. The ± term is one standard deviation. For the *Mathematica* program in Fig. 19.2, likewise with

$X \equiv X_Q \cong 0.3265386$ and 10 runs of 1 million trials each, the results were $24\,982.0 \pm 140.3$. Each program in Fig. 19.2 was easily modified to test for numbers of events falling *below* the 95% CI. The results were $25\,033.3 \pm 108.7$ (*Light-Stone/ExtendSim*) and $24\,954.1 \pm 165.4$ (*Mathematica*). In all cases, the obtained results were statistically the same as the theoretical expectation value.

19.4 EXPERIMENT

The laser-excited molecular fluorescence experimental system was exactly the same one as used previously [14, 15], with the same laser dye analyte: rhodamine 6G tetrafluoroborate (R6G) in ethanol. The standards were either the exact same solutions previously prepared, used [14], and then carefully stored for possible future use, or were prepared from them by one-step dilutions using digital pipettes and pure ethanol. With the exception of using more standards, all other conditions were as previously reported [15], with noise attenuator setting of 1.00, that is, 10% of full scale added noise. Each analyte was measured 12 001 times, so that data were collected at a rate of 1000 samples per second for 12 s.

The analyte solution concentrations (X_i), in mg of R6G/100 mL ethanol, are reported in Table 19.1, along with their mean responses ($\bar{y}_i$), standard deviations, and approximate RMEs, that is, RME/$\eta^{1/2}$. The mean of the noise standard deviations was 0.0329589 V, which was taken as the estimate, $\hat{\sigma}_0$, of the population standard deviation of the homoscedastic noise. Then the approximate RMEs were computed, ignoring the $\eta^{1/2}$ factor in eqn 19.3, as $1.959964\hat{\sigma}_0/(\bar{y}_i - a)$, where $a = -0.037961$ V.

Table 19.1 Concentrations and Experimental Response Data

Concentration (mg/100 mL)	Response (V)	Noise (V)	RME/ $\eta^{1/2}$
0.03392	0.1025	0.03302	0.4599
0.04240	0.1322	0.03318	0.3797
0.08480	0.2835	0.03291	0.2009
0.1272	0.4411	0.03265	0.1348
0.2120	0.8386	0.03282	0.07369
0.2968	1.1062	0.03293	0.05646
0.3604	1.3606	0.03319	0.04619
0.4240	1.6272	0.03304	0.03879
0.01696	0.0434	0.03338	0.7939
0.02120	0.0564	0.03317	0.6844
0.04240	0.1294	0.03273	0.3859
0.06360	0.2037	0.03282	0.2673
0.1060	0.3797	0.03313	0.1547
0.1484	0.5008	0.03283	0.1199
0.1802	0.6382	0.03298	0.09554
0.2120	0.7717	0.03257	0.07978

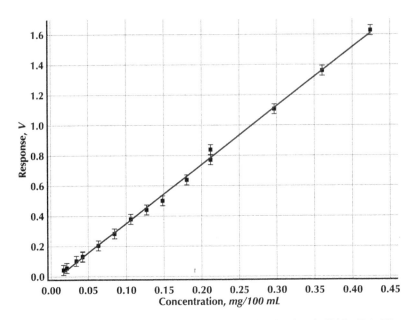

Figure 19.3 Molecular fluorescence calibration curve for the data in Table 19.1. The error bars are ±1 standard deviation.

Figure 19.3 shows a plot of the mean fluorescence responses $(\bar{y}_i)$, versus R6G concentrations (X_i). The ordinary least squares fit yields $a = -0.037961$ V, $b = 3.88777$ V/(mg/100 mL), $s_a = 0.00835$ V, $s_b = 0.0436$ V/(mg/100 mL), and adjusted $R^2 = 0.9981$. The $\eta^{1/2}$ blank subtraction factor is estimated by pooling $\hat{\sigma}_0$ and s_a:

$$\eta^{1/2} = [(\hat{\sigma}_0^2 + s_a^2)/\hat{\sigma}_0^2]^{1/2} \cong 1.0316 \tag{19.12}$$

so the approximate RMEs in Table 19.1 are about 3% low.

Figure 19.4 shows a plot of the RME, with the above $\eta^{1/2}$ factor included, *versus X*. Also shown is an inset plot of the inverse relationship curve fit between RME and X:

$$\text{RME} = 0.017179/X \tag{19.13}$$

so that $\text{RME} \equiv 0.05$ implies $X_Q = 0.344$ mg R6G/100 mL ethanol. The predicted values of X_C and X_Q for the experiment were $X_C = 1.644854 \times 1.0316 \times 0.0329589/3.88777 \cong 0.0144$ mg/100 mL, from eqn 19.5, and, from eqn 19.6, $X_Q \equiv (20 \times 1.959964/1.644854)X_C \cong 0.343$ mg/100 mL. Note that from eqn 19.3, with β estimated by b, $z_{0.025} \times \eta^{1/2} \times \hat{\sigma}_0/b = 0.017141$ mg/100 mL, in reasonable agreement with the obtained curve fit value of 0.017179 mg/100 mL.

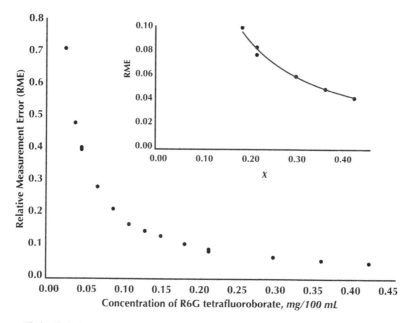

Figure 19.4 Relative measurement error (RME) *versus* analyte concentration (*X*) for the data in Table 19.1. Inset shows the power law fit of the indicated six data.

19.5 DISCUSSION AND CONCLUSION

The LOQ methodology suggested above has several advantages. First, it assures the user that there is 95% probability, over many repetitions of the process, that analyte present at the LOQ will be quantifiable with *at least one significant figure*. The detection limit, in contrast, is essentially a qualitative assurance that, at high probability, analyte is present and has not escaped detection; it is *not usable* for quantitation purposes. Second, if false negatives are of no concern, then the decision levels in Fig. 19.1 also serve as detection limits, and the expressions with subscripts "*D*" are not used. This is the pre-Currie detection limit methodology underlying obsolete expressions such as $LOD = ks_{blank}/slope$. Third, since the LOQ is related to the decision level, which depends on noise on the blank rather than noise at X values above 0, extension to heteroscedastic noise is straightforward. For example, if the noise were linearly heteroscedastic, that is, $\sigma(\beta X) \equiv \sigma_0 + \mu \beta X$, then

$$X_Q = \frac{z_{0.025}}{z_p(0.05 - z_{0.025}\eta^{1/2}\mu)}X_C \tag{19.14}$$

where μ is the heteroscedasticity, that is, asymptotic relative standard deviation (RSD), and the related expressions for Y_Q, x_Q, and y_Q are trivially found. Note that if $\mu = 0$, the system is homoscedastic and eqn 19.14 reduces to eqn 19.6. Equation 19.14 also shows that, for X_Q to exist, μ has an upper limit: $\mu < 0.05/z_{0.025}\eta^{1/2}$. Therefore,

μ must be less than about 2.5% for a linearly heteroscedastic system if $RME(X_Q)$ is defined as 0.05.

Mermet was right that "any LOQ can be selected" [6], but the authors believe that LOQ may never step out from under the shadow of the detection limit unless it can be seen as fulfilling a fundamentally important and, at present, largely unmet need.

ACKNOWLEDGMENTS

This research is dedicated to the memory Dr. Elena Nikolova Dodova.

REFERENCES

1. L. A. Currie, "Limits for qualitative and quantitative determination – application to radio-chemistry", *Anal. Chem.* **40** (1968) 586–593.

2. P. B. Adams, W. O. Passmore, D. E. Campbell, Paper 14, *Symposium on Trace Characterization – Chemical and Physical*, National Bureau of Standards, Oct. 1966.

3. ACS Committee on Environmental Improvement, "Guidelines for data acquisition and data quality evaluation in environmental chemistry", *Anal. Chem.* **52** (1980) 2242–2249.

4. H. Kaiser, "Part II quantitation in elemental analysis", *Anal. Chem.* **42** (1970) 26A–59A.

5. C. G. Fraga, A. M. Melville, B. W. Wright, "ROC-curve approach for determining the detection limit of a field chemical sensor", *Analyst* **132** (2007) 230–236.

6. J.-M. Mermet, "Limit of quantitation in atomic spectrometry: An unambiguous concept?", *Spectrochim. Acta, B* **63** (2008) 166–182.

7. J.-M. Mermet, G. Granier, P. Fichet, "A logical way through the limits of quantitation in inductively coupled plasma spectrochemistry", *Spectrochim. Acta, B* **76** (2012) 221–225.

8. D. Coleman, J. Auses, N. Grams, "Regulation – From an industry perspective *or* Relationships between detection limits, quantitation limits, and significant digits", *Chemom. Intell. Lab. Syst.* **37** (1997) 71–80.

9. A. Hubaux, G. Vos, "Decision and detection limits for linear calibration curves", *Anal. Chem.* **42** (1970) 849–855.

10. E. Voigtman, "Limits of detection and decision. Part 1", *Spectrochim. Acta, B* **63** (2008) 115–128.

11. E. Voigtman, "Limits of detection and decision. Part 2", *Spectrochim. Acta, B* **63** (2008) 129–141.

12. E. Voigtman, "Limits of detection and decision. Part 3", *Spectrochim. Acta, B* **63** (2008) 142–153.

13. E. Voigtman, "Limits of detection and decision. Part 4", *Spectrochim. Acta, B* **63** (2008) 154–165.

14. E. Voigtman, K. T. Abraham, "Statistical behavior of ten million experimental detection limits", *Spectrochim. Acta, B* **66** (2011) 105–113.

15. E. Voigtman, K. T. Abraham, "True detection limits in an experimental linearly heteroscedastic system. Part 1", *Spectrochim. Acta, B* **66** (2011) 822–827.

16. E. Voigtman, K. T. Abraham, "True detection limits in an experimental linearly heteroscedastic system. Part 2", *Spectrochim. Acta, B* **66** (2011) 828–833.

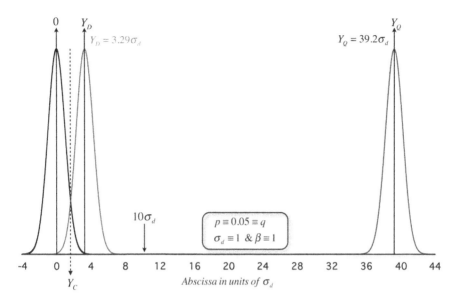

Figure 19.5 The Gaussian distributions centered on 0, the detection limit and the quantitation limit. The figure is to scale, with the abscissa parsed in units of σ_d.

19.6 POSTSCRIPT

Suppose $z'_p \equiv z_{0.025} \cong 1.959964$ and $p \equiv 0.05 \equiv q$, so that $z_p = z_q \cong 1.644854$. Then, in the theoretical domain, eqn 19.4 shows that $X_Q \simeq 23.8X_C = 11.9X_D$ and $Y_Q \simeq 23.8Y_C = 11.9Y_D$. Compared to the 10σ LOQ definition, these are 3.92 times larger due to the definitional change from 10% RSD, that is, 10σ, to 5% RME. The wide gap between Y_D and Y_Q is shown in Fig. 19.5.

Obviously, there is negligible probability of false negatives when analyte is present at the LOQ. More importantly, however, there is 95% probability that analyte at X_Q can be quantified with one or more significant digits of measurement precision. Since RME is inversely proportional to X, as per eqn 19.3, and $X_Q \simeq 11.9X_D$, there are *zero* significant digits of measurement precision, at *higher* than 95% confidence, at X_D. Exactly the same applies to Y_D. For the y_Q and x_Q expressions in the experimental quadrants of Fig. 19.1, ratios of critical t values are involved, rather than ratios of critical z values. Hence, the value of ≈ 11.9 is replaced by values that are dependent upon ν.

As a direct consequence of the above, *when analyte content is in the vicinity of the detection limit, or lower, it is unusable for quantitation purposes.* In particular, when analyte is present in the range between X_C and X_D, detection power is very poor: the 10σ LOQ is actually in the semiquantitation regime, as discussed in Chapter 24.

Typically, LOQs are reported much less often than decision levels and detection limits simply because LOQs are unpalatably large. Since the proposed LOQ definition is even higher, by a factor of 3.92 or more, this will certainly work against its

Table 19.2   Limits of Quantitation Expressions Associated with Chapters 7–14

Chapter	Response Domain	Net Response Domain	Content Domain
7	$R_Q = \alpha + \dfrac{20 z_{0.025}}{z_p} Y_C$	$Y_Q = \dfrac{20 z_{0.025}}{z_p} Y_C$	$X_Q = \dfrac{20 z_{0.025}}{z_p} X_C$
8	$r_Q = \hat{\alpha} + \dfrac{20 z_{0.025}}{z_p} Y_C$	$Y_Q = \dfrac{20 z_{0.025}}{z_p} Y_C$	$X_Q = \dfrac{20 z_{0.025}}{z_p} X_C$
9	$R_Q = \alpha + \dfrac{20 z_{0.025}}{z_p} Y_C$	$Y_Q = \dfrac{20 z_{0.025}}{z_p} Y_C$	$x_Q = \dfrac{20 z_{0.025}}{z_p} x_C$
10	$r_Q = \hat{\alpha} + \dfrac{20 z_{0.025}}{z_p} Y_C$	$Y_Q = \dfrac{20 z_{0.025}}{z_p} Y_C$	$x_Q = \dfrac{20 z_{0.025}}{z_p} x_C$
11	$r_Q = \alpha + \dfrac{20 t_{0.025}}{t_p} y_C$	$y_Q = \dfrac{20 t_{0.025}}{t_p} y_C$	$x_Q = \dfrac{20 t_{0.025}}{t_p} x_C$
12	$r_Q = \hat{\alpha} + \dfrac{20 t_{0.025}}{t_p} y_C$	$y_Q = \dfrac{20 t_{0.025}}{t_p} y_C$	$x_Q = \dfrac{20 t_{0.025}}{t_p} x_C$
13	$r_Q = \alpha + \dfrac{20 t_{0.025}}{t_p} y_C$	$y_Q = \dfrac{20 t_{0.025}}{t_p} y_C$	$x_Q = \dfrac{20 t_{0.025}}{t_p} x_C$
14	$r_Q = \hat{\alpha} + \dfrac{20 t_{0.025}}{t_p} y_C$	$y_Q = \dfrac{20 t_{0.025}}{t_p} y_C$	$x_Q = \dfrac{20 t_{0.025}}{t_p} x_C$

usage. Indeed, Badocco *et al.* [46 in Bibliography] concluded that, in comparison with Currie's LOQ definitions in the signal and concentration domains, "the LOQ based on the significant digit, although conceptually sensible, is often unsuitable because it is too large." This was a serious, but gentle, understatement. However, while analysts ponder the current motley farrago of LOQ definitions [6, 7], Table 19.2 summarizes the LOQ expressions relevant to Chapters 7–14.

19.7   CHAPTER HIGHLIGHTS

In this chapter, the LOQ was defined as a multiple of the decision level, *not* the detection limit. The definition was tested, via both real laser-excited molecular fluorescence experiments and via computer simulations, and found to be in excellent agreement with the derived theory. A significant disadvantage, however, is that limits of quantitation tend to be well above detection limits, a factor that often militates against their use.

20

THE SAMPLED STEP FUNCTION

20.1 INTRODUCTION

In the previous chapters, time, as a variable, appeared only because random noise is intrinsically time dependent, while the chemical measurement system (CMS), itself, was assumed to be time invariant. With regard to the various measurements performed on a given specimen under test (SUT), and among the various different SUTs, there were neither explicitly stated nor implied temporal relationships. In particular, there was no required temporal order for the individual measurements and no required temporal spacing between measurements.

However, as noted in Chapter 2, there are many widely used CMSs wherein the response is time dependent even if the analyte content, x, is constant. Three such CMS types are schematically illustrated in Fig. 20.1.

In gas chromatography (GC) or high-performance liquid chromatography (HPLC), the SUT is a mixture of chemical substances that is injected into the instrument. Typically, each constituent in the SUT has constant chemical content and the best case scenario is that the separation process results in well-separated temporal concentration peaks for each constituent. Subsequently, an appropriate detector converts the concentration peaks into temporal response peaks. Unfortunately, there are many possibilities for nonideal separation and detection. For example, peaks may be incompletely resolved or interferences may occur. As well, there is no universal consensus with regard to the functional form of the temporal response; in a 2001 survey of "almost 200" chromatography publications, Di Marco and Bombi found "about 90" different mathematical functions used to model $r(t)$ [1]. Furthermore, it is usually the case that response peaks must be integrated, to obtain peak areas, since chromatographic signals are typically due to the amount of analyte, not concentration [2].

Limits of Detection in Chemical Analysis, First Edition. Edward Voigtman.
© 2017 John Wiley & Sons, Inc. Published 2017 by John Wiley & Sons, Inc.
Companion Website: www.wiley.com/go/Voigtman/Limits_of_Detection_in_Chemical_Analysis

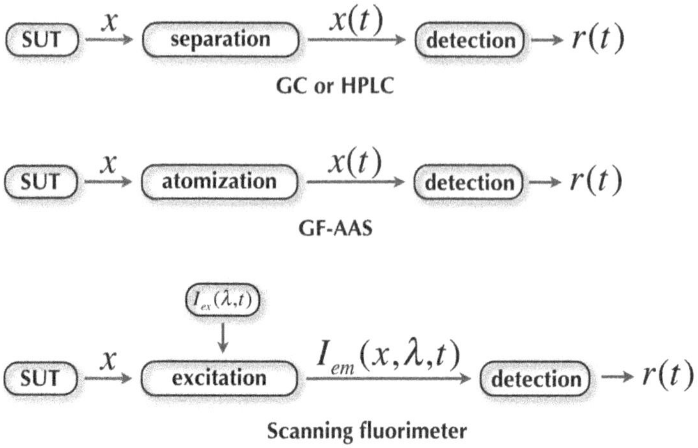

Figure 20.1 Three common CMS types that produce transient signals.

For a graphite furnace-atomic absorption spectroscopy (GF-AAS) instrument, the SUT is usually a liquid solution, of several microliters volume, placed into the graphite furnace. Then, the furnace operational protocol involves drying, ashing, atomization, and detection. Consequently, the time-dependent analyte number density in the furnace is a complicated function of time, three spatial coordinates in the furnace, and a variety of furnace conditions, pretreatments, and even operational history. The temporal response function is even more complicated since it necessarily involves integration over multiple optical paths, prefiltering effects, noise from stray light, and so on [3, 4]. A great deal of research has been devoted to understanding GF-AAS responses [5–7], yet the matter is far from resolved. One certainty, though, is that $r(t, x, y, z) \neq x(t, x, y, z)$.

For the scanning fluorimeter schematically shown in Fig. 20.1, the SUT is typically a liquid solution of the fluorescent analyte, for example, rhodamine 6G tetrafluoroborate in ethanol, as in Chapter 15. Then, even if x is a constant fluorophore concentration in a sealed sample cell, and the excitation wavelength is fixed, the emission intensity, $I_{em}(x, \lambda, t)$, will be time dependent if the excitation intensity, $I_{ex}(\lambda, t)$, is time dependent. This will always be true if a pulsed light source, for example, a pulsed laser or filtered Xe flashlamp, is used to excite the fluorescence. The photodetector processes $I_{em}(x, \lambda, t)$, yielding a time-dependent photocurrent, which is then integrated. Under optimum conditions, the result is a fluorescence signal directly proportional to x.

These three examples suffice to demonstrate that, even if x is constant, the CMS may cause $x(t)$ to be complicated, as in the above examples. Then, $r(t)$ is almost certainly even more complicated. Furthermore, the ultimate signal, $r(x)$, may be a function of $r(t)$, for example, the integral or average of $r(t)$ over a specified period of time. Yet, despite the myriad possibilities for defining usable signals for CMSs, it is almost always feasible to implement Currie's detection limit schema for transient signals in additive Gaussian white noise (AGWN).

Perhaps the most common way to proceed simply involves processing CMS outputs as required, for example, via boxcar averager, lock-in amplifier, computerized data acquisition, and processing, or the like, resulting in a compound measurement value, $r(x)$, for each x input used. Then calibration curves of $r(x)$ versus x are prepared, and analysis proceeds as in Chapters 7–14. The signal processing often results in the noise on $r(x)$ being nonwhite, so the model defined in Chapter 7 is technically inapplicable. In practice, this problem is usually ignored, and little harm results, particularly when the CMS response is periodic or repetitive and the signal processing involves *averaging* over many periods or individual signal pulses. The scanning fluorimeter system, with pulsed excitation light source and boxcar averager signal processing, is an example of a system that is in this category. What ultimately saves the day is the ability to repeatedly and nondestructively evoke a series of signal responses from a single SUT and then *average* the many responses. However, the chromatographic and GF-AAS systems are typically "one-shot" CMSs: a single SUT, which is destroyed in the process, yields a transient response that must be acquired and processed to yield an $r(x)$ value.

Dealing with "one-shot" CMSs involves extending the material presented in Chapters 7–14 by restricting attention to an analytical blank plus one standard, and assumption of temporal structure, particularly with regard to data acquisition. As seen below, this results in a sampled step function response, that is, blank followed by analyte. Then, the next step is to consider more complicated single signal response temporal waveshapes. The steps involved are discussed in this chapter and the following three chapters.

20.2   A NOISY STEP FUNCTION TEMPORAL RESPONSE

Consider the CMS model schematically shown in Fig. 20.2. The ideal flow injection switch has zero response time: it can switch instantaneously from one SUT to another. The spectrophotometer similarly has zero response time, that is, no memory or time constant, so it can respond instantaneously to input changes. As shown in Fig. 20.2, the ideal flow injection switch first allows the blank SUT, where $x \equiv X_0 \equiv 0$, to pass through the input flow cell of the spectrophotometer. This results in a noisy baseline,

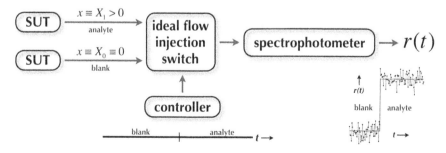

Figure 20.2   A CMS with instantaneous switching from blank SUT to analyte SUT.

that is, "background," output response. The input is then switched to the analyte SUT, where $x \equiv X_1 > 0$, causing the spectrophotometer's output to jump up to a higher average level. This assumes, as in the previous chapters, that the true sensitivity, β, is greater than zero. Consequently, $r(t)$ is a noisy step function, as shown at the lower right of Fig. 20.2.

20.3 SIGNAL PROCESSING PRELIMINARIES

From the outset, it must be realized that the field of signal processing is vast: there are tens of thousands of published papers on the topic, plus hundreds of books, at all levels of expertise. Even the far smaller subset concerned with data acquisition issues, *per se*, is large. For present purposes, it suffices to restrict attention to a small number of necessary definitions and utilitarian concepts. However, those interested in more detail, or greater depth of treatment, will easily be able to find it [8–16].

Decades ago, $r(t)$ might have been recorded on a strip chart recorder, thereby eliminating almost every sensible data processing option. Not surprisingly, more recent practice is to use a computer-controlled data acquisition system (DAS) in place of the chart recorder. This provides many benefits, but at the initial expense of requiring an understanding of how analog temporal responses are converted into digital records. The following discussion focuses on the bare essentials of what is needed to validly digitize analog responses.

The key components of a typical DAS are, in signal pathway order, (i) a fast sample-and-hold circuit, (ii) an antialiasing filter, and (iii) an analog-to-digital converter (ADC) having high resolution (many "bits") and fast sampling frequency, f_s [12]. The sampling frequency, also known as the digitization rate, is usually a user-specified constant, for example, 1000 Hz or samples/s. The antialiasing filter suppresses the input frequency content above the Nyquist sampling frequency, $f_{Nyquist}$, where $f_{Nyquist} \equiv f_s/2$. This is required because it is impossible for the digitized response to have legitimate frequency content above $f_{Nyquist}$ [13]. The digitization process adds quantization noise, also known as digitization error, but this added white noise may be minimized by the use of high-resolution ADCs. As well, many DASs have programmable gain preamplifiers before the ADC itself, in order to better utilize the input range of the ADC and minimize quantization noise effects [12].

When f_s is constant, the time between sampled data points, Δt, is also constant and is equal to $1/f_s$. Hence, if N consecutive data points are collected, they span a time interval, τ, of $N \times \Delta t$. The bilateral white noise power spectral density, $\eta_{bilateral}$, is simply the digital record's white noise variance, evenly distributed between $-f_{Nyquist}$ and $f_{Nyquist}$. Hence,

$$\eta_{bilateral} \equiv \sigma_0^2 / 2f_{Nyquist} \equiv \sigma_0^2 / f_s = \sigma_0^2 \, \Delta t \tag{20.1}$$

For the unilateral white noise power spectral density, $\eta_{unilateral}$, the white noise variance is evenly distributed between 0 and $f_{Nyquist}$. Thus,

$$\eta_{\text{unilateral}} \equiv \sigma_0^2/f_{\text{Nyquist}} = 2\eta_{\text{bilateral}} \tag{20.2}$$

The units are (signal units)^2/Hz for both power spectral densities (PSDs).

20.4 PROCESSING THE SAMPLED STEP FUNCTION RESPONSE

Now consider Fig. 20.3, which is a detailed elaboration of the response plot shown at the lower right of Fig. 20.2. The noise is assumed to be homoscedastic AGWN, with σ_0 population standard deviation. The true value of the blank response is α and $N_{\text{blank}}(\equiv M_0)$ *i.i.d.* samples are taken of the blank's portion of the noisy response waveform. From the N_{blank} samples, a sample mean, $\bar{r}_{\text{blank}}$, and sample standard deviation, s_{blank}, may be calculated, with $v_{\text{blank}}(\equiv N_{\text{blank}} - 1)$ degrees of freedom. Similarly, the true value of the analyte response is μ_{analyte} and $N_{\text{analyte}}(\equiv M)$ *i.i.d.* samples are taken of the analyte's portion of the noisy response waveform. From the N_{analyte} samples, a sample mean, $\bar{r}_{\text{analyte}}$, and sample standard deviation, s_{analyte}, may be calculated, with $v_{\text{analyte}}(\equiv N_{\text{analyte}} - 1)$ degrees of freedom. The true signal is the true net response, that is, $\mu_{\text{analyte}} - \alpha$, so a *possible* true (amplitude) signal-to-noise ratio, denoted SNR_{true}, is $(\mu_{\text{analyte}} - \alpha)/\sigma_0$. Note that SNR_{true} does not involve any use of averaging: σ_0 is independent from N_{blank} and N_{analyte}.

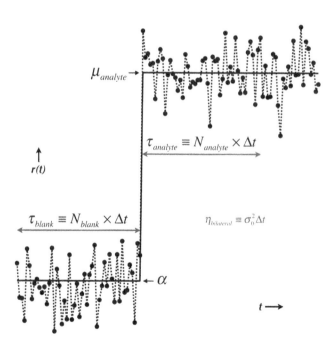

Figure 20.3   A detailed version of the sampled noisy step function in Fig. 20.2.

20.5   THE STANDARD t-TEST FOR TWO SAMPLE MEANS WHEN THE VARIANCE IS CONSTANT

Since the noise is homoscedastic, it is informative to compare $\bar{r}_{analyte}$ and $\bar{r}_{blank}$ via the standard t test for two sample means when the variance is constant. The t test statistic is

$$t \equiv \frac{\bar{r}_{analyte} - \bar{r}_{blank}}{s_p} = \frac{\bar{r}_{analyte} - \bar{r}_{blank}}{s_{pooled}\eta^{1/2}} \tag{20.3}$$

where

$$s_{pooled} \equiv \left(\frac{v_{blank} s_{blank}^2 + v_{analyte} s_{analyte}^2}{v_{blank} + v_{analyte}} \right)^{1/2} \tag{20.4}$$

and

$$\eta^{1/2} \equiv \left(\frac{1}{N_{analyte}} + \frac{1}{N_{blank}} \right)^{1/2} \equiv \left(\frac{1}{M} + \frac{1}{M_0} \right)^{1/2} \tag{20.5}$$

with s_{pooled} and s_p being the point estimate test statistics for σ_0 and $\sigma_p \equiv \sigma_0 \eta^{1/2}$, respectively. Note that $M (\equiv N_{analyte})$ is the number of independent future analyte measurement data for which their sample mean is to be compared with the decision level. In the previous chapters, the canonical and best choice for M was unity because only a single future blank measurement was to be compared with R_C or r_C. Henceforth, however, M will be greater than one, to realize the benefit of averaging, for measurements of either future blanks or analytes.

The t test statistic in eqn 20.3 has a noncentral t distribution:

$$p_t(t) = t(t | v_{total}, \delta) \tag{20.6}$$

where $\delta \equiv (\mu_{analyte} - \alpha)/\sigma_p$ and the total degrees of freedom, v_{total}, is simply $v_{blank} + v_{analyte}$. The noncentrality parameter, δ, is the true signal-to-noise ratio (SNR) in this scenario. It intrinsically involves averaging, since $\sigma_p \equiv \sigma_0 \eta^{1/2}$, so $\delta = SNR_{true}/\eta^{1/2}$. From eqn 20.3, t is the point estimate test statistic of δ, that is, t is an experimental SNR [17]. It is positively biased because it is the ratio of an unbiased estimate $(\bar{r}_{analyte} - \bar{r}_{blank})$ divided by the negatively biased s_p variate. The expectation value of t, for $v \equiv v_{total} > 1$, is [18]

$$E[t] = \delta \times (v/2)^{1/2} \Gamma((v-1)/2) / \Gamma(v/2) \tag{20.7}$$

and the variance of t, for $v \equiv v_{total} > 2$, is [18]

$$var[t] = [v/(v-2)] + \delta^2 \{ [v/(v-2)] - (v/2)\Gamma^2((v-1)/2)/\Gamma^2(v/2) \} \tag{20.8}$$

20.6 RESPONSE DOMAIN DECISION LEVEL AND DETECTION LIMIT

Since σ_0 and α are assumed to be unknown, the equations for the response domain decision level, r_C, and detection limit, r_D, are taken from Table 14.1:

$$r_C = \bar{r}_{\text{blank}} + t_p \eta^{1/2} s_{\text{pooled}} \qquad (20.9)$$

and

$$r_D = \bar{r}_{\text{blank}} + (\eta^{1/2} t_p + t_q / M^{1/2}) s_{\text{pooled}} \qquad (20.10)$$

with $\bar{r}_{\text{blank}} \equiv \hat{\alpha}$, $s_{\text{pooled}} \equiv s_0$, $\nu \equiv \nu_{\text{total}}$ and $\eta^{1/2}$ as in eqn 20.5. Less efficiently, s_{pooled} may be replaced with s_{blank} in eqns 20.9 and 20.10, in which case $\nu \equiv \nu_{\text{blank}}$.

20.7 HYPOTHESIS TESTING

The t value computed via eqn 20.3 may be used in hypothesis testing in the customary manner. Thus, if $t < t_p$, the null hypothesis (that is, $\mu_{\text{analyte}} - \alpha = 0$) is not rejected, $\bar{r}_{\text{analyte}}$ is declared to be a blank response, and no detection has occurred. However, if $t > t_p$, the null hypothesis is rejected, the one-sided alternative hypothesis (that is, $\mu_{\text{analyte}} - \alpha > 0$) is not rejected, $\bar{r}_{\text{analyte}}$ is declared to be an analyte response, and detection has occurred.

20.8 IS THERE ANY ADVANTAGE TO INCREASING N_{analyte}?

As Fig. 20.3 shows, the step function continues past τ_{analyte}, so there is nothing fundamental to prevent increasing N_{analyte} by increasing τ_{analyte}. Likewise, N_{blank} may be increased by collecting more data for the blank. If M was equal to unity, the net effect would be to refine the various test statistics at the cost of significantly increasing the measurement time. In the limit, the test statistics approach the population parameters for which they are the relevant point estimates, for example, r_C and r_D would become R_C and R_D, respectively. The expressions for R_C and R_D are given in Table 7.1; Fig. 20.4 shows a response domain step function where $\alpha \equiv 1$, $\sigma_0 \equiv 0.1$, $M \equiv 1$, $p \equiv 0.05 \equiv q$, $R_C \cong 1.1645$, and $R_D \cong 1.3290$. On *average*, there is only a 5% probability that a sample value from the blank's half period may exceed R_C and only 5% probability that a sample value from the analyte's half period may fall below R_C.

Using the same equations and parameters as in Fig. 20.4, except with $M = 50$, results in $R_C \cong 1.0233$ and $R_D \cong 1.0465$. This is shown in Fig. 20.5. Thus, there is only a 5% probability that the sample mean of 50 *i.i.d.* samples from the blank's half period may exceed R_C and only 5% probability that the sample mean of 50 *i.i.d.* samples from the analyte's half period may fall below R_C. Note that unguided visual detection of the step is unreliable. It is also clear that increasing both N_{blank} and N_{analyte}, as may be feasible, is beneficial.

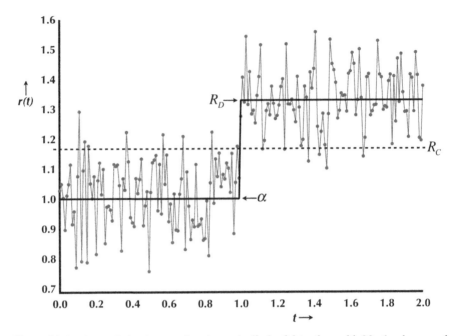

Figure 20.4 A sampled noisy step function at the limit of detection, with $M = 1$, when α and σ_0 are known. Both N_{blank} and N_{analyte} equal 50.

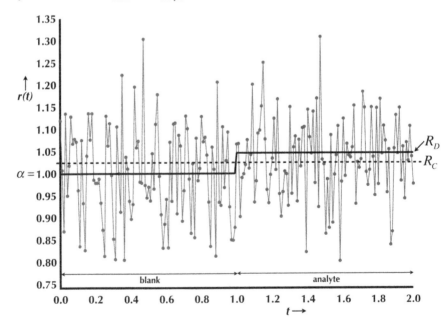

Figure 20.5 A sampled noisy step function at the limit of detection, when α and σ_0 are known. Note that $M = N_{\text{blank}} = N_{\text{analyte}} = 50$.

20.9   NET RESPONSE DOMAIN DECISION LEVEL AND DETECTION LIMIT

The equations for the net response domain decision level, y_C, and detection limit, y_D, are directly from Table 14.1:

$$y_C = t_p \eta^{1/2} s_{\text{pooled}} = r_C - \bar{r}_{\text{blank}} \tag{20.11}$$

and

$$y_D = (t_p + t_q)\eta^{1/2} s_{\text{pooled}} = y_C + t_q \eta^{1/2} s_{\text{pooled}} \tag{20.12}$$

with $s_{\text{pooled}} \equiv s_0$, $v \equiv v_{\text{total}}$ and $\eta^{1/2}$ as in eqn 20.5. With less efficiency, s_{pooled} may be replaced with s_{blank}, with $v \equiv v_{\text{blank}}$. Note that the y_D expression is symmetric with regard to the $\eta^{1/2}$ factor and, if $p = q$, then $y_D = 2y_C$.

20.10   NET RESPONSE DOMAIN SNRs

The SNR at y_C, denoted by SNR_{yC}, follows immediately from eqn 20.11:

$$\text{SNR}_{yC} \equiv y_C / \eta^{1/2} s_{\text{pooled}} = t_p \tag{20.13}$$

Similarly, the SNR at y_D, denoted by SNR_{yD}, follows from eqn 20.12:

$$\text{SNR}_{yD} \equiv y_D / \eta^{1/2} s_{\text{pooled}} = t_p + t_q \tag{20.14}$$

Given that SNR_{yD} depends upon degrees of freedom for t_p and t_q, *the limit of detection cannot be defined as having a single, universal numerical value.* In particular, *the limit of detection can never sensibly be defined as having an SNR of 3 and the same applies to the decision level.*

20.11   CONTENT DOMAIN DECISION LEVEL AND DETECTION LIMIT

In the highly unlikely event that β was known, despite both μ_{analyte} and α being unknown, the content domain decision level, x_C, and detection limit, x_D, are obtained from eqns 20.13 and 20.14, respectively: $x_C = y_C/\beta$ and $x_D = y_D/\beta$. This is as in Table 12.1, with s_0 being replaced by s_{pooled}. If β is unknown, then its estimate is

$$b \equiv \hat{\beta} \equiv (\bar{r}_{\text{analyte}} - \bar{r}_{\text{blank}})/(X_1 - X_0) = (\bar{r}_{\text{analyte}} - \bar{r}_{\text{blank}})/X_1 \tag{20.15}$$

since $X_0 \equiv 0$. The error in b, s_b, is a standard error [17], rather than a standard deviation:

$$s_b \equiv s_p/X_1 = s_{\text{pooled}}\eta^{1/2}/X_1 \tag{20.16}$$

The multiple measurements made in determining $\bar{r}_{\text{analyte}}$ and $\bar{r}_{\text{blank}}$ merely serve to refine their difference; they do not result in a systematic increase or decrease in the estimate of β. Thus, the expressions for x_C and x_D are obtained from eqns 20.13 and 20.14, respectively, by dividing by b. This is as in Table 14.1, with s_0 being replaced by s_{pooled}.

20.12   THE RSDB–BEC METHOD

The Relative Standard Deviation of the Background–Background Equivalent Concentration (RSDB–BEC) method was devised by Boumans [19–21], to facilitate "examination or instrumental optimization of the separately measured RSDB and BEC." [17] The simplest and most common variant uses only a background SUT, measured $N_{\text{background}}$ times, and an analyte SUT, measured N_{analyte} times. However, the original method neglects false negatives and uses a defined value of 3 in place of $t_p \eta^{1/2}$, so that it neither properly specifies a probability of false positives, p, nor controls whatever the *de facto* value of p may happen to be. Furthermore, the original RSDB–BEC method uses standard deviations, rather than standard errors, so it takes no direct advantage of averaging, that is, there is no $\eta^{1/2}$ factor. Consequently, it yields needlessly conservative detection limits, albeit ignoring false negatives.

These problems are relatively easy to correct. First, using eqn 20.12 and $x_D = y_D/b$,

$$x_D = (t_p + t_q)\eta^{1/2} \times s_{\text{background}} \times \frac{1}{b} \tag{20.17}$$

where s_{pooled} has been replaced by $s_{\text{background}}$. Then eqn 20.17 is partitioned thusly:

$$x_D = (t_p + t_q)\eta^{1/2} \times \frac{s_{\text{background}}}{\bar{r}_{\text{background}}} \times \frac{\bar{r}_{\text{background}}}{b} \equiv (t_p + t_q)\eta^{1/2} \times \text{RSDB} \times \text{BEC} \tag{20.18}$$

Correction of previously published RSDB–BEC detection limits is then simply a matter of multiplying them by $(t_p + t_q)\eta^{1/2}/c$, where c is whatever arbitrary value was originally used in place of $t_p \eta^{1/2}$, for example, 3. Note that $(t_p + t_q)\eta^{1/2}/c$ is typically well below unity.

The standard t test, *per se*, cannot be used with this method. However, a fix is readily available by defining t' as $t_p \eta^{1/2}$. Via eqn 20.3, t' is seen to be an alternative (lower) experimental SNR. It has a noncentral t distribution given by

$$p_{t'}(t') = \eta^{-1/2} t(\eta^{-1/2} t' | \nu_{\text{total}}, \delta) \tag{20.19}$$

with δ and σ_p as defined above. Then, t testing would proceed as above, with t' in place of t.

20.13 CONCLUSION

As demonstrated, analysis of a step function temporal response built naturally on the results was presented in Chapters 7–14. Aside from introducing a few necessary temporal domain concepts, such as sampling frequency and noise power spectral density, the extension was straightforward. It was found that the standard t test for two sample means, when variance is constant, was useful in determining whether or not detection had occurred. As discussed above, the original RSDB–BEC method [19–21] cannot be recommended for calculation of detection limits. However, it has some practical utility when used for preliminary instrumental or method optimization purposes, particularly with solid SUTs [22].

The following chapter focuses on the rectangular pulse temporal waveform, which is just a step function that abruptly returns to baseline. Indeed, a rectangular pulse is defined as the difference between a pair of temporally displaced step functions [23]. However, the rectangular pulse waveform has a well-defined temporal duration (that is, width) and this simple fact has important ramifications. Furthermore, it is perhaps the easiest pulse waveform to analyze, is commonly encountered in practice, at least approximately, and is a useful "building block" in constructing more complex pulse waveforms.

20.14 CHAPTER HIGHLIGHTS

The step function is, perhaps, the simplest temporal waveform: it has only a "before" and "after" the step transition. Accordingly, this chapter focused on the signal processing concepts relevant to the detection of a step function response from a CMS. It was shown that the standard t test for sample means was applicable, the decision level and detection limit expressions were given, and the beneficial effect of having $M \gg 1$ was demonstrated. Lastly, the original RSDB–BEC method was shown to be problematic. For further discussion of the method, see Section 2.2 in Ref. [17].

REFERENCES

1. V.B. Di Marco, G.G. Bombi, "Mathematical functions for the representation of chromatographic peaks", *J. Chromatogr. A* **931** (2001) 1–30.

2. J. Foley, J. Dorsey, "Clarification of the limit of detection in chromatography", *Chromatographia* **18** (1984) 503–511.

3. J.B. Dawson, R.J. Duffield, P.R. King, M. Hajizadeh-Saffar, G.W. Fisher, "Signal processing in electrothermal atomization atomic absorption spectroscopy (ETA-AAS)", *Spectrochim. Acta* **43B** (1988) 1133–1140.

4. E. Voigtman, A.I. Yuzefovsky, R.G. Michel, "Stray light effects in Zeeman atomic absorption spectrometry", *Spectrochim. Acta* **49B** (1994) 1629–1641.

5. Honorary issue dedicated to J.A. Holcombe, *Spectrochim. Acta* **105B** (2015).

6. B.V. L'vov, "Fifty years of atomic absorption spectrometry", *J. Anal. Chem.* **60** (2005) 382–392.

7. G. Schlemmer, B. Radziuk, *Analytical Graphite Furnace Atomic Absorption Spectrometry – A Laboratory Guide*, 1st Ed., Birkhäuser Verlag, Basel, Switzerland, ©1999.

8. T.H. Wilmshurst, *Signal Recovery from Noise in Electronic Instrumentation*, 2nd Ed., Taylor and Francis, NY, ©1990.

9. A.B. Carlson, *Communication Systems*, 2nd Ed., McGraw-Hill, New York, ©1975.

10. A.D. Whalen, *Detection of Signals in Noise*, 1st Ed., Academic Press, New York, ©1971.

11. H.A. Blinchikoff, A.I. Zverev, *Filtering in the Time and Frequency Domains*, 1st Ed., John Wiley and Sons, New York, ©1976.

12. A.F. Arbel, *Analog Signal Processing and Instrumentation*, 1st Ed., Cambridge University Press, Cambridge, ©1980.

13. E.O. Brigham, *The Fast Fourier Transform and its Applications*, 1st Ed., Prentice Hall, Englewood Cliffs, NJ, ©1988, p. 83.

14. T.A. Schonhoff, A.A. Giordano, *Detection and Estimation Theory and Its Applications*, 1st Ed., Pearson, Prentice Hall, Upper Saddle River, NJ, ©2006.

15. M. Schwartz, *Information Transmission, Modulation, and Noise*, 1st Ed., McGraw-Hill, New York, ©1980.

16. C.D. McGillem, G.R. Cooper, *Continuous and Discrete Signal and System Analysis*, 2nd Ed., Holt, Rinehart and Winston, New York, ©1984.

17. E. Voigtman, "Limits of detection and decision. Part 4", *Spectrochim. Acta* **63B** (2008) 154–165.

18. D.B. Owen. "A survey of properties and applications of the noncentral t-distribution", *Technometrics* **10** (1968) 445–478.

19. P.W.J.M. Boumans, J.J.A.M. Vrakking, "Detection limit including selectivity as a criterion for line selection in trace analysis using inductively coupled plasma-atomic emission spectrometry (ICP-AES) – a tutorial treatment of a fundamental problem of AES", *Spectrochim. Acta* **42B** (1987) 819–840.

20. P.W.J.M. Boumans, "Atomic emission detection limits: more than incidental analytical figures of merit! – a tutorial discussion of the differences and links between two complementary approaches", *Spectrochim. Acta* **46B** (1991) 917–939.

21. P.W.J.M. Boumans, "Detection limits and spectral interferences in atomic emission spectrometry", *Anal. Chem.* **66** (1994) 459A–467A.

22. A.B. Anfone, R.K. Marcus, "Radio frequency glow discharge optical emission spectrometry (rf-GD-OES) analysis of solid glass samples", *J. Anal. At. Spectrom.* **16** (2001) 506–513.

23. R.N. Bracewell, *The Fourier Transform and Its Applications*, 2nd Ed., Revised, McGraw-Hill, New York, ©1986.

21

THE SAMPLED RECTANGULAR PULSE

21.1 INTRODUCTION

The rectangular pulse shape is one of the simplest possible signal pulse shapes, making it relatively easy to analyze. Even so, several technical issues arise that must be considered before limits of detection may be deduced. First, the entire pulse must be taken into consideration because the measurement is a compound one: the measurement is a function of $r(t)$ rather than simply a nearly instantaneous sample of it. This requires consideration of the noise bandwidth (see Section 21.7) of the system that processes the output of the CMS, that is, that implements the function of $r(t)$, to yield the desired compound measurement. Second, the relationship between temporal integration over an aperture period τ_a, and temporal averaging over τ_a, must be established, and, in this regard, the integration time constant, τ_i, must also be considered. Third, the simplest digital integration approximation, that is, the rectangular approximation, must be introduced and its connection to averaging established.

21.2 THE SAMPLED RECTANGULAR PULSE RESPONSE

A sampled rectangular pulse response is shown in Fig. 21.1. Obviously, it would be highly inefficient (and absurd) to sample just one value during the time interval $\tau_{analyte}$. It is well known [1] that both matched filtering and cross-correlation provide optimum signal-to-noise ratios (SNRs) for the detection of rectangular signal pulses in white noise. Matched filters, in particular, were originally developed for radar detection purposes (the "North" filter) during World War II and are not restricted to either rectangular signal pulses or white noise. Indeed, the matched filter

recognizes a specified signal shape in the presence of noise (white or colored) and accordingly yields a higher output peak signal-to-mean noise power ratio (SNR) for this signal shape than for any other signal shape with the same energy. If the noise is not Gaussian, the matched filter is the optimum linear filter. If the noise is also Gaussian,

Limits of Detection in Chemical Analysis, First Edition. Edward Voigtman.
© 2017 John Wiley & Sons, Inc. Published 2017 by John Wiley & Sons, Inc.
Companion Website: www.wiley.com/go/Voigtman/Limits_of_Detection_in_Chemical_Analysis

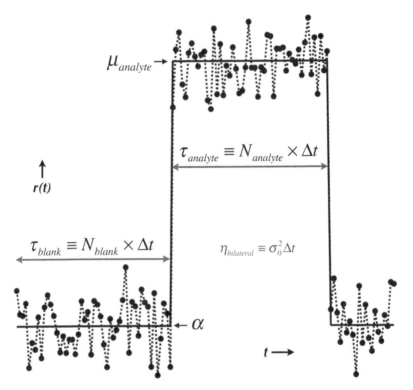

Figure 21.1 A sampled rectangular pulse, starting at t_0, with pulse width τ_{analyte}. As shown, N_{blank} and N_{analyte} are each 50 data samples.

> this filter is the optimum of all time-invariant filters, linear or nonlinear, and provides
> an optimum solution to the signal detection problem. [1, p. 312]

Furthermore, gated integration, with integration aperture perfectly synchronized with the rectangular pulse, is equally optimal in terms of SNR, as shown below. From Fig. 21.1, it is also apparent that N_{blank} data may be sampled before the pulse and N_{analyte} data during τ_{analyte}. Once the close connection between integration and averaging has been established, it will be seen that these data are ultimately what are used in the computation of the limit of detection.

21.3 INTEGRATING THE SAMPLED RECTANGULAR PULSE RESPONSE

Fig. 21.1 shows that N_{analyte} samples have been acquired during τ_{analyte}. The spacing between samples, Δt, is constant and equal to the reciprocal of the constant sampling frequency, f_s. Therefore, the integrated area of the rectangular pulse may be approximated by considering that each of the N_{analyte} sampled values, shown as markers in

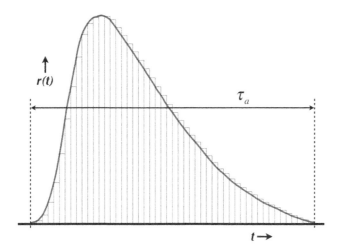

Figure 21.2 The rectangular integration approximation for integration.

the figure, is the height of a rectangle having width Δt, and then summing the areas of the rectangles. Thus,

$$\text{Area of rectangular pulse} \cong \Delta t \times \sum_{i=1}^{N_{\text{analyte}}} r\left((i-1)\Delta t + t_0\right) \qquad (21.1)$$

where t_0 is the moment the rectangular pulse starts. This is, without doubt, the crudest integration approximation [2]. Yet, with large numbers of sufficiently narrow rectangles, and modern data acquisition systems and computers, it is fit for purpose. This integration approximation is illustrated in Fig. 21.2, which shows an arbitrary unipolar pulse shape and a deliberately coarse set of area-approximating rectangles.

Clearly, the rectangles in Fig. 21.2 have greater area error after the peak of the waveform, so the net result is overestimation of the peak's area. But narrowing the rectangles reduces both the underestimation before the peak and the overestimation after it, so the total error is reduced. Again, it must be noted that the rectangular integration approximation is by far the crudest and least efficient numerical integration procedure: more than a hundred generations ago, the great Archimedes is said, perhaps apocryphally, to have spurned this approximation. Yet, it clearly provides the connection between integration and averaging over τ_{analyte}:

$$\text{Average of rectangular pulse} \equiv \frac{\text{Area of rectangular pulse}}{\tau_{\text{analyte}}}$$

$$\cong \frac{1}{N_{\text{analyte}}} \times \sum_{i=1}^{N_{\text{analyte}}} r\left((i-1)\Delta t + t_0\right) \qquad (21.2)$$

since $\Delta t/\tau_{analyte} = 1/N_{analyte}$. So the average value of the rectangular pulse is obtained in the customary way: sum the $N_{analyte}$ sample values and divide the sum by $N_{analyte}$. Each of the $N_{analyte}$ *i.i.d.* samples is from $N{:}\mu_{analyte}, \sigma_0$, so the sample mean of $N_{analyte}$ data is itself a sample from $N{:}\mu_{analyte}, \sigma_0/N_{analyte}^{1/2}$. This is how averaging reduces white noise and it is clearly due to the use of the standard error, that is, $\sigma_0/N_{analyte}^{1/2}$, in this compound measurement.

21.4 RELATIONSHIP BETWEEN DIGITAL INTEGRATION AND AVERAGING

For an analog integrator [3], with $r(t)$ as its input waveform, the temporal integral over τ_a is defined by eqn 21.3:

$$\text{Integral of } r(t) \text{ over } \tau_a \equiv \frac{1}{\tau_i} \int_0^{\tau_a} r(t)dt \qquad (21.3)$$

where τ_i is the integration time constant, in seconds. For a simple operational amplifier integrator, τ_i is the product of the input resistance, in ohms, and the feedback capacitance, in farads [3]. Thus, the integral has the same units as $r(t)$, that is, volts if $r(t)$ is a voltage waveform. For true mathematical integration, τ_i is simply dropped, so the integral has units of seconds times those of $r(t)$. For digital integration, there are three commonly used τ_i options:

- Drop τ_i from eqn 21.3, so the units of the integral are those of $r(t)$ times seconds.
- Define τ_i as 1 s, so the units of the integral are those of $r(t)$.
- Define τ_i as equal to τ_a, so the units of the integral are those of $r(t)$.

With the third option, the DC gain is unity, that is, if the input is constant, then the output, at the end of the integration period, will equal the constant input. In this case, integration and averaging become functionally equivalent, as shown by eqn 21.4:

$$\text{Average of } r(t) \text{ over } \tau_a \equiv \frac{1}{\tau_a} \times \int_0^{\tau_a} r(t)dt = \text{Integral of } r(t) \text{ over } \tau_a \qquad (21.4)$$

The unity DC gain condition will be assumed henceforth, unless stated otherwise. As a consequence, either averaging or integration may be used, as convenient.

A summary of the signal processing terms is provided in Table 21.1. These will be used in the remainder of this chapter and the following two.

Table 21.1   Summary of Signal Processing Definitions

Term	Definition
A	Constant net signal level (signal units)
σ_0	Population standard deviation of white noise (signal units)
τ_a	Integration aperture or "gate" (s)
τ_i	Integration time constant (s)
τ_a/τ_i	DC (zero frequency) gain (unitless)
$(\tau_a/\tau_i)^2$	DC power gain (unitless)
B_n	Noise bandwidth (Hz)
f_s	Constant sampling frequency (Hz)
$\Delta t \equiv f_s^{-1}$	Constant point spacing between sampled data (s)
$N \equiv \tau/\Delta t$	Number of sampled data points in τ (unitless)
$\eta_{\text{bilateral}} \equiv \sigma_0^2/f_s$	Bilateral white noise power spectral density, ((signal units)^2/Hz)
$\eta_{\text{unilateral}} \equiv 2\eta_{\text{bilateral}}$	Unilateral white noise power spectral density ((signal units)^2/Hz)

21.5   WHAT IS THE SIGNAL IN THE SAMPLED RECTANGULAR PULSE?

The sampled rectangular pulse consists of an ideal noiseless rectangular pulse plus additive Gaussian white noise (AGWN). These are independent, so they may be treated separately. Therefore, the net signal is given by

$$\text{Net signal} \equiv \frac{1}{\tau_i}\int_{t_0}^{t_0+\tau_a} \mu_{\text{analyte}}dt - \frac{1}{\tau_i}\int_{t_0-\tau_a}^{t_0} \alpha\, dt = \mu_{\text{analyte}} - \alpha \qquad (21.5)$$

where t_0 is the start of the rectangular pulse and $\tau_{\text{blank}} \equiv \tau_{\text{analyte}} \equiv \tau_a \equiv \tau_i$. Thus, $N_{\text{blank}} = N_{\text{analyte}}$. For convenience, A is defined as the true net signal, that is, $\mu_{\text{analyte}} - \alpha$.

21.6   WHAT IS THE NOISE IN THE SAMPLED RECTANGULAR PULSE?

When white noise is integrated over a period τ_a, the resulting variance at the end of the integration period is [4]

$$\sigma^2 = \int_0^\infty \eta_{\text{unilateral}}(\tau_a/\tau_i)^2\text{sinc}^2(\tau_a f)df = \eta_{\text{unilateral}}(\tau_a/\tau_i)^2\int_0^\infty \text{sinc}^2(\tau_a f)df \qquad (21.6)$$

where $\text{sinc}(\lambda) \equiv \sin(\pi\lambda)/\pi\lambda$ and $\eta_{\text{unilateral}}$ is constant, as per Table 21.1. The power transfer function is $|H(f)|^2 = (\tau_a/\tau_i)^2\text{sinc}^2(\tau_a f)$ [1] and the zero frequency power gain, that is, the DC power gain, is $|H(0)|^2 = (\tau_a/\tau_i)^2$. Since the integral in eqn 21.6 is independent of $\eta_{\text{unilateral}}$, it may be evaluated in advance, thereby simplifying subsequent white noise calculations. This leads to the concept of noise bandwidth.

21.7   THE NOISE BANDWIDTH

The noise bandwidth, B_n, is the bandwidth of a particular ideal (in the frequency domain) low-pass filter (LPF) [1]. It arises in the following way: consider a real LPF with power transfer function $|H(f)|^2$ [1]. Then imagine a fictitious ideal (in the frequency domain) LPF that satisfies two conditions: (1) its power transfer function has the same area as that of the real LPF and (2) its DC power gain equals the maximum value of $|H(f)|^2$ for the real LPF [1, p. 147]. Since the power transfer function of the fictitious ideal LPF is a rectangle, and both the area and height of the rectangle are specified, the resulting bandwidth is determined as well:

$$B_n \equiv \frac{1}{|H(0)|^2} \int_0^\infty |H(f)|^2 df = \frac{1}{\text{height}} \times \text{area} = \text{width} \qquad (21.7)$$

If white noise, with $\eta_{\text{unilateral}}$ power spectral density (PSD), were present at the inputs of both the real LPF and the fictitious ideal LPF, the resulting outputs would have equal variances. Thus, B_n simplifies white noise calculations for LPFs: with white noise input, the output variance is

$$\sigma^2 = \eta_{\text{unilateral}} \times B_n \times |H(0)|^2 \qquad (21.8)$$

For the gated integrator, the noise bandwidth is

$$B_n \equiv \frac{1}{(\tau_a/\tau_i)^2} \int_0^\infty (\tau_a/\tau_i)^2 \text{sinc}^2(\tau_a f)\, df = \int_0^\infty \text{sinc}^2(\tau_a f)\, df = \frac{1}{2\tau_a} \qquad (21.9)$$

Hence, with $\tau_a = \tau_i$, $|H(0)|^2 = (\tau_a/\tau_i)^2 = 1$ and

$$\sigma^2 = \eta_{\text{unilateral}} B_n = 2\sigma_0^2 \Delta t / 2\tau_a = \sigma_0^2 / N_{\text{analyte}} \qquad (21.10)$$

since $\tau_a = \tau_{\text{analyte}}$. Therefore, $\sigma = \sigma_0 / N_{\text{analyte}}^{1/2}$, as expected. A summary of the ideal signal and noise expressions is provided in Table 21.2, where $\tau_{\text{blank}} \equiv \tau_{\text{analyte}} \equiv \tau_a \equiv \tau_i$, $N \equiv N_{\text{analyte}} = N_{\text{blank}}$, and $A \equiv \mu_{\text{analyte}} - \alpha$. If $\tau_a \neq \tau_i$, then eqn 21.8 yields $\sigma^2 = \sigma_0^2 (\tau_a/\tau_i)^2 / N_{\text{analyte}}$.

The noise and SNR expressions in Table 21.2 tacitly assume that the offset, α, is either zero or is known and therefore may be subtracted without increasing the noise and decreasing the SNR. If α is nonzero and unknown, as is almost always

Table 21.2   Ideal Signal, Noise, and SNR Expressions

	Integration	Averaging (with $\tau_i \equiv \tau_a$)	Peak Detection
Signal	$A\tau_a/\tau_i$	A	A
Noise	$\sigma_0 \tau_a / N^{1/2}\tau_i (= \sigma_0 \Delta t N^{1/2}/\tau_i)$	$\sigma_0 / N^{1/2}$	σ_0
SNR	$A N^{1/2}/\sigma_0$	$A N^{1/2}/\sigma_0$	A/σ_0

the case, it must be estimated and this is typically accomplished by averaging the N_{blank} data to yield an estimate, $\hat{\alpha}$. With the typical assumption that $N_{analyte} = N_{blank}$, subtraction of $\hat{\alpha}$, from the average of the $N_{analyte}$ data, doubles the noise power without changing the signal. Therefore, the integration and averaging noise expressions in Table 21.2 increase by a factor of $\sqrt{2}$ and their SNR expressions decrease by a factor of $2^{-1/2}$. The "Peak Detection" column remains unchanged and is always inferior to both integration and averaging.

21.8   THE SNR WITH MATCHED FILTER DETECTION OF THE RECTANGULAR PULSE

A particularly nice feature of matched filters is that their power SNR is easy to compute: it is the signal pulse energy, E_p, divided by $\eta_{bilateral}$. Thus, for the sampled rectangular pulse in Fig. 21.1, with $\alpha = 0$ or α known and subtracted,

$$E_p \equiv \int_{-\infty}^{\infty} (\mu_{analyte} - \alpha)^2 dt = \int_{t_0}^{t_0+\tau_a} (\mu_{analyte} - \alpha)^2 dt = (\mu_{analyte} - \alpha)^2 \tau_{analyte}$$

$$(21.11)$$

since $\tau_a = \tau_{analyte}$. Therefore, the power SNR is

$$\text{Power SNR} = \frac{E_p}{\eta_{bilateral}} = \frac{(\mu_{analyte} - \alpha)^2 \tau_{analyte}}{\sigma_0^2 \Delta t} = \frac{(\mu_{analyte} - \alpha)^2}{\sigma_0^2} \times N_{analyte} \quad (21.12)$$

so that the customary amplitude SNR is

$$\text{SNR} = \frac{\mu_{analyte} - \alpha}{\sigma_0} N_{analyte}^{1/2} = \frac{A}{\sigma_0} N^{1/2} \quad (21.13)$$

as in Table 21.2. Note that the discussion following Table 21.2, concerning what happens if α must be estimated, is applicable here in exactly the same way. Thus, the gated integrator, with τ_a perfectly synchronized with $\tau_{analyte}$, achieves the optimal SNR of all time-invariant filters.

21.9   THE DECISION LEVEL AND DETECTION LIMIT

If both σ_0 and α are known, then Table 7.2 shows, with $M > 1$, that

$$R_C = \alpha \tau_a / \tau_i + z_p \sigma_0 \tau_a / M^{1/2} \tau_i = \alpha \tau_a / \tau_i + Y_C \quad (21.14)$$

and

$$R_D = \alpha \tau_a / \tau_i + (z_p + z_q) \sigma_0 \tau_a / M^{1/2} \tau_i = \alpha \tau_a / \tau_i + Y_D \quad (21.15)$$

where M is the number of future blanks for which their sample mean is to be compared with R_C and $\tau_i \equiv \tau_a$, for averaging over τ_a. The same table provides the base expressions for Y_C, Y_D, and, if β is known, X_C and X_D. In the previous chapters, M was best defined as unity because, for elementary measurements, only a single future blank should be compared to R_C. For pulse signals, however, M will necessarily be greater than one.

If α is unknown, then it is estimated as described above and the relevant expressions are given in Table 8.1, with σ_0 staying unchanged if $\tau_i \equiv \tau_a$. This is because $\eta^{1/2}$, given below, already accounts for the standard error due to M:

$$\eta^{1/2} = \left(\frac{1}{M} + \frac{1}{N_{\text{blank}}} \right)^{1/2} \tag{21.16}$$

See also eqn 21.15 in Appendix C, with $M_0 = N_{\text{blank}}$. Similar substitutions must be made if σ_0 or β is unknown and, in every case, the fundamental starting expressions are to be found in the summary tables in Chapters 7–14. If σ_0 is estimated by s_0, the degrees of freedom are those associated with the determination of s_0.

21.10 A SQUARE WAVE AT THE DETECTION LIMIT

For the sampled rectangular pulse in Fig. 21.1, the true parameter values were $\mu_{\text{analyte}} \equiv 1.5$, $\alpha \equiv 0.5$, $\sigma_0 \equiv 0.1$, $M \equiv N_{\text{blank}} \equiv 50$, $N_{\text{analyte}} \equiv 50$, $\Delta t \equiv 0.01$ s, and $p \equiv 0.05 \equiv q$. Hence, $R_C = 0.523$ and $R_D = 0.546$. From Table 21.2, the pulse's SNR is $1 \times 50^{1/2}/0.1 = 70.7$, while the SNR at the detection limit is 3.29, that is, $z_p + z_q$. If the rectangular pulse were to be periodically repeated, the result would be a square wave, since N_{blank} and N_{analyte} are equal. This is illustrated in Fig. 21.3, with $\mu_{\text{analyte}} \equiv R_D$, $\alpha \equiv 1$, $\sigma_0 \equiv 0.1$, $M \equiv N_{\text{blank}} \equiv 50$, $N_{\text{analyte}} \equiv 50$, $\Delta t \equiv 0.01$ s, and $p \equiv 0.05 \equiv q$, so that $R_C = 1.0233$ and $R_D = 1.0465$. Restricting attention to a single period of 1 s duration, the SNR is 3.29. Even though the square wave is quite noisy, it is deliberately such that there is only a 5% probability that the sample mean of 50 i.i.d. data from any "α" half period may exceed R_C and only 5% probability that the sample mean of 50 i.i.d. data from any "R_D" half period may fall below R_C.

If the $N^{1/2}$ factor were discarded, the SNR would only be 0.465, that is, the peak detection SNR in Table 21.2. Even a cursory examination of Fig. 21.3 shows that single point peak detection cannot be effective. Collecting only one point per 1 s period, and averaging, would certainly help. However, it is grossly inefficient; in Fig. 21.3, each 1 s period has 50 equally good "collectable" points.

If the square wave in Fig. 21.3 were the actual signal waveform, rather than a single rectangular pulse and baseline as in Fig. 21.1, then averaging over multiple periods, and using all of the data in each period, can greatly improve upon the single-period SNR. This signal processing is commonly done with phase synchronous detection, for example, using a lock-in amplifier [5]. But it must be remembered that this possibility is not available unless the CMS produces a periodic response for a given specimen under test (SUT). As noted in the previous chapter, this is not the case with "one-shot" CMSs.

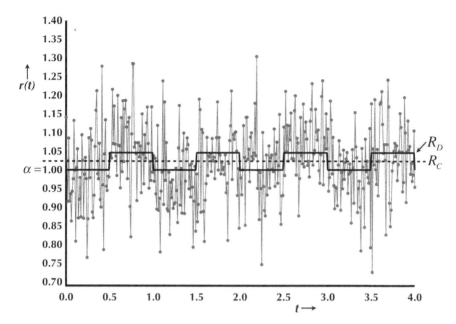

Figure 21.3 A 1 Hz square wave with α offset and net peak amplitude equal to Y_D, that is, $R_D - \alpha$.

21.11 EFFECT OF SAMPLING FREQUENCY

Provided that the noise is white, increasing f_s increases both N_{blank} and $N_{analyte}$ proportionately and decreases the noise, as discussed in Chapter 16. As a result, SNRs increase and both decision levels and detection limits decrease. However, if the noise is nonwhite, the use of noise-whitening filters, ensemble averaging, modulation, or other such techniques, may become necessary [6, 7]. Discussion of the troublesome issues that arise when the dominant noise is nonwhite is beyond the scope of this text, but Wilmshurst [6] provides an accessible and particularly relevant introduction to the topic.

21.12 EFFECT OF AREA FRACTION INTEGRATED

A rectangular pulse is optimally integrated when the integration aperture is perfectly matched to the pulse: it has the same duration as the pulse and is synchronized with it in the obvious way; they start and stop at the same time. If the aperture starts before the pulse or ends after it, extra noise is acquired and the SNR declines from the optimum. If the aperture is shorter than the rectangular pulse, and entirely within it, then the SNR also declines because only a fraction of the available signal area is acquired.

In order for the *a priori* specified values of p and q to obtain, the average waveform value over τ_a, denoted AWV for brevity, must equal Y_D. Optimal integration of the entire rectangular pulse gives $AWV = A_{max}$, that is,

$$AWV \equiv \frac{\text{Area integrated}}{\tau_a} = \frac{A_{max} \times \tau_a}{\tau_a} = A_{max} \qquad (21.17)$$

where A_{max} is the maximum net peak height of the rectangular pulse. For the rectangular pulse, changing τ_a has no effect on AWV so long as the integration does not extend outside the pulse.

Assuming σ_0 and α are known, as in Fig. 21.3, then

$$A_{max} = Y_D = (z_p + z_q)\sigma_0/N_{analyte}^{1/2} \qquad (21.18)$$

where $N_{analyte}$ is the number of data values acquired during τ_a, and τ_i is defined as 0.5 s, which is the maximum possible value of τ_a in Fig. 21.3. Note that $\tau_a \neq \tau_i$ in this example because τ_a may vary while τ_i is constant.

Reducing the fraction of the total area integrated concomitantly also reduces $N_{analyte}$, linearly for the rectangular pulse shape, but nonlinearly for other pulse shapes. For an optimally integrated rectangular pulse, the integrated area fraction is unity, and $A_{max} = R_D - \alpha$, as shown in Fig. 21.3. If only 50% of the possible signal area were integrated, then $N_{analyte}$ would be halved and, as eqn 21.18 shows, A_{max} and Y_D would have to increase by a factor of $\sqrt{2}$ in order for p and q to obtain. This will be important in the following two chapters, when nonrectangular signal pulses are considered.

21.13   AN ALTERNATIVE LIMIT OF DETECTION POSSIBILITY

In the previous chapter, it was noted that the standard t test for two sample means, when the variance was constant, was useful in determining whether or not detection had occurred. The same applies here since a rectangular pulse is just a truncated step function. Accordingly, the relevant decision level and detection limit equations in the previous chapter may be used, with appropriate degrees of freedom. A caution is that this should only be done if it is certain that the noise is homoscedastic and that no additional noise, for example, pulse-to-pulse (p–p) amplitude fluctuations, exists.

21.14   PULSE-TO-PULSE FLUCTUATIONS

It often happens that signal pulses may not be constant in amplitude even though analyte content is constant. In every case where p–p fluctuation noises exist, it is essential to study the relevant published literature, which is beyond the scope of this text. Due to the diverse and abundant possibilities that exist for these noises, the discussion that follows is only intended as a minimal introduction to a decidedly complex topic.

A commonly encountered CMS with p–p fluctuation noises is that of the scanning fluorimeter in Fig. 20.1: even with constant fluorophore content, p–p fluctuations in the excitation light source cause p–p fluctuations in the signal response, but not necessarily the background response. The total output noise variance, σ^2_{total}, is then the sum of two independent variances: one due to the AGWN and the other due to the p–p fluctuation noise.

The variance due to integration of the AGWN has already been exhibited above:

$$\sigma^2_{\text{AGWN}} = \sigma^2_0 (\tau_a/\tau_i)^2 / M \qquad (21.19)$$

where $M \equiv N_{\text{analyte}}$. The additional variance due to the p–p noise is given by

$$\sigma^2_{\text{fluctuations}} = \sigma^2_{p\text{-}p} \times (\tau_a/\tau_i)^2 \qquad (21.20)$$

where $\sigma^2_{p\text{-}p}$ is the population variance of the p–p noise. It is assumed that integration does not uselessly extend outside the pulse. Thus,

$$\sigma^2_{\text{total}} = \sigma^2_{\text{AGWN}} + \sigma^2_{\text{fluctuations}} = \left(\frac{\sigma^2_0}{M} + \sigma^2_{p\text{-}p} \right) (\tau_a/\tau_i)^2 \qquad (21.21)$$

The total output noise will be non-Gaussian unless the p–p noise is Gaussian. Assuming Gaussian p–p noise, the expression for R_D is modified accordingly:

$$R_D = \alpha(\tau_a/\tau_i) + z_p \sigma_{\text{AGWN}} + z_q \sigma_{\text{total}} \qquad (21.22)$$

If the noise is non-Gaussian, then a more complicated modification of R_D would be necessary, since neither z_q nor t_q critical values would be appropriate. This would likely require empirical information about the actual PDF of the total noise.

Note that the noise is heteroscedastic, but the noise precision model has not yet been specified. It may be that the p–p fluctuation noise is directly proportional to the true net peak height of the rectangular pulse, that is, $\sigma_{p\text{-}p} \propto A_{\max}$, so that the noise precision model (NPM) is either of the "hockey stick" variety or possibly linear, as discussed in Chapter 18. Thus, at sufficiently low analyte content, AGWN dominates, while at higher analyte content, especially when M is large, the p–p noise dominates. Hence, the relative standard deviation (RSD) is approximately constant at higher analyte content, while the standard deviation is approximately constant near the detection limit.

21.15 CONCLUSION

Rectangular signal pulses in additive white noise are among the easiest signal pulses to detect: they are open invitations to take advantage of the significant SNR improvements that are provided by averaging multiple *i.i.d.* samples. Assuming DC unity gain, the relevant decision levels and detection limits are simple modifications of the expressions given in Chapters 7–14.

21.16   CHAPTER HIGHLIGHTS

The rectangular pulse shape is the simplest pulsed waveform, and a useful "building block" for more complex waveform, that is, other shapes may be approximated as a contiguous series of "narrow enough" rectangular pulses. Digital integration and averaging are discussed, relevant signal processing definitions are presented, and noise bandwidth is defined and used. Optimum gated integration is shown to equal matched filtering, in terms of SNR, and the expressions for R_C and R_D are given. The effects of sampling frequency, area fraction integrated, and pulse-to-pulse fluctuations are discussed.

REFERENCES

1. H.A. Blinchikoff, A.I. Zverev, *Filtering in the Time and Frequency Domains*, 1st Ed., John Wiley and Sons, New York, ©1976.
2. W.H. Press, B.P. Flannery, S.A.Teukolsky, W.T. Vetterling, *Numerical Recipes in C*, 1st Ed., Cambridge University Press, Cambridge, ©1988.
3. P. Horowitz, W. Hill, *The Art of Electronics*, 2nd Ed., Cambridge University Press, Cambridge, ©1989.
4. E. Voigtman, J.D. Winefordner, "Low-pass filters for signal averaging", *Rev. of Sci. Inst.* **57** (1986) 957–966.
5. E. Voigtman, J.D. Winefordner, "Time variant filters for analytical measurements – electronic measurement systems", *Prog. Anal. Spectrosc.* **9** (1986) 7–143.
6. T.H. Wilmshurst, *Signal Recovery from Noise in Electronic Instrumentation*, 2nd Ed., Taylor and Francis, NY, ©1990.
7. A.F. Arbel, *Analog Signal Processing and Instrumentation*, 1st Ed., Cambridge University Press, Cambridge, ©1980.

22

THE SAMPLED TRIANGULAR PULSE

22.1 INTRODUCTION

The rectangular signal pulse in the previous chapter was relatively easy to analyze because of its simple geometry and due to the fact that, for it, optimum gated integration, matched filtering, and cross-correlation are equally optimum in terms of pulse detection signal-to-noise ratios (SNRs). Indeed, for rectangular pulses, they are functionally equivalent and easily implemented in either software or hardware [1]. It was also clear that peak detection would be a poor option for rectangular signal pulses due to its obvious inefficiency relative to simple averaging.

However, for unipolar nonrectangular signal pulses, for example, triangular and Gaussian pulses, optimum gated integration, with its conventional equally weighted integration aperture, has inferior SNR to matched filtering and cross-correlation. It also does not acquire the total area of the signal pulse. If the signal pulse is time-limited, that is, it is zero everywhere outside a finite duration period, the area deficit issue may be addressed by integration of the entire nonrectangular signal pulse. This results in all of the area being acquired, but the resulting SNR is always lower than that of optimum gated integration. Furthermore, signal pulses with tails, for example, Gaussian signal pulses, generally result in mild compromises; a small amount of "far tail" area is sacrificed in order to reduce damage to the detection SNR.

22.2 A SIMPLE TRIANGULAR PULSE SHAPE

The isosceles triangular signal pulse is a useful "zero-order" unipolar signal shape that is symmetric about its single peak. Similar to the rectangular pulse shape, it has simple geometry and is time-limited. An example of this signal shape is shown in

Limits of Detection in Chemical Analysis, First Edition. Edward Voigtman.
© 2017 John Wiley & Sons, Inc. Published 2017 by John Wiley & Sons, Inc.
Companion Website: www.wiley.com/go/Voigtman/Limits_of_Detection_in_Chemical_Analysis

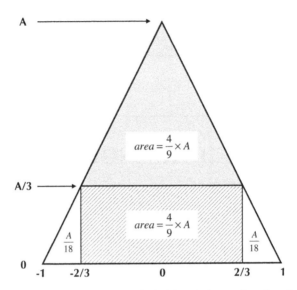

Figure 22.1 An isosceles triangle with height $= A$, base $= 2$, and total area $= A$.

Fig. 22.1, where the peak height is A, the base is 2, and the total area is also A. The full width at half maximum height (FWHM) is unity. The energy of the peak, E_p, is $2A^2/3$, so the matched filter (amplitude) SNR is $(2/3)^{1/2}A/\eta_{\text{bilateral}}^{1/2}$, assuming additive white with $\eta_{\text{bilateral}}$ power spectral density (PSD). If the entire peak is integrated, the signal is simply A/τ_i, while the noise is

$$\sigma^2 = 2\eta_{\text{bilateral}} \times B_n \times |H(0)|^2 = 2\eta_{\text{bilateral}} \times \frac{1}{2\tau_a} \times \left(\frac{\tau_a}{\tau_i}\right)^2 = \frac{2\eta_{\text{bilateral}}}{\tau_i^2} \quad (22.1)$$

since $\tau_a = 2$, that is, the base of the triangle. Therefore, $\sigma = 2^{1/2}\eta_{\text{bilateral}}^{1/2}/\tau_i$ and the SNR is $(1/2)^{1/2}A/\eta_{\text{bilateral}}^{1/2}$, that is, 86.60% of the matched filter optimum SNR.

Optimum gated integration is between $\pm 2/3$, as shown in Fig. 22.1 [2]. The resulting SNR is

$$\text{SNR} = (4/3^{3/2}) \times A/\eta_{\text{bilateral}}^{1/2} \quad (22.2)$$

which is 94.28% of the matched filter optimum. Note that the optimum integration aperture is symmetric about the peak and is such that the area of the rectangle in Fig. 22.1 is exactly equal to the area of the triangle above it. This is generally true for unipolar pulse shapes in white noise, as Wilmshurst has elegantly shown [3]. Results for several other common pulse shapes, including the Gaussian pulse shape, are given elsewhere [2].

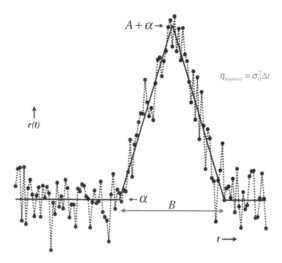

$$A + \alpha \rightarrow$$

$$\eta_{bilateral} \equiv \sigma_0^2 \Delta t$$

$$\uparrow \\ r(t)$$

$$\leftarrow \alpha$$

$$B$$

$$t \longrightarrow$$

Figure 22.2 A noisy sampled isosceles triangular pulse with $A = 1$ V, $B = 0.5$ s, total triangle area $= 0.25$ V s, $\sigma_0 = 0.1$ V, and $\Delta t = 0.01$ s.

22.3   PROCESSING THE SAMPLED TRIANGULAR PULSE RESPONSE

Now consider the sampled noisy isosceles triangular signal pulse shown in Fig. 22.2, with $r(t)$ arbitrarily assumed to have voltage units. Since integration of the entire signal results in a relatively small reduction in SNR relative to both optimum integration and matched filtering, assume that $\tau_a = B = \tau_{analyte} = \tau_i$. If α is zero or is known and subtracted, the SNR is 35.36.

As before, the SNR will be reduced by a factor of $2^{-1/2}$ if α must be estimated by averaging over B seconds of the baseline. Assuming that $\tau_{baseline} = B$, the parameter values given in the caption of Fig. 22.2 would result in SNR $= 25.00$. In contrast, dividing the net peak height by σ_0, and then accounting for the α estimation noise, yields an SNR of about 7.07 because integration adventitiously averaged over $N_{analyte} = 50$ sample values in the triangular pulse. A summary of SNRs relevant to Fig. 22.2 is given in Table 22.1. Relative to Fig. 22.1, SNRs are half as large because B is one-fourth as large: see "Table 22.1 calculations.pdf" at the companion website.

Table 22.1   Summary of Numerical SNRs Relevant to Fig. 22.2

Signal Processing	SNR (α known)	SNR (α estimated; $\tau_{baseline} = B$)
Matched filter	40.825	28.868
Optimum gated integration ($\tau_a = \pm B/3$)	38.490	27.217
Full area gated integration ($\tau_a = B$)	35.355	25.000

22.4 THE DECISION LEVEL AND DETECTION LIMIT

The equations for the decision level and detection limit are exactly the same as that in the previous chapter, for example, eqns 21.14 and 21.15. Suppose σ_0 and α are known, $\tau_i \equiv \tau_a \equiv B = 0.5\,\text{s}$, $p \equiv 0.05 \equiv q$ and assume the entire signal peak in Fig. 22.2 is integrated. Then the integrated area is $AB/2$ and the AWV is simply $A/2$, that is, Y_D. Thus,

$$A = 2Y_D = 2(z_p + z_q)\sigma_0/N_{\text{analyte}}^{1/2} \cong 0.0930\,\text{V} \qquad (22.3)$$

where $N_{\text{analyte}} = 50$ and $\sigma_0 = 0.1$ V.

For optimum gated integration, the AWV is $2A/3$ and $N_{\text{analyte}} = (2/3) \times B/\Delta t \simeq 33$, so A could be reduced to about 0.0859 V ($=3Y_D/2$) while keeping $p \equiv 0.05 \equiv q$. In each case, the net response domain decision level, Y_C, is simply $Y_D/2$. It is interesting that even $A = 0.0930$ V is less than σ_0, so, as for the train of rectangular pulses in the previous chapter, triangular pulses may be at the limit of detection and yet be all but indiscernible to the unguided eye.

22.5 DETECTION LIMIT FOR A SIMULATED CHROMATOGRAPHIC PEAK

Chromatography, in its myriad varieties, is ubiquitous in modern chemical analysis and, quite simply, it is impossible to overstate its importance; the capacity to separate complex mixtures, and quantify the amounts of the various resolved constituents, is absolutely indispensable. There is no difficulty in digitizing chromatograms, that is, the temporal response outputs from chromatographs, because even the fastest chromatographs, yielding the narrowest peaks, are slow. Indeed, if a chromatographic temporal peak were Gaussian, with σ_t as its standard deviation in seconds, it would have FWHM $= (8\ln 2)^{1/2}\sigma_t \cong 2.355\sigma_t$ and almost all of its frequency content would be below $\approx 1.5/\text{FWHM}$, in Hz. Thus, with $f_{\text{Nyquist}} \equiv 1.5/\text{FWHM}$, the minimum sampling frequency would be $f_{s,\min} \equiv 2f_{\text{Nyquist}} \equiv 3/\text{FWHM}$ and, as a practical matter, f_s would be set about 8–10 times higher than $f_{s,\min}$. With FHWM $= 1$ s, this results in $f_s \approx 24$–$30\,\text{Hz}$, that is, 24–30 digitized samples per second.

Clearly, there is no sampling rate problem. There is also no problem regarding potentially large file sizes or their processing and storage. Rather, chromatography is particularly challenging because chromatograms, by reason of being inherently low-frequency temporal waveforms, are afflicted with low-frequency noises, drift, offsets, occasional sporadic noise events ("bubbles"), system artifacts ("pump noises") and, of course, peaks are very often not completely resolved. This is not even a complete list of things that make life difficult for chromatographers; the determination of fundamentally well-defined chromatographic limits of detection has been curiously unsettled for decades [4], *even for the simple case of a single chromatographic peak in additive Gaussian white noise (AGWN)*. This latter issue is addressed as follows.

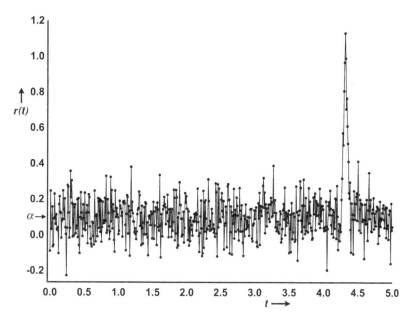

Figure 22.3   A simulated chromatographic peak, with time in minutes.

Consider, therefore, the simulated chromatographic peak shown in Fig. 22.3. For simplicity, it is modeled as a symmetric triangular signal pulse, with AGWN and α baseline offset. Note that using a more realistic peak model [5] would only change the mathematical details in an inessential way, that is, the specific numerical values would change rather little. Furthermore, it formerly was common, when chromatograms were recorded on paper, to approximate peaks as triangles; peak areas were then easily, if not accurately, calculated.

The simulation parameters were $\sigma_0 = 0.1$ V, $\alpha = 0.1$ V, $f_s = 2$ points/s, $\Delta t = 0.5$ s, retention time $= 4.3125$ min, and $B = 7.5$ s. Hence

$$Y_D = (z_p + z_q)\sigma_0 / N_{\text{analyte}}^{1/2} \cong 0.085\,\text{V} \tag{22.4}$$

since $N_{\text{analyte}} \equiv B/\Delta t \equiv Bf_s = 15$ and $\tau_i \equiv \tau_a \equiv B$. Then the content domain limit of detection, X_D, would be obtained by dividing Y_D by the sensitivity, that is, the response per unit amount of the analyte that would have generated the peak, had it been real. For example, if the sensitivity was 1 V/μg of analyte, then X_D would be approximately 85 ng.

Suppose that σ_0, α, and B are unknown, as is almost always the case with real data. Then the digitized file would have to be processed in order to estimate them. Using a spreadsheet program, the first 4 min of the file, comprising 481 points, were found to have a sample mean, $\hat{a}$, of 0.096158 V and a sample standard deviation, s_0, of 0.10004 V. To estimate the FWHM of the triangular signal peak, a Gaussian curve fit was performed on the 49 data between 4.1 and 4.5 min. The estimated σ_0 value, $\hat{\sigma}_0$,

was found to be 1.6394 s. For a Gaussian pulse shape, FWHM $= (8 \ln 2)^{1/2} \sigma_0$, so the
FWHM of the triangular signal peak was estimated as 3.86 s, that is, $(8 \ln 2)^{1/2} \hat{\sigma}_0$ for
the Gaussian fit. Then, since $B = 2 \times$ FWHM for a symmetric triangular peak, and
$N_{analyte} = B/\Delta t$, their estimates were found to be 7.72 s and 15.4, respectively. Thus,
with $\hat{N}_{analyte} = 15.4$,

$$y_D = (t_p + t_q)s_0/N_{analyte}^{1/2} \cong 0.084 \text{ V} \tag{22.5}$$

with $t_p = t_q = 1.64803$ for $p = 0.05 = q$ and $v = 480$. Of course, $N_{analyte}$ must be an
integer, so $\hat{N}_{analyte} = 15$ yields $y_D \cong 0.085 \text{ V}$ and $\hat{N}_{analyte} = 16$ yields $y_D \cong 0.082 \text{ V}$.
Thus, $x_D \cong 0.084 \, \mu g$.

In the above example, the sampling frequency was only 2 Hz, but $30/$FWHM $\approx$
7.8 Hz. If f_s were increased to 8 Hz, then $N_{analyte} = Bf_s = 60$, $\hat{N}_{analyte}$ would increase
by a factor of about 4 and both Y_D and y_D would decrease by a factor of about $1/\sqrt{4}$,
as per eqns 22.4 and 22.5, respectively. The reason for the decrease is that *averaging*
over B, not simply integration, is taking place.

22.6   WHAT SHOULD NOT BE DONE?

First, primary data should never be recorded solely on paper; it should be acquired
with a suitable data acquisition system (DAS), the properties of which are understood
[6, 7]. Then the digitized data may be processed subsequently, that is, processed with
noise-reducing digital [8] or analog filters [9], as appropriate. Assuming sampling has
been adequately fast, the digitized data may even be reprocessed, if future developments should warrant. Old chromatograms, available only as strip chart paper records,
are problematic; if it is desired to process such a chromatogram, it should be scanned
at high resolution, converted to a digital record, and then analyzed as above.

Second, techniques using peak heights as proxies for peak areas should be spurned
for detection limit purposes; they yield unnecessarily high limits of detection because
peak detection SNRs are always decidedly inferior to peak area SNRs, *ceteris paribus*
[2]. Third, estimation of the standard deviation of the baseline, as one-fifth of the maximum peak-to-peak range of the baseline noise, is often questionable: best practice is
to avoid having to resort to this estimation procedure. Finally, note that it is impossible to simultaneously maximize detection SNR while minimizing the distortion of
signal peaks. This is discussed in every communications theory text, for example,
Refs [10, 11].

22.7   A BAD PLAY, IN THREE ACTS

To illustrate the above issues, suppose that the simulated chromatogram in Fig. 22.3
was available only as a paper copy and that the sensitivity is 1 V/μg of analyte, as
above. Then what is the effect, upon the limit of detection (LOD), of techniques that

violate the recommendations stated above? To answer this question, note that Boqué and Vander Heyden have briefly (and neutrally) summarized three *ad hoc* empirical approaches, each based on peak heights [4]. These were then applied, to the extent possible, to the paper copy of Fig. 22.3.

The first method involves making measurements on a series of standard solutions until "a peak is found whose height is three times taller than the maximum height of the baseline (measured at both sides of the chromatographic peak). The concentration corresponding to that peak is taken as the LOD" [4]. For the paper copy of Fig. 22.3, the peak is about 1.03 V, corresponding to 1.03 µg of analyte. It is also about 2.5 times the height of the noise spike at 4.5 min. A plausible guess is that the LOD, with false negatives neglected, is about 1.2 µg.

The second method computes the LOD as $2z_{1-\alpha}h_{noise}/R$, where $z_{1-\alpha} \equiv z_p$, R is the responsivity, that is, the sensitivity, and h_{noise} is "half of the maximum amplitude of the noise" [4]. Hence, with $p \equiv 0.05$, $R = 1$ V/µg, and $h_{noise} \simeq 0.31$ V, the LOD is about 1 µg.

The third method defines an SNR as $2H/h$, where H is the net peak height and h is "the range (maximum amplitude) of the background noise" [4]. With $H \simeq 1.03$ V and $h \simeq 0.62$ V, SNR $\simeq 3.3$. If the LOD is defined as the analyte mass at SNR $= 3$, then the LOD is about 0.9 µg. This is the lowest LOD of the three, yet it is 10 times higher than $x_D \cong 0.084$ µg, determined above with $f_s = 2$ Hz.

The sample calculations for these three methods reveal a variety of defects:

- General loss of information due to nonelectronic primary data recording
- Use of peak heights as proxies for peak areas
- Crude estimation of baseline standard deviations
- Neglect of false negatives
- Definition of the LOD as having an arbitrary fixed SNR, for example, 3
- Detection limits are unnecessarily high, that is, quite conservative.

These methods, and similar variants, may have some practical utility for preliminary optimization purposes. However, they should not be used to estimate detection limits unless it is infeasible to perform fundamentally sound detection limit calculations.

If a real chromatogram were to consist of a series of fully resolved peaks, with known peak shapes, constant baseline offset and only AGWN, there would be no difficulty in properly estimating the detection limits associated with individual peaks. Unfortunately, as alluded to above, real chromatograms are often remarkably complex, that is, the desired analyte peak might be small, mired in a poorly resolved tangle of large peaks, and the baseline might be far from benign. Then an obvious option is to calculate detection limits using the chromatographic system's supplied software or other third-party software, assuming the calculation methodology is clearly documented. Alternatively, "quick and dirty" methods, such as the three criticized above, may have to suffice. In all such cases, the method used, and the specific details involved, should be disclosed in as much detail as possible. As well, such detection limits should be treated with caution, particularly if they are to be used for comparison purposes between laboratories.

22.8 PULSE-TO-PULSE FLUCTUATIONS

As in the previous chapter, the total output noise variance is the sum of two independent variances: one due to the AGWN, exhibited above in eqn 22.4, and the other given by

$$\sigma^2_{\text{fluctuations}} = \sigma^2_{p-p} \times (\tau_a/\tau_i)^2 \times (\text{AWV}/A_{\text{max}})^2 \tag{22.6}$$

where σ^2_{p-p} is the population variance of the p–p noise and A_{max} is the maximum net peak height. For rectangular pulses, the average waveform value (AWV) is always A_{max}, so it did not appear in eqn 21.20. However, for nonrectangular pulses, the AWV depends on the shape of the waveform, the specific portion integrated and the maximum net peak height. With reference to Fig. 22.1, where A is both the area (A_{area}) and the maximum net peak height (A_{max}), optimal integration results in AWV $= 2A_{\text{max}}/3$, that is,

$$\text{AWV} \equiv \frac{\text{area integrated}}{\tau_a} = \frac{(8/9) \times A_{\text{area}}}{4/3} = \frac{2}{3}A_{\text{max}} \tag{22.7}$$

Alternatively, integration of the entire triangular pulse gives AWV $= A_{\text{max}}/2$, that is,

$$\text{AWV} \equiv \frac{\text{area integrated}}{\tau_a} = \frac{(1) \times A_{\text{area}}}{2} = \frac{1}{2}A_{\text{max}} \tag{22.8}$$

Thus, the total noise variance is

$$\sigma^2_{\text{total}} = \sigma^2_{\text{AGWN}} + \sigma^2_{\text{fluctuations}} = \left(\frac{\sigma^2_0}{N_{\text{analyte}}} + \sigma^2_{p-p}(\text{AWV}/A_{\text{max}})^2 \right)(\tau_a/\tau_i)^2 \tag{22.9}$$

The discussion following eqn 21.20 in the previous chapter (*quo vide*) applies here without change.

22.9 CONCLUSION

The symmetric triangular signal pulse shape is, in some sense, the next step up in complexity from the rectangular pulse shape. For triangular signal pulses, it has been shown that gated integration detection, even when optimized, is inferior to matched filtering in terms of SNR. However, the somewhat reduced SNR (see Table 22.1) is of little practical significance in the large majority of circumstances; it results in less than a factor of two increase in the limit of detection.

On the positive side, the ease with which triangular signal pulse models may be analyzed makes it abundantly clear that peak area detection is always superior to peak height detection and limits of detection should be based on the former, not the latter. This was clearly demonstrated in the chromatographic peak example. For real chromatograms, it may sometimes be infeasible to calculate fundamentally sound detection limits.

22.10   CHAPTER HIGHLIGHTS

The symmetric triangular pulse shape is the second simplest pulsed waveform: it has finite duration and trivially simple geometry, with a clear peak. Optimum gated integration is shown to be inferior to matched filtering, and integration of the entire triangular pulse is worse still, yet the latter is quite acceptable in practice. Expressions for Y_D and Y_C were presented and applied to an example of a simulated chromatographic peak. With reference to chromatographic peak detection, a few recommendations were made and three empirical chromatographic detection limit methods were shown to provide relatively poor performance relative to the extension of Currie's schema. The effects of pulse-to-pulse fluctuations were also discussed.

REFERENCES

1. H.A. Blinchikoff, A.I. Zverev, *Filtering in the Time and Frequency Domains*, 1st Ed., John Wiley and Sons, New York, ©1976.

2. E. Voigtman, "Gated peak integration versus peak detection in white noise", *Appl. Spectrosc.* **45** (1991) 237–241.

3. T.H. Wilmshurst, *Signal Recovery from Noise in Electronic Instrumentation*, 2nd Ed., Taylor & Francis, NY, ©1990.

4. R. Boqué, Y. Vander Heyden, "The limit of detection", *LCGC Eur.* **22** (2009) 4 pages.

5. V.B. Di Marco, G.G. Bombi, "Mathematical functions for the representation of chromatographic peaks", *J. Chromatogr. A* **931** (2001) 1–30.

6. A. Felinger, A. Kilár, B. Boros, "The myth of data acquisition rate", *Anal. Chim. Acta* **854** (2015) 178–182.

7. M.F. Wahab, P.K. Dasgupta, A.F. Kadjo, D.W. Armstrong, "Sampling frequency, response times and embedded signal filtration in fast, high efficiency liquid chromatography: a tutorial", *Anal. Chim. Acta* **907** (2016) 31–44.

8. C.D. McGillem, G.R. Cooper, *Continuous and Discrete Signal and System Analysis*, 2nd Ed., Holt, Rinehart and Winston, NY ©1984.

9. E. Voigtman, J.D. Winefordner, "Low-pass filters for signal averaging", *Rev. Sci. Instrum.* **57** (1986) 957–966.

10. A.B. Carlson, *Introduction to Communication Systems*, 2nd Ed., McGraw-Hill, NY ©1975.

11. M. Schwartz, *Information Transmission, Modulation, and Noise*, 3rd Ed., McGraw-Hill, NY ©1980.

23

THE SAMPLED GAUSSIAN PULSE

23.1 INTRODUCTION

The Gaussian shape is ubiquitous throughout science, engineering, and technology. One reason for this is that the hidden hand of the central limit theorem naturally causes Gaussians to arise as limiting functions. Another is simply that the Gaussian function has relatively simple and highly desirable mathematical properties, making it particularly useful for modeling purposes. As a signal peak shape model, it is especially convenient because it is symmetric about its single peak, it has tails that die relatively quickly, and it is fully specified by its location and scale parameters. Single Gaussian peaks, and linear combinations of them, are often used as "zero-order" approximations of more complicated signal shapes, for example, idealized chromatographic peaks. Furthermore, Gaussians may serve as starting shapes from which to construct more complicated waveforms, for example, an exponentially modified Gaussian (EMG) chromatographic peak [1, 2]. As a practical matter, convolving a Gaussian with an exponential function involves nothing more than having the Gaussian be the input to an RC low-pass filter: the filter's output is the EMG waveform. Of course, this is easy to approximate with software.

However, there are several minor issues that must be considered. First, optimum gated integration of a Gaussian peak shape, with conventional equally weighted integration aperture, has inferior signal-to-noise ratio (SNR) to matched filtering and cross-correlation [3]. Second, integration does not acquire the total area of a Gaussian signal pulse unless the limits of integration are infinite. Typically, the integration limits are restricted to no more than $\pm 5\sigma$ about the location parameter, resulting in a negligible area deficit: less than 0.58 ppm. For $\pm 4\sigma$ and $\pm 3\sigma$, the area deficits are less than 64 ppm and 0.27%, respectively.

Limits of Detection in Chemical Analysis, First Edition. Edward Voigtman.
© 2017 John Wiley & Sons, Inc. Published 2017 by John Wiley & Sons, Inc.
Companion Website: www.wiley.com/go/Voigtman/Limits_of_Detection_in_Chemical_Analysis

23.2 PROCESSING THE SAMPLED GAUSSIAN PULSE RESPONSE

Consider the sampled noisy Gaussian signal pulse shown in Fig. 23.1. This may represent a symmetric chromatographic peak, a spectral absorbance peak, or any of dozens of other possibilities. The net peak amplitude is A, the scale parameter is σ_w, the location parameter (not shown) is μ_{location}, and the area of the peak is $A\sigma_w(2\pi)^{1/2}$. Matched filter detection provides the maximum (amplitude) SNR [3]:

$$\text{SNR}_{\text{MF}} = \pi^{1/4}A\left(\frac{\sigma_w}{\eta_{\text{bilateral}}}\right)^{1/2} \cong \left(\frac{\pi^{1/2}N}{10}\right)^{1/2}\frac{A}{\sigma_0} \qquad (23.1)$$

where $\tau_a \equiv N\Delta t \equiv 10\sigma_w$. In Fig. 23.1, $\sigma_w = 0.1$ s, $\sigma_0 = 0.1$ V, and $\Delta t = 0.01$ s, so $\text{SNR}_{\text{MF}} \cong 42.10 \times A$.

Optimum gated integration is in the interval $\pm 1.40\sigma_w$, centered on the peak, resulting in 94.35% of the matched filter SNR [3] and 83.85% of the total area acquired. Integration in the interval $\pm 5\sigma_w$ has negligible area deficit, but the SNR is lower:

$$\text{SNR} = \frac{A(2\pi)^{1/2}\sigma_w/\tau_i}{\sigma_0\Delta t N^{1/2}/\tau_i} = \left(\frac{\pi N}{50}\right)^{1/2}\frac{A}{\sigma_0} = \left(\frac{\pi^{1/2}}{5}\right)^{1/2} \times \text{SNR}_{\text{MF}} \qquad (23.2)$$

where $\tau_a \equiv N\Delta t \equiv 10\sigma_w$. Hence, the SNR is only 59.54% of the matched filter SNR [3]. Even in this case, however, the SNR is $25.07 \times A$. Note that, so long as the noise is white, and τ_a is constant, both SNR_{MF} and SNR increase as $N^{1/2}$. Equivalently,

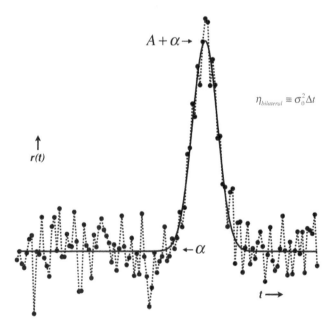

Figure 23.1   A noisy sampled Gaussian pulse.

Table 23.1   Summary of SNRs Relevant to Fig. 23.1, with A = 1

Signal Processing	SNR (α Known)	SNR (α Estimated; $\tau_{\text{baseline}} = \tau_a$)
Matched filter	42.10	29.77
Optimum gated integration ($\tau_a = \pm 1.40\sigma_w$)	39.72	28.09
Full area gated integration ($\tau_a = \pm 5\sigma_w$)	25.07	17.72

they increase as $f_s^{1/2}$. In contrast, the SNR for peak detection is fixed at A/σ_0, that is, $10 \times A$, if $\sigma_0 = 0.1$ V.

As in the previous chapter, SNRs are reduced by a factor of $2^{-1/2}$ if α must be estimated by averaging over τ_a seconds of the baseline. Assuming that $\tau_{\text{baseline}} = \tau_a$, if α must be estimated, Table 23.1 provides a summary of the Gaussian signal detection SNRs relevant to Fig. 23.1 (with A = 1 assumed).

23.3   THE DECISION LEVEL AND DETECTION LIMIT

The equations for the decision level and detection limit are exactly the same as in the previous chapter. Suppose σ_0 and α are known, $\tau_i \equiv \tau_a \equiv 10\sigma_w = 1$ s, $p \equiv 0.05 \equiv q$ and assume the entire signal peak in Fig. 23.1 is integrated. Then the integrated area is $A\sigma_w(2\pi)^{1/2}$ and the average waveform value (AWV) is $A(2\pi)^{1/2}/10$. Thus, $Y_D = A(2\pi)^{1/2}/10 \cong 0.25066A$ and

$$A = \frac{Y_D}{(2\pi)^{1/2}/10} = \frac{(z_p + z_q)\sigma_0/N_{\text{analyte}}^{1/2}}{(2\pi)^{1/2}/10} \cong 0.131 \text{ V} \qquad (23.3)$$

where $N_{\text{analyte}} = 100$, $\sigma_0 = 0.1$ V, and $\Delta t = 0.01$ s.

For optimum gated integration, AWV $\cong 0.75063A$ (see eqn 23.5) and $N_{\text{analyte}} = 2.8\sigma_w/\Delta t = 28$, so A could be reduced to 0.0828 V ($= Y_D/0.75063$) while keeping $p \equiv 0.05 \equiv q$. As before, the net response domain decision level, Y_C, is simply $Y_D/2$.

23.4   PULSE-TO-PULSE FLUCTUATIONS

Once again, the total output noise variance is the sum of two independent variances: one due to the AGWN and the other given by

$$\sigma_{\text{fluctuations}}^2 = \sigma_{p-p}^2 \times (\tau_a/\tau_i)^2 \times (\text{AWV}/A_{\text{max}})^2 \qquad (23.4)$$

where σ_{p-p}^2 is the population variance of the $p-p$ noise and A_{max} ($= A$) is the maximum net peak height. For Gaussian peaks, optimal integration results in

$AWV \cong 0.75063A_{max}$, that is,

$$AWV \equiv \frac{\text{area integrated}}{\tau_a} = \frac{\left[(2 \times 0.41924) \times A_{max}\sigma_w(2\pi)^{1/2}\right]}{(2 \times 1.40\sigma_w)} \cong 0.75063A_{max}$$

(23.5)

Alternatively, integration of the entire Gaussian pulse gives $AWV \cong 0.25066A_{max}$, that is,

$$AWV \equiv \frac{\text{area integrated}}{\tau_a} \cong \frac{1 \times A_{max}\sigma_w(2\pi)^{1/2}}{2 \times 5\sigma_w} \cong 0.25066A_{max} \qquad (23.6)$$

Thus, the total noise variance is

$$\sigma_{total}^2 = \sigma_{AGWN}^2 + \sigma_{fluctuations}^2 = \left(\frac{\sigma_0^2}{N_{analyte}} + \sigma_{p-p}^2(AWV/A_{max})^2\right)(\tau_a/\tau_i)^2 \quad (23.7)$$

The discussion following eqn 21.20 (*quo vide*) applies here without change.

23.5   CONCLUSION

Mathematical details aside, detection of Gaussian and triangular signal pulse shapes is similar and the relevant expressions for decision levels and detection limits are the same. More complicated unipolar pulse shapes, for example, those produced by GF-AAS instruments, may be dealt with by the same methods covered in this chapter and the previous one. Fundamentally, all that really matters is that averaging of multiple *i.i.d.* signal waveform samples must be feasible; the specific techniques, and instruments from which data are acquired, are of secondary importance. *It is the properties of the CMS under consideration, not its identity, that determine its figures of merit, including its limits of detection.*

23.6   CHAPTER HIGHLIGHTS

The Gaussian pulse shape is ubiquitously used as a model pulsed waveform, for many reasons. As for the triangular pulse, optimum gated integration is shown to be inferior to matched filtering, and integration of the entire Gaussian pulse is worse still, but acceptable in practice. Expressions for Y_D and Y_C are presented and the effects of pulse-to-pulse fluctuations are also discussed.

REFERENCES

1. J. Foley, J. Dorsey, "Clarification of the limit of detection in chromatography", *Chromatographia* **18** (1984) 503–511.

2. V.B. Di Marco, G.G. Bombi, "Mathematical functions for the representation of chromatographic peaks", *J. Chromatogr. A* **931** (2001) 1–30.

3. E. Voigtman, "Gated peak integration versus peak detection in white noise", *Appl. Spectrosc.* **45** (1991) 237–241.

24

PARTING CONSIDERATIONS

24.1 INTRODUCTION

This chapter begins by addressing issues that have significantly impeded progress in the formal study of limits of detection. Next, several recent distractions are discussed, and then the chapter concludes with a brief examination of a few promising research avenues.

Issue 1: Conflating Sample Test Statistics and Population Parameters

It is still quite common for publications, even recent ones, to ignore the fact that sample test statistics are *not* the same as the population parameters they estimate. This issue has probably caused more damage to analytical detection limit theory and practice than everything else combined. In particular, $s \neq \sigma$, and best practice is to avoid, to the maximum extent possible, use of Greek letters for sample test statistics. Related issues include the following:

- Carelessly combining critical z values with sample test statistics
- Assuming $SNR = 3$ at the detection limit
- Ignoring the error involved in estimation of the blank's true value
- Using long obsolete, and repeatedly supplanted, IUPAC guidance from the 1970s
- Defining "k" as 3, failing to recognize that it is a composite factor, for example, $k = t_p \eta^{1/2}$
- Using the irrelevant δ_{critical} method (see Chapter 17)

Limits of Detection in Chemical Analysis, First Edition. Edward Voigtman.
© 2017 John Wiley & Sons, Inc. Published 2017 by John Wiley & Sons, Inc.
Companion Website: www.wiley.com/go/Voigtman/Limits_of_Detection_in_Chemical_Analysis

- Failing to perform even rudimentary Monte Carlo computer simulation testing
- Uncritically accepting results, in the refereed literature, as factually correct
- Anchoring to oversimplified or long obsolete publications
- Using inadequately quantified experimental systems to "validate" preferred theories.

Issue 2: Measurements Are *Not* Made in the Content Domain

In Chapters 2 and 3, it was noted that *measurements are made in the response domain* and, by subtraction of either α or $\hat{\alpha}$, net response measurements result. In contrast, no measurements are made in the content domain because *it is impossible to make measurements in the content domain*. Thus, if X_{unknown} is an unknown, but constant, input to a chemical measurement system (CMS), then the best that can be achieved in terms of determining X_{unknown} is to obtain an *inferred* value x_{unknown}, plus range limits bracketing an *a priori* specified confidence interval in which X_{unknown} *might* be found. Ideally, x_{unknown} should be an unbiased estimate of X_{unknown} and the confidence interval should be narrow.

One issue, then, is that of referring to measurements as being made in the content domain. This abuse of language is not innocuous: it leads directly to the misconception, as detailed in Chapter 3, that *inferred* distributions of x variates must be truncated or "corrected" in order to avoid unphysical negative content values. In fact, all such proposed manipulations are the result of failure to understand what is fundamentally involved in the detection process: it is a mapping between distinct domains, that is, from the content domain to the response domain. The following discussion, addressing this matter, is largely based on the work of Coleman *et al.* [1].

Consider the possible detection regimes in the response, net response, and content domains, as shown in Fig. 24.1. Now suppose the analyte content is unknown, but constant, that is, $x = X_{\text{unknown}}$, and that this results in the response measurement $r = r_{\text{unknown}}$. Since analyte content is of primary interest to analysts, the *inferred* result, x_{unknown}, may be back-calculated from r_{unknown}. Then x_{unknown} is the estimate of X_{unknown}. Regardless of the detection regime in which x_{unknown} falls, its numerical value should be *both* recorded *and* reported [2]. As Thompson has noted [2], this would be necessary for any subsequent statistical analysis and might well facilitate error correction, if a mistake were to be discovered later. Thus, *no censoring of numerical values should be performed*, and this includes both negative inferred x_{unknown} values and entirely negative confidence intervals. In general, *all numerical values should be recorded and reported*, even if they appear to be "unphysical."

Accordingly, if x_{unknown} is in region (a), its value should be reported, along with the decision level and the statement that analyte was *not* detected. It is indistinguishable from a true blank. If x_{unknown} is in region (b), then its value should

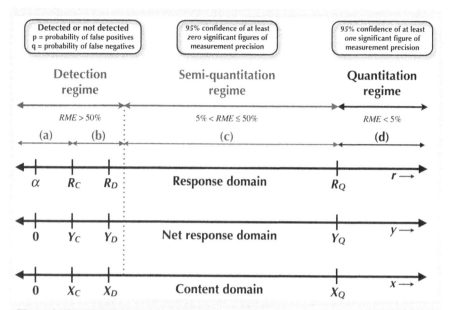

Figure 24.1 The possible detection regimes in all three domains. Note that (a) includes negative values and the boundary between detection and semiquantitation is *not* at the detection limit.

be reported, along with *both* the decision level *and* the detection limit, and the statement that analyte was detected, but detection power was poor. Provision of an uncertainty estimate is possible, but not particularly useful; at 95% confidence, the result has less than zero significant figure of measurement precision [1].

In the semiquantitation regime, the number of significant figures of measurement precision is between zero and one. Hence, if $x_{unknown}$ is in region (c), its value would be reported, along with the detection limit and the statement that analyte was detected, but with poor to moderate precision. Provision of a confidence interval is strongly recommended. If $x_{unknown}$ is in region (d), then its value would be reported, with a confidence interval. At 95% confidence, RME < 5%, so the measurement has at least one significant figure of measurement precision.

24.2 THE MEASURAND DICHOTOMY DISTRACTION

Thompson, in advocating for his "uncertainty function" (a noise precision model, NPM) as a superior alternative to detection limits, states that [3, p. 1174]

the detection limit c_L ... artificially dichotomises the concentration axis in an analytical system. ... This tendency encourages analysts and their customers to believe – incorrectly – that a result just below the detection limit (say at $0.9c_L$) has a qualitatively different status to one just above the limit at $1.1c_L$. Analysts have consequently become unwilling to report a result that falls below such a limit, replacing the numerical result, originally on a rational or interval scale, with an ordinal form such as 'not detected' or 'less than c_L'. Such reporting is detrimental because it renders much more difficult an unbiased statistical treatment of the resulting dataset.

There is, of course, a fundamental dichotomy involved in Neyman–Pearson hypothesis testing: measured values must be compared with the decision level. However, the issue raised by Thompson is artificial. As demonstrated in Chapter 19, there is 95% confidence that the RME does not exceed 5% at the limit of quantitation, X_Q. It is known that RME $\propto 1/x$, as per Fig. 19.4, and

$$X_Q = 20\frac{z'_p}{z_p}X_C \cong 20\frac{1.959964}{1.644854}X_C \simeq 23.83X_C \tag{24.1}$$

from eqn 19.6. With homoscedastic additive Gaussian white noise (AGWN) and $p \equiv q$, $X_D = 2X_C$. Therefore, $X_Q \simeq 11.92X_D$. Consequently, at $x \equiv X_D$, RME $\simeq$ 59.6%, that is, $11.92 \times 5\%$. As well, eqn 7.9 yields the same result:

$$\text{RME at } X_D \equiv \frac{z_{0.025}\sigma_0/\beta}{X_D} \equiv \frac{z_{0.025}\sigma_0/\beta}{(z_p + z_q)\sigma_0/\beta} = \frac{z_{0.025}}{z_p + z_q} \cong 0.596 \quad \text{if } p \equiv 0.05 \equiv q \tag{24.2}$$

Hence, as shown in Fig. 24.1, X_D is *below* the boundary between the detection and semiquantitation regimes.

For Thompson's $0.9c_L$ and $1.1c_L$ values, the RME values are 66.1% and 54.1%, respectively. Obviously, there is no dichotomy and no meaningful qualitative difference between his example values: both are roughly 60%. For heteroscedastic AGWN, the situation is precisely similar; Fig. 24.2 shows the distributions for a CMS with $\mu \simeq 6.1\%$ heteroscedasticity [4, Fig. 2]. In this case, RME $\simeq 0.596 \times 45.6 \text{ mV}/37.8 \text{ mV} \simeq 0.719$. Again, nothing unexpected would happen at $0.9Y_D$ relative to $1.1Y_D$.

One of the *actual* issues that arise with experimental detection limits is simply that they are samples from distributions that may be relatively wide, particularly at low degrees of freedom. This is illustrated in Fig. 24.3, which is based on Fig. 15.9 [5]. The 10 million total real experimental detection limits, that is, x_D variates, were binned into 400 histogram bins, in the range 0–3.2. Clearly, almost all the x_D variates were in the lower 200 bins, in the range 0–1.6. Furthermore, the histogram of experimental detection limits is fitted all but perfectly by eqn 15.11:

$$p_{x_D}(x_D) = (7.10331/x_D^2)t(7.10331/x_D|2, 12.838) \tag{24.3}$$

As noted in Chapter 15, *the 99% CI for X_D is $0.251_6 \pm 0.001_0$ mg R6G/100 mL ethanol*. Hence, $\hat{X}_D \cong 0.252$ mg R6G/100 mL ethanol and, for all *practical* purposes, this is X_D.

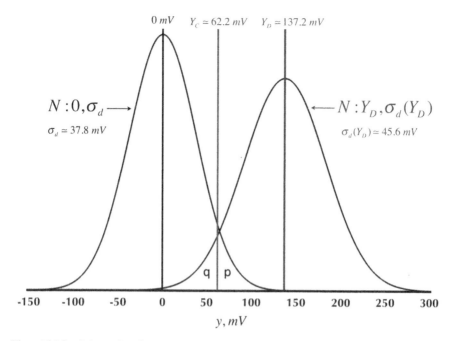

Figure 24.2   Schematic of Y_C and Y_D and associated normal distribution PDFs for the net blank response and the net analyte response at Y_D. Voigtman, 2011 [4]. Reproduced with permission of Elsevier.

Given eqn 24.3, it is easy to compute, numerically, any desired confidence interval for x_D in Fig. 24.3. However, Fig. 24.3 clearly shows that distributions of experimental x_D variates may be relatively wide; the upper limit of the 95% CI is almost 13 times the lower limit. As degrees of freedom increase, the ratio drops, as is apparent in Fig. 15.10. Nevertheless, the distributions are relatively wide even for the narrowest histogram in Fig. 15.10. This means that it is as fruitless to compare $0.9x_D$ with $1.1x_D$, as it was to suppose that $0.9X_D$ was qualitatively different from $1.1X_D$.

The histogram of 10^7 experimental x_D variates in Fig. 24.3, ignoring units, had $\bar{x}_D \cong 0.4932$ and $s_{x_D} \cong 0.2619$, while the histogram of 10^7 simulation x_D variates, also in Fig. 24.3, had $\bar{x}_D \cong 0.4934$ and $s_{x_D} \cong 0.2617$. As for theory, Badocco et al. [6] applied propagation of errors (POEs) to the x_D expression as follows:

$$x_D = (t_p + t_q)\eta^{1/2}\sqrt{s_0^2/b} \tag{24.4}$$

resulting in an approximate expression for s_{x_D}:

$$s_{x_D} \cong (t_p + t_q)\eta^{1/2}\frac{s_0}{b}\left[\frac{1}{2v} + \frac{s_0^2}{b^2 S_{XX}}\right]^{1/2} = x_D\left[\frac{1}{2v} + \frac{s_0^2}{b^2 S_{XX}}\right]^{1/2} \tag{24.5}$$

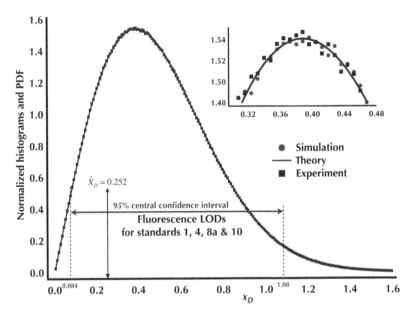

Figure 24.3 The overplotted theoretical (solid line) detection limit PDF for x_D (eqn 15.10), together with the normalized 10 million event histograms of experimental (filled squares, uppermost) and simulation (filled circles, farthest back) detection limits for calibration with four standards. The units of x_D are mg R6G/100 mL ethanol. Voigtman, 2011 [5]. Reproduced with permission of Elsevier.

Using $\bar{x}_D$ and average values for b and s_0^2, this yields $s_{x_D} \cong 0.280$. The result is somewhat high due to the relative error in $\sqrt{s_0^2}$ not being sufficiently "small," as required by the first-order Taylor-series expansion derivation of the customary POE formula for quotients [7, Table 1].

A more accurate result is readily obtained as follows. First, the estimated population mean for eqn 24.3 is computed by numerical integration, as shown in Fig. 24.4. The result is $E[u] = \hat{\mu}_{x_D} \cong 0.4934$, as shown in Fig. 24.4 [8, Fig. 7]. The estimated value for δ, and the values for k and the "scale factor," that is, $S_{XX}^{1/2}$, are given in Table 15.3. Then, the estimated population standard deviation of eqn 24.3 is also obtained by numerical integration, as shown in Fig. 24.5.

The result is $\hat{\sigma}_{x_D} \cong 0.2617$, as shown in Fig. 24.5. Thus, as expected, experiment, theory, and simulation are in excellent agreement and the percent relative standard deviation is 53%. This rather large value corroborates the fact that it is pointless to compare $0.9x_D$ with $1.1x_D$ and see a dichotomy of any sort. All of the above then leads to the realization that there is no real dichotomization of the measurand space and analysts should simply follow the recommendations made in regard to Issue 2.

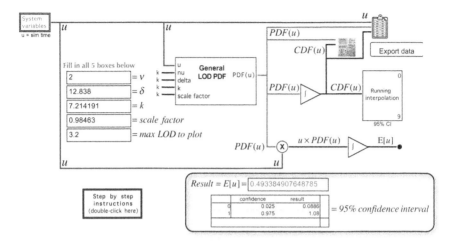

Figure 24.4 *LightStone* numerical estimation of the first moment, that is, the population mean, of eqn 24.3. Voigtman, 2011 [8]. Reproduced with permission of Elsevier.

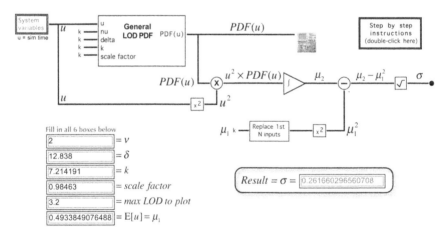

Figure 24.5 *LightStone* numerical estimation of the square root of the second central moment, that is, the population standard deviation, of eqn 24.3. Note that $E[u]$ is from the simulation in Fig. 24.4.

24.3 A "NEW DEFINITION OF LOD" DISTRACTION

Fonollosa *et al.* have suggested [9] a "new definition of LOD based on the amount of information that the chemical system can extract from the presented stimulus." They state [9] as follows:

All methodologies based on the probability of Type I and Type II errors assume implicitly that the two states representing analyte/no-analyte exposure are presented with the same probability, i.e., the same a priori probability for both classes. However, in most of the applications one of the possible states is expected to be found more often than the other one, thereby leaning the probability of the system towards one of the classes. In this paper we show that the amount of information that can be extracted about the sample from the analytical system depends on the prior probability of presence or absence of the analyte.

Basically, they have proposed a methodology mingling information theory, Bayesian statistics, linear systems theory, and receiver operating characteristic (ROC) curves. While certainly original, their synthesis is ultimately baffling; all that matters is whether a CMS, and the detection protocol, satisfies *a priori* specifications of p and q. In other words, if a blank is presented to the CMS, is the probability of a false positive equal to p? Similarly, if the CMS is presented with analyte content at the detection limit, is the probability of a false negative equal to q?

There is no binary distribution of blanks versus analytes and no "prior probabilities of presenting the analyte or a blank sample" [9]. Knowing that the inputs to a CMS are usually blanks, or usually analytes, would have *no effect whatsoever* upon the decision level or the detection limit. Taken to the extreme, if it was known that the inputs were *invariably* blanks, there would be no need for the CMS at all. Similarly, suppose that the inputs are *always* analytes, but always at levels in region (a) in Fig. 24.1. In this case, the information would be used to set a *new* decision level as low as possible, so that there was zero probability of a false negative, even though the analyte would always be *below* the original decision level. Once again, there would be no need for the CMS at all. Accordingly, there is no "leaning the probability of the system towards one of the classes" [9].

24.4   POTENTIALLY IMPORTANT RESEARCH PROSPECTS

In contrast to the above issues and distractions, a number of promising research avenues have recently been identified. Some of these are briefly discussed as follows.

24.4.1   Extension to Method Detection Limits

The preceding chapters were devoted to showing how Currie's detection limit schema could be rigorously instantiated for the model CMS defined in Chapter 7. Extension to heteroscedastic noises was discussed in Chapter 18. However, the discussion was limited to instrumental detection limits and no mention was made of the next level up, that is, method detection limits. This was deliberate; instrumental detection limits are the foundation, upon which all higher levels must be based, and it would be counterproductive to construct an elaborate hierarchical edifice on a faulty foundation.

Method detection limits are worse than instrumental detection limits because of various noninstrumental error sources. These additional noises are specific to the method, for example, method "A" may be highly susceptible to interferences while method "B" is not. Furthermore, it is often the case that methods are beset with several different noninstrumental error sources and it may be that some of these noises are unrecognized, poorly understood, or difficult to individually quantify. In such cases, it may be beneficial to employ a partitioning strategy; the method's total noise is the sum of the strictly instrumental noise plus the mingled total of all the other noises.

This is basically what Badocco *et al.* [10] accomplished in their detailed study of sources of error in the trace analysis of 41 elements by ICP-MS. Their analysis demonstrated that the noninstrumental error source was predominantly due to reagent contamination via environmental urban air particulates. They specifically state that the "obtained results clearly indicated the need of using the two-component variance regression approach for the calibration of all the elements usually present in the environment at significant concentration levels" [10].

More generally, they demonstrated the practical value of the partitioning strategy, thereby providing a clue as to how a rigorous method detection limit schema might be formulated. Overall, this is an exciting development and it would be well worth pursuing further.

24.4.2 Confidence Intervals in the Content Domain

As discussed in Chapter 14, it is currently unknown if there is a general way to construct confidence intervals for x_D and X_D, if only a small amount of data is available. Repeated attempts to use the method of Holstein *et al.* [11] resulted in uninformative failures, but perhaps this was simply the present author's fault. In any event, if their method is valid, and can be shown to work for linear CMSs, it would be an important advance, with considerable practical utility. Further investigation is desirable, and will necessarily involve extensive computer simulations. It is hoped that the challenge will be taken up.

24.4.3 Noises Other Than AGWN

When the CMS noise is not Gaussian, everything becomes considerably more complicated. The present text almost exclusively considers AGWN, thereby resulting in the simplest possible model system. Shot noise, with its discrete Poissonian probability density function (PDF), is of enormous practical importance since it is involved (1) whenever particles, for example, photons or ions, must be detected and (2) their average arrival rate is low. But there is already an immense body of published information pertaining to detection in the shot noise limit, so it was not considered herein, and the freely available MARLAP publications [12] are recommended for further information on the topic.

In Appendix C, simulations are used to show what would happen if a CMS is afflicted with additive uniformly distributed white noise (AUWN). With α as the blank's true value, σ_0 as the population standard deviation, $M = 1, p < 0.5$, and $q < 0.5$, it is shown that

$$R_D \equiv \alpha + Y_D = \alpha + [(1 - 2p) + (1 - 2q)]3^{1/2}\sigma_0 \qquad (24.6)$$

Table 24.1 Bias Factors for AUWN

ν	Bias Factor
1	0.8164
2	0.9118
3	0.9464
4	0.9626
5	0.9721
6	0.9778
7	0.9816
8	0.9844
9	0.9864

With $p \equiv 0.05 \equiv q$, $R_D \cong \alpha + 3.1177\sigma_0$, and $Y_D \cong 3.1177\sigma_0$. From Chapter 7, with AGWN and the same values of p, q, and M, $R_D = \alpha + (z_p + z_q)\sigma_0 \cong \alpha + 3.2897\sigma_0$. Hence, $Y_D \cong 3.2897\sigma_0$ and, Y_D depends upon the noise PDF. In particular, $Y_D \approx 3.3\sigma_0$ only for Gaussian noise, with $M = 1$ and $p \equiv 0.05 \equiv q$.

If σ_0 is unknown, and must be estimated by s_0, then $r_D \simeq \alpha + [(1 - 2p) + (1 - 2q)]3^{1/2}s_0$, but simulations show that s_0 is a negatively biased estimator of σ_0 and the analytical expression for its bias factor, analogous to the $c_4(\nu)$ factor for AGWN, is unknown to the present author. Table 24.1 shows the first nine bias factors for AUWN, as estimated via simulations.

As with $c_4(\nu)$, these bias factors are only properly used when they are divided into the *average* of *many i.i.d.* s_0 variates. If only a single s_0 variate is divided by its relevant bias factor, the bias is altered but not eliminated. Hence, $r_D \simeq \alpha + [(1 - 2p) + (1 - 2q)]3^{1/2}s_0$ is negatively biased and its bias cannot be eliminated simply by dividing a single s_0 by its bias factor. Unlike the situation with AGWN, there appears to be nothing analogous to the central t distribution and critical t values, so the bias may be uncorrectable.

With AGWN, $R_D = \alpha + (z_p + z_q)\sigma_0$ and $r_D = \alpha + (t_p + t_q)s_0$. This simple change, replacing critical z values by corresponding critical t values, when σ_0 is estimated by s_0, is perhaps be unique to the Gaussian distribution case. But it would be useful to know if this is generally true or not.

24.5 SUMMARY

The main purpose of this text was to explore and illustrate the theory and practice of limits of detection as they exist in analytical chemistry and chemical analysis. The first five chapters provided a general development of detection limits, with an emphasis on understanding the purposes they are intended to serve and how they are crafted to achieve those purposes. These chapters necessarily introduced important concepts and terminologies, and dispelled several myths, for example, the need to correct back-calculated negative analyte content values. Chapter 6 provided a brief introduction to ROC curves and demonstrated their versatility in situations where relatively little is known about the CMS or its NPM. In particular, it was shown to be possible to instantiate Currie's detection limit schema without any use of calibration curves.

Then Chapters 7–14 were devoted to a detailed analysis of Currie's detection limit schema as applied to the simplest CMS model, with the most benign noise: homoscedastic AGWN. A total of 48 decision level and detection limit equations were presented in the summary figures in Chapters 7–14, and their performance, as judged by extensive Monte Carlo simulations, was uniformly excellent. Chapter 15 demonstrated how Currie's detection limit schema applied to a real laser-excited molecular fluorescence experiment and yielded an important insight: with appropriate use of bootstrapping, it is possible to obtain a highly accurate estimate of the fluorophore analyte's true detection limit, X_D.

Chapter 16 presented a discussion of important issues raised in the preceding nine chapters, while Chapter 17 demonstrated that two detection limit standards, that is, ISO 11843-2 [13] and IUPAC 1995 [14], are unusably biased, due to their use of the $\delta_{critical}$ method, and should not be used. Then the correct extension of Currie's schema to heteroscedastic noise was demonstrated in Chapter 18 and Chapter 19 carefully examined a metrologically sound redefinition of the limit of quantitation. Chapters 20–23 discussed how Currie's detection limit schema would be extended to time-variant responses, such as step responses and isolated chromatographic peaks. These four chapters were far from exhaustive; indeed, how many real chromatograms have only a single peak, with only homoscedastic AGWN? Finally, Chapter 24 deals with sundry issues, distractions, and research prospects identified along the way.

REFERENCES

1. D. Coleman, J. Auses, N. Grams, "Regulation – from an industry perspective or relationships between detection limits, quantitation limits, and significant digits", *Chemom. Intell. Lab. Syst.* **37** (1997) 71–80.

2. Analytical Methods Committee of the Royal Society of Chemistry, "What should be done with results below the detection limit? Mentioning the unmentionable," 5 (April 2001) 2 pp. (Fig. 2).

3. M. Thompson, "Uncertainty functions, a compact way of summarising or specifying the behavior of analytical systems", *Trends Anal. Chem.* **30** (2011) 1168–1175.

4. E. Voigtman, K.T. Abraham, "True detection limits in an experimental linearly heteroscedastic system. Part 1", *Spectrochim. Acta, B* **66** (2011) 822–827.

5. E. Voigtman, K.T. Abraham, "Statistical behavior of ten million experimental detection limits", *Spectrochim. Acta, B* **66** (2011) 105–113.

6. D. Badocco, I. Lavagnini, A. Mondin, P. Pastore, "Estimation of the uncertainty of the quantification limit", *Spectrochim. Acta, B* **96** (2014) 8–11.

7. H.H. Ku, "Notes on the use of propagation of error formulas", *J. Res. Natl. Bur. Stand. – C. Eng. Instrum.* **70C** (1966) 263–273.

8. E. Voigtman, K.T. Abraham, "True detection limits in an experimental linearly heteroscedastic system. Part 2", *Spectrochim. Acta, B* **66** (2011) 828–833.

9. J. Fonollosa, A. Vergara, R. Huerta, S. Marco, "Estimation of the limit of detection using information theory measures", *Anal. Chim. Acta* **810** (2014) 1–9.

10. D. Badocco, I. Lavagnini, A. Mondin, A. Tapparo, P. Pastore, "Limit of detection in the presence of instrumental and non-instrumental errors: study of the possible sources of error and application to the analysis of 41 elements at trace levels by ICP-MS technique", *Spectrochim. Acta, B* **107** (2015) 178–184.

11. C.A. Holstein, M. Griffin, J. Hong, P.D. Sampson, "Statistical method for determining and comparing limits of detection of bioassays", *Anal. Chem.* **87** (2015) 9795–9801.

12. Multi-Agency Radiological Laboratory Analytical Protocols (MARLAP) Manual Volume III, Section 20A.3.1, 20-54–20-58, 2004.

13. ISO 11843-2, *"Capability of Detection – Part 2: Methodology in the Linear Calibration Case"*, ISO, Genève, 2000.

14. L.A. Currie, for IUPAC, "Nomenclature in evaluation of analytical methods including detection and quantification capabilities" *Pure Appl. Chem.* **67** (1995) 1699–1723. IUPAC ©1995.

APPENDIX A

STATISTICAL BARE NECESSITIES

This appendix is intended to provide a concise presentation of useful items, from elementary statistics, that are used in the text. These include basic concepts, definitions, probability density functions (PDFs), formulas, and nomenclature. This appendix is absolutely *not* intended as a review of basic statistics, and it contains only a very tiny percentage of the vast body of statistics that exists. Rather, it is deliberately restricted so as to include only the essential statistical "machinery" relevant to limits of detection. Readers with a significant knowledge of statistics should look over this appendix, if for no other reason than to avoid nomenclature surprises or confusion. It also serves as a reference for a number of equations and expressions that are used in the text.

A.1 PRIOR KNOWLEDGE AND REFERENCES

The level of statistics knowledge required for a solid understanding of limits of detection is quite modest; nothing more than a reasonable undergraduate course or two in statistics. Maybe not even that. The same applies in regard to mathematics: only basic integral calculus is required to obtain expectation values and the like, and these are already in hand.

Five tables of useful statistical values are present in this appendix. Almost every topic in this appendix is *not* the original concept of the author, the major exception being the *LightStone* software. In some cases, equations (for example, eqn A.28) were derived from first principles, mostly because they were too hard, at the time, to track back to their first published occurrences. For clarity of presentation, numerical values, such as critical values, may be shown as rounded to several decimal places. However, all computations use full double-precision accuracy.

Limits of Detection in Chemical Analysis, First Edition. Edward Voigtman.
© 2017 John Wiley & Sons, Inc. Published 2017 by John Wiley & Sons, Inc.
Companion Website: www.wiley.com/go/Voigtman/Limits_of_Detection_in_Chemical_Analysis

A.2   MONTE CARLO COMPUTER SIMULATIONS

Cogent results from computer simulations will be presented where they can facilitate understanding or corroborate theoretical predictions. The main simulation software used will be the author's freely available and fully commented *LightStone* libraries of add-on blocks for the *ExtendSim* (*v. 9.2*) simulation program. To a far lesser extent, *Mathematica* (*v. 9*) will also be used. All of the simulation executable files, that is, "programs," are freely available. Both *ExtendSim* and *Mathematica* are commercial programs and each has freely available demonstration versions that may be used to perform the simulations that will appear in the book. To obtain the demo versions of *ExtendSim* or *Mathematica*, respectively, go to http://www.extendsim.com or http://www.wolfram.com/mathematica. The free *LightStone* libraries of add-on blocks and all other software models used in this book are available at the URL given in Appendix B.

A.3   THE GAUSSIAN DISTRIBUTION

The Gaussian distribution, also known as the normal distribution, is one of the fundamental concepts in statistics and in all experimental sciences [1, 2]. In common parlance, it is *the* "Bell curve." Mathematically, the Gaussian distribution is described by its PDF, which is given by

$$p_x(x) \equiv \frac{1}{\sqrt{2\pi}\sigma} e^{-(x-\mu)^2/2\sigma^2} \tag{A.1}$$

where x is the Gaussian distributed random variate, μ is the population mean, and σ is the population standard deviation. Note that $\sigma > 0$ and the x subscript on the PDF will be omitted henceforth, unless necessary to avoid confusion between or among PDFs.

A.4   PROBABILITIES FROM PDFs

The PDF of a random variate determines the probability that the variate may be found between a pair of possible bounding values. For example, if x has the Gaussian PDF in eqn A.1, and a and b are two constants, with $a < b$, then the probability that x is between a and b is

$$\Pr(a \le x \le b) = \int_a^b p(x)dx \tag{A.2}$$

Figure A.1 shows a graph of eqn A.1, with $\mu \equiv 0$ and $\sigma \equiv 1$.

A.4.1   Normal Distribution Notation and z "Scores"

It is convenient to represent the Gaussian PDF by a short form notation such as $N{:}\mu, \sigma$ or $N(\mu, \sigma^2)$. We will use the former, although the latter is more commonly found. If the

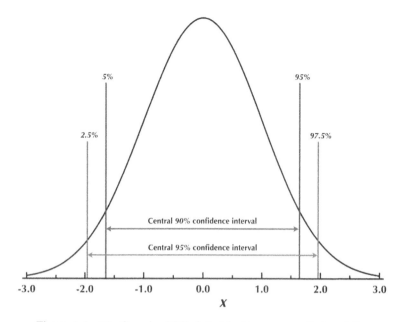

Figure A.1   The Gaussian PDF plotted for the range $\mu - 3\sigma$ to $\mu + 3\sigma$.

variate x is distributed as $N{:}\mu, \sigma$, this will be denoted by $x \sim N{:}\mu, \sigma$. In Fig. A.1, the abscissa is x, $\mu \equiv 0$ and $\sigma \equiv 1$, so the Gaussian PDF is that of the *standard unit normal variate*. Then the Gaussian PDF in Fig. A.1 is simply $N{:}0, 1$. It is often convenient to use the *standard normal variate z*:

$$z \equiv \frac{x - \mu}{\sigma} \tag{A.3}$$

so that any particular x variate is simply a scaled and shifted z variate, that is, $x = z\sigma + \mu$. Note that, in Fig. A.1, $z = x$. Frequently, z variates are referred to as z "*scores*."

A.4.2   Central Confidence Intervals and Critical Values

Aside from the unit normal PDF, Fig. A.1 also depicts the central 90% and 95% confidence intervals (CIs). All PDFs are normalized to unit area, since their total integrated area is 100%, that is, the variate must be *somewhere* in its allowed range. Accordingly,

$$1 = \int_{-\infty}^{\infty} p(x)dx \tag{A.4}$$

Any *central* confidence interval, for the Gaussian distribution, will simply be the interval that symmetrically spans μ and contains whatever area is defined. Thus, the central 90% confidence interval in Fig. A.1 is from $\mu - 1.645\sigma$ to $\mu + 1.645\sigma$, or equivalently, from $z = -1.645$ to $z = +1.645$. In equation form, these are

$$0.90 = \int_{\mu-1.645\sigma}^{\mu+1.645\sigma} p(x)dx = \int_{-1.645}^{1.645} (2\pi)^{-1/2} e^{-z^2/2} dz \qquad (A.5)$$

Similarly, the central 95% confidence interval in Fig. A.1 is from $\mu - 1.960\sigma$ to $\mu + 1.960\sigma$, or equivalently, from $z = -1.960$ to $z = +1.960$. In equation form, these are

$$0.95 = \int_{\mu-1.960\sigma}^{\mu+1.960\sigma} p(x)dx = \int_{-1.960}^{1.960} (2\pi)^{-1/2} e^{-z^2/2} dz \qquad (A.6)$$

The symmetric z values that specify central confidence intervals are referred to as *critical z values* and they are usually given subscripts indicating the area of *either* of the two equal tails. Thus, $z_{0.05} \simeq 1.645$ and $z_{0.025} \simeq 1.960$. Table A.1 provides a few of the commonly encountered critical z values, with more digits, together with the areas of either tail. Extensive tables are widely available, but integration of eqn A.1 is easy and critical z values are readily available in widely used software, for example, in *Excel*, $z_p = -\text{NORMSINV}(p)$, while in *Mathematica*, $z_p = \text{InverseCDF[NormalDistribution[], } 1 - p]$.

If a random selection were to be made from the Gaussian PDF in Fig. A.1, the probability would be 95% that it would fall within $\mu \pm 1.960\sigma$ and there would be 90% probability that it would fall within $\mu \pm 1.645\sigma$. In terms of z scores, the random selection would have 95% probability of having $|z| \leq 1.960$ and 90% probability of having $|z| \leq 1.645$. Suppose N *independent and identically distributed (i.i.d.)* such selections, denoted $x_1, x_2, \ldots, x_N$, were made and their sample mean was computed in the conventional way:

$$\bar{x} = \frac{1}{N} \sum_{i=1}^{N} x_i \qquad (A.7)$$

Since each x_i is normally distributed, and any linear combination of normal variates is normally distributed, x is normally distributed. It is a random sample from the *sampling distribution of the mean* and its PDF differs from the parent PDF in eqn A.1 only by replacing σ by the population standard error, σ_e. This latter parameter is defined as $\sigma_e \equiv \sigma/\sqrt{N}$. Hence, $\bar{x}$ is distributed as $N{:}\mu, \sigma_e$, that is, $\bar{x} \sim N{:}\mu, \sigma_e$.

The central confidence intervals shown in Fig. A.1 are conventionally expressed in the following notation:

$$(1 - 2p) \times 100\% \text{ CI for } x \equiv \mu \pm z_p\sigma \qquad (A.8)$$

and

$$(1 - 2p) \times 100\% \text{ CI for } \bar{x} \equiv \mu \pm z_p\sigma_e \qquad (A.9)$$

where p is the area fraction of one tail and $2p$ is the total area of the two tails. Each of these CIs may be *inverted* to provide corresponding CIs for μ when σ is known and either x or $\bar{x}$ is available. The results are

Table A.1   Critical z Values

Upper Tail Area, p	Critical z Value, z_p
0.50000	0
0.45000	0.125661346855074
0.40000	0.253347103135800
0.35000	0.385320466407568
0.30000	0.524400512708041
0.25000	0.674489750196082
0.20000	0.841621233572914
0.15866	1.00000000000000
0.15000	1.03643338949379
0.10000	1.28155156554460
0.09000	1.34075503369022
0.08000	1.40507156030963
0.07000	1.47579102817917
0.06000	1.55477359459685
0.05000	1.64485362695147
0.04500	1.69539771027214
0.04000	1.75068607125217
0.03500	1.81191067295260
0.03000	1.88079360815125
0.02500	1.95996398454005
0.02275	2.00000000000000
0.02000	2.05374891063182
0.01500	2.17009037758456
0.01000	2.32634787404085
0.00500	2.57582930354892
0.00400	2.65206980790220
0.00300	2.74778138544498
0.00234	2.82842712474625
0.00200	2.87816173909550
0.00135	2.99999999999995
0.00100	3.09023230616779
0.00050	3.29052673149177
0.00025	3.48075640434619
0.00010	3.71901648545737

$$(1 - 2p) \times 100\% \text{ CI for } \mu \equiv x \pm z_p \sigma \tag{A.10}$$

and

$$(1 - 2p) \times 100\% \text{ CI for } \mu \equiv \bar{x} \pm z_p \sigma_e \tag{A.11}$$

The rationale for eqn A.10 is that eqn A.8 shows that x can be no further away from μ than $z_p \sigma$ while remaining *not outside* the CI given by eqn A.8. Therefore, μ can be no further away from x than $z_p\sigma$ while remaining *not outside* the CI given by eqn A.10. Similar considerations apply to eqns A.11 and A.9.

A.5 SAMPLE STANDARD DEVIATIONS

If σ is unknown, then it is estimated by an appropriate sample standard deviation, denoted by s and defined by

$$s \equiv + \left[\frac{1}{N-1} \sum_{i=1}^{N} (x_i - \bar{x})^2 \right]^{1/2} \tag{A.12}$$

where N is the number of *i.i.d.* samples from $N: \mu, \sigma$. If μ were known, eqn A.12 would be modified by replacing $N-1$ by N and x by μ. Assume μ and σ are both unknown, with N as immediately above. Then the CIs in eqns A.10 and A.11 become, respectively

$$(1 - 2p) \times 100\% \text{ CI for } \mu \equiv x \pm t_p s \tag{A.13}$$

and

$$(1 - 2p) \times 100\% \text{ CI for } \mu \equiv \bar{x} \pm t_p s_e \tag{A.14}$$

where s_e is the sample standard error of the mean, defined by $s_e \equiv s/\sqrt{N}$, the number of degrees of freedom, denoted by v, is $N-1$, and t_p is the critical t value, for p and v. If necessary for clarity, t_p may be denoted by $t_{p,v}$. Note that eqn A.13 is a nonsensical construct because N cannot equal 1.

Table A.2 provides a few of the commonly encountered critical t values, as functions of v and p. As for critical z values, extensive critical t tables are widely available, but again software evaluation is very convenient: in *Excel*, $t_p = \text{TINV}(2p, v)$, while in *Mathematica*, $t_p = \text{InverseCDF[StudentTDistribution}[v], 1 - p]$. The t distribution is discussed immediately following the next section.

A.6 THE χ DISTRIBUTION

Unlike x, s is not Gaussian distributed. Rather, it is χ distributed, with PDF given by

$$p(s) = \frac{v^{v/2} s^{v-1} e^{-vs^2/2\sigma^2}}{\sigma^v 2^{(v-2)/2} \Gamma(v/2)} U(s) \tag{A.15}$$

where $U(s)$ is the unit step function: $U(s) = 1$ for $s \geq 0$ and 0 for $s < 0$. The infinite family of s PDFs is indexed by v, where $v \equiv N - 1$. An example, with $\sigma \equiv 1$, is shown in Fig. A.2.

The χ PDF in Fig. A.2 is clearly asymmetric. As v increases, the χ PDFs become more like Gaussians, that is, more symmetric about σ. In the limit, as $v \to \infty$, the

Table A.2   Central (Two-Sided) Critical *t* Values

Confidence Levels			
ν	90%	95%	98%	99%
1	6.31375151357386	12.7062047339870	31.8205159483141	63.6567411519546
2	2.91998558009756	4.30265272954454	6.96455673396344	9.92484320047470
3	2.35336343453313	3.18244630488688	4.54070285842150	5.84090930943221
4	2.13184678190398	2.77644510504380	3.74694738775648	4.60409487123225
5	2.01504837208812	2.57058183469754	3.36492999735038	4.03214298334391
6	1.94318027429198	2.44691184643268	3.14266840313005	3.70742802038721
7	1.89457860365580	2.36462425094932	2.99795156635778	3.49948329725447
8	1.85954803330183	2.30600413329912	2.89645944621375	3.35538733113484
9	1.83311292255007	2.26215715817358	2.82143792141052	3.24983554112748
10	1.81246110219722	2.22813884242587	2.76376945778846	3.16927267160917
11	1.79588481423219	2.20098515872184	2.71807918317643	3.10580651358217
12	1.78228754760568	2.17881282716507	2.68099799196004	3.05453958595050
13	1.77093338264828	2.16036865224854	2.65030883595298	3.01227583313491
14	1.76131011506196	2.14478668128208	2.62449406449589	2.97684273395330
15	1.75305032520786	2.13144953567595	2.60248029039023	2.94671288283488
16	1.74588366894289	2.11990528516258	2.58348717869037	2.92078162148262
17	1.73960671564883	2.10981555859266	2.56693397470020	2.89823051834251
18	1.73406359230939	2.10092203686118	2.55237961821875	2.87844047091164
19	1.72913279247219	2.09302404985486	2.53948318919090	2.86093460403877
20	1.72471821821380	2.08596344129554	2.52797700085489	2.84533970664782
21	1.72074287148535	2.07961383708272	2.51764801361881	2.83135955405598
22	1.71714433543983	2.07387305831561	2.50832454984430	2.81875605568542
23	1.71387151707496	2.06865759861054	2.49986673571863	2.80733567778810
24	1.71088206673347	2.06389854731807	2.49215946856631	2.79693949760654
25	1.70814074523276	2.05953853565859	2.48510716990898	2.78743580520601
26	1.70561790054927	2.05552941848069	2.47862981708430	2.77871452344142
27	1.70328842296808	2.05183049297067	2.47265990434998	2.77068294570595
28	1.70113090761181	2.04840711466289	2.46714008916997	2.76326244241061
29	1.69912699562287	2.04522961110855	2.46202135007114	2.75638590209806
30	1.69726085107213	2.04227244936679	2.45726153095181	2.74999565175574
31	1.69551874206184	2.03951343844151	2.45282418049957	2.74404191722513
32	1.69388870259190	2.03693333440703	2.44867761923314	2.73848147966702
33	1.69236025759198	2.03451528722141	2.44479418375496	2.73327663971165
34	1.69092419777125	2.03224449783959	2.44114961016523	2.72839436412004
35	1.68957243954679	2.03010791544831	2.43772252764334	2.72380558592897
40	1.68385101380743	2.02107536985045	2.42325677441038	2.70445926227922
45	1.67942739312867	2.01410335926697	2.41211586850486	2.68958501201957
50	1.67590502564271	2.00855907214326	2.40327190688862	2.67779326114136
60	1.67064886538840	2.00029780432954	2.39011945702846	2.66028301372293
70	1.66691447954213	1.99443708586968	2.38080746044128	2.64790460304919
80	1.66412457907751	1.99006338664240	2.37386824476334	2.63869059127950
90	1.66196108451881	1.98667449723875	2.36849744179118	2.63156515925826
100	1.66023432657454	1.98397146629437	2.36421735599704	2.62589051386574

(*continued*)

Table A.2 (*Continued*)

Confidence Levels			
v 90%	95%	98%	99%
200 1.65250810140191	1.97189617759775	2.34513705783638	2.60063441900884
300 1.64994867442820	1.96790294718438	2.33884189096349	2.59231638816454
400 1.64867194195541	1.96591226750121	2.33570637516545	2.58817605394922
500 1.64790685442018	1.96471975370631	2.33382891367766	2.58569780676637
600 1.64739719225081	1.96392553243662	2.33257887286075	2.58404811685245
800 1.64676055986494	1.96293369654506	2.33101811937767	2.58198878390535
1000 1.64637881777427	1.96233903613052	2.33008262484126	2.58075466446738
1E+04 1.64500601854625	1.96020118477496	2.32672081798068	2.57632100752206
1E+05 1.64486886490831	1.95998765128930	2.32638514430343	2.57587845811086
1E+06 1.64485514965029	1.95996630070791	2.32635158203039	2.57583420852098

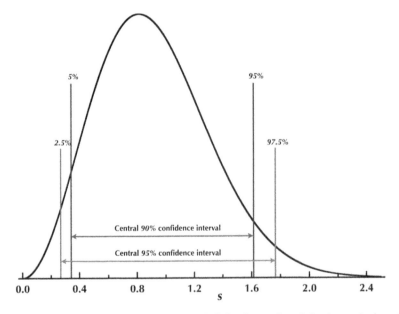

Figure A.2 The χ PDF plotted for the range 0–2.5σ, for $v = 3$ and abscissa units in σ. The 95% CI for s is 0.268–1.765s and the 90% CI for s is 0.342–1.614s.

PDF would simply become a so-called "delta function," located at σ. As was the case for CIs for Gaussian PDFs, CIs also exist for χ PDFs. The central 90% and 95% CIs for s, with $v = 3$ and given $\sigma \equiv 1$, are shown in Fig. A.2 and its caption. These may be inverted in a similar manner to those for Gaussian CIs. Thus, the 95% CI for σ, for the example in Fig. A.2, is (1/1.765)s to (1/0.268)s, that is, from 0.5665s to 3.729s. More generally, Table A.3 provides the 90% CIs for s, given σ, and the 90% CIs for σ, given s. Similarly, Table A.4 provides the 95% CIs for s, given σ, and the 95% CIs

Table A.3   Central 90% Confidence Intervals for s and σ

ν	90% CI for s		90% CI for σ	
	Lower Limit $\times \sigma$	Upper Limit $\times \sigma$	Lower Limit $\times s$	Upper Limit $\times s$
1	0.062704	1.959896	0.5102	15.9479
2	0.226479	1.730801	0.5778	4.4154
3	0.342463	1.613966	0.6196	2.9200
4	0.421521	1.540104	0.6493	2.3724
5	0.478638	1.487983	0.6721	2.0893
6	0.522075	1.448652	0.6903	1.9154
7	0.556435	1.417599	0.7054	1.7972
8	0.584447	1.392268	0.7183	1.7110
9	0.607829	1.371088	0.7293	1.6452
10	0.627717	1.353034	0.7391	1.5931
11	0.644896	1.337403	0.7477	1.5506
12	0.659925	1.323695	0.7555	1.5153
13	0.673215	1.311546	0.7625	1.4854
14	0.685076	1.300680	0.7688	1.4597
15	0.695745	1.290885	0.7747	1.4373
16	0.705409	1.281995	0.7800	1.4176
17	0.714214	1.273879	0.7850	1.4001
18	0.722281	1.266431	0.7896	1.3845
19	0.729707	1.259563	0.7939	1.3704
20	0.736572	1.253203	0.7980	1.3576
21	0.742944	1.247293	0.8017	1.3460
22	0.748877	1.241780	0.8053	1.3353
23	0.754421	1.236622	0.8087	1.3255
24	0.759616	1.231783	0.8118	1.3165
25	0.764496	1.227231	0.8148	1.3081
26	0.769094	1.222939	0.8177	1.3002
27	0.773433	1.218883	0.8204	1.2929
28	0.777538	1.215041	0.8230	1.2861
29	0.781430	1.211396	0.8255	1.2797
30	0.785125	1.207931	0.8279	1.2737
31	0.788640	1.204632	0.8301	1.2680
32	0.791989	1.201486	0.8323	1.2626
33	0.795184	1.198482	0.8344	1.2576
34	0.798238	1.195609	0.8364	1.2528
35	0.801159	1.192857	0.8383	1.2482
40	0.814083	1.180661	0.8470	1.2284
45	0.824786	1.170529	0.8543	1.2124
50	0.833837	1.161936	0.8606	1.1993
60	0.848409	1.148055	0.8710	1.1787
80	0.868845	1.128491	0.8861	1.1510
100	0.882776	1.115087	0.8968	1.1328

Table A.4   Central 95% Confidence Intervals for s and σ

ν	95% CI for s		95% CI for σ	
	Lower Limit $\times \sigma$	Upper Limit $\times \sigma$	Lower Limit $\times s$	Upper Limit $\times s$
1	0.031336	2.241278	0.4462	31.9127
2	0.159114	1.920614	0.5207	6.2848
3	0.268200	1.765245	0.5665	3.7286
4	0.348000	1.669072	0.5991	2.8736
5	0.407727	1.602026	0.6242	2.4526
6	0.454118	1.551845	0.6444	2.2021
7	0.491334	1.512459	0.6612	2.0353
8	0.521982	1.480478	0.6755	1.9158
9	0.547761	1.453835	0.6878	1.8256
10	0.569821	1.431193	0.6987	1.7549
11	0.588969	1.411640	0.7084	1.6979
12	0.605790	1.394532	0.7171	1.6507
13	0.620715	1.379397	0.7250	1.6110
14	0.634074	1.365883	0.7321	1.5771
15	0.646123	1.353720	0.7387	1.5477
16	0.657060	1.342696	0.7448	1.5219
17	0.667046	1.332644	0.7504	1.4991
18	0.676212	1.323428	0.7556	1.4788
19	0.684662	1.314940	0.7605	1.4606
20	0.692486	1.307087	0.7651	1.4441
21	0.699757	1.299795	0.7694	1.4291
22	0.706537	1.293000	0.7734	1.4154
23	0.712879	1.286647	0.7772	1.4028
24	0.718828	1.280690	0.7808	1.3912
25	0.724422	1.275091	0.7843	1.3804
26	0.729696	1.269814	0.7875	1.3704
27	0.734680	1.264830	0.7906	1.3611
28	0.739397	1.260113	0.7936	1.3525
29	0.743873	1.255639	0.7964	1.3443
30	0.748125	1.251389	0.7991	1.3367
31	0.752173	1.247344	0.8017	1.3295
32	0.756033	1.243488	0.8042	1.3227
33	0.759717	1.239807	0.8066	1.3163
34	0.763240	1.236289	0.8089	1.3102
35	0.766613	1.232921	0.8111	1.3044
40	0.781553	1.218007	0.8210	1.2795
45	0.793951	1.205636	0.8294	1.2595
50	0.804454	1.195158	0.8367	1.2431
60	0.821398	1.178259	0.8487	1.2174
80	0.845230	1.154494	0.8662	1.1831
100	0.861521	1.138249	0.8785	1.1607

for σ, given s. In each table, $v \leq 100$, and it is usually more important to have the CIs for σ.

A.7   THE CENTRAL AND NONCENTRAL t DISTRIBUTIONS

Unlike critical z values, critical t values always depend on v, though this is often suppressed, for purposes of clarity, in the notational designation of critical t values. The v dependence is a consequence of the fact that s is a negatively biased point estimate of σ, and the bias is a function of v, as explicitly shown in the following section. The central t PDF, with dummy independent variable x, is

$$p(x|v) = \frac{(1 + x^2/v)^{-(v+1)/2}\Gamma((v+1)/2)}{\sqrt{v\pi} \times \Gamma(v/2)} \equiv p(x) \qquad (A.16)$$

and Fig. A.3 shows the PDFs for $v = 1, 2, 4, 8, 15, 30$ and the Gaussian limit at infinity.

The critical t values are obtained from their respective central t distributions by the same sort of area finding process used to find critical z values for the Gaussian PDF. For example, if $v = 4$, then $t_{0.05} \simeq 2.132$ and $t_{0.025} \simeq 2.776$ since

$$0.90 = \int_{-2.132}^{2.132} p(x|4)dx \qquad (A.17)$$

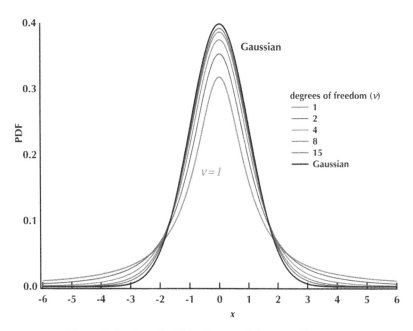

Figure A.3   Central t PDFs for several degrees of freedom.

and

$$0.95 = \int_{-2.776}^{2.776} p(x|4)dx \tag{A.18}$$

As a practical matter, the cumulative distribution function (CDF) simplifies the process since CDFs are just the running area integrals of associated PDFs. Thus, the CDF of a given PDF is of the form

$$CDF(x) = \int_{-\infty}^{x} p(x)dx \tag{A.19}$$

The noncentral t distribution is the generalization of the central t distribution. It is a built-in function in various software programs such as *Mathematica* and *R* and is also available in the author's *LightStone* software. Of note here is that the noncentral t PDF contains two controlling parameters, v and δ, where δ is the *noncentrality parameter*. The central or "Student" t distribution is the $\delta = 0$ special case of the noncentral t distribution.

Consider the quotient of two independent random variates, x and y, where $x \sim N{:}\delta, 1$, $y \sim (\chi^2/v)^{1/2}$, "$\sim$" means "is distributed as," v is degrees of freedom, and δ the noncentrality parameter. Thus, x is a Gaussian variate, y is a χ distributed variate, and, by definition, $u \equiv x/y$ is distributed as the noncentral t distribution [3, p. 509]:

$$u \sim t(u|v, \delta) = C(1 + u^2/v)^{-(v+1)/2} e^{-v\delta^2/2(v+u^2)} Hh(k) \tag{A.20}$$

where

$$C \equiv \frac{v!}{2^{(v-1)/2}\Gamma(v/2)(\pi v)^{1/2}}, \quad k \equiv \frac{-u\delta}{(v+u^2)^{1/2}}, \quad Hh(k) \equiv \int_0^{\infty} \left(\frac{u^v}{v!}\right) e^{-(u+k)^2/2} du$$

and $Hh(k)$ is the Airey integral [1, p. 515].

As v goes to infinity, both the central and noncentral t distributions asymptotically become Gaussian [3, p. 519]. Owen [4] gives the expectation value of u, for $v > 1$,

$$E[u] = \delta \times \left(\frac{v}{2}\right)^{1/2} \frac{\Gamma[(v-1)/2]}{\Gamma(v/2)} \tag{A.21}$$

and the variance of u, for $v > 2$,

$$\text{var}[u] = \frac{v}{v-2} + \delta^2 \left\{ \frac{v}{v-2} - \left(\frac{v}{2}\right) \frac{\Gamma^2[(v-1)/2]}{\Gamma^2(v/2)} \right\} \tag{A.22}$$

Confidence intervals are computed in the customary manner by inverting the CDF of the noncentral t distribution at the desired bounding values. The CDF is simply

$$CDF(u) \equiv \int_{-\infty}^{u} t(u'|v, \delta) du' \tag{A.23}$$

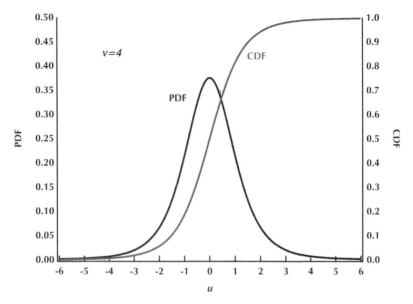

Figure A.4   Central t PDF and CDF for $v = 4$ degrees of freedom.

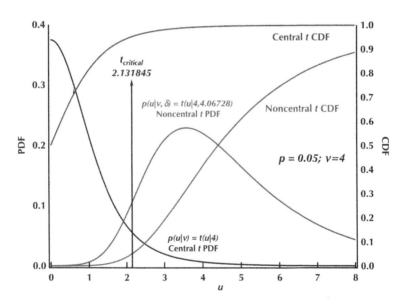

Figure A.5   Central and noncentral t PDFs, and their CDFs, for 4 degrees of freedom.

Figure A.4 shows the central t PDF and CDF, with dummy independent variable u, for $v = 4$. Note that the central t PDF is symmetric about zero. Figure A.5 shows an overplot of the PDFs and CDFs for both the central and noncentral t distributions when $v = 4$. It is useful for understanding the "statistical power" of the t test. The figure clearly shows $t_{0.05} \simeq 2.131845$ and the corresponding critical δ value of 4.06728 that would result in the noncentral t PDF having 5% of its area *below* $t_{0.05}$ for $v = 4$.

A.8   LIMITING VALUES OF THE CRITICAL VALUES

For all finite values of p, q, and v

$$(z_p + z_q) < \delta_{p,q,v} < (t_{p,v} + t_{q,v}) \tag{A.24}$$

In the limit, as $v \to \infty$, $s \to \sigma$, $t_{p,v} \to z_p$, and $\delta_{p,q,v} \to z_p + z_q$.

A.9   EXPECTATION VALUES

If x is a variate with continuous PDF $p(x)$, then the expectation value of x, denoted by $E[x]$, is defined as

$$E[x] \equiv \int_{-\infty}^{\infty} x p(x) dx \tag{A.25}$$

assuming the integral exists, that is, does not diverge. The nth moment is defined similarly

$$E[x^n] \equiv \int_{-\infty}^{\infty} x^n p(x) dx \tag{A.26}$$

if the integral exists and the nth central moment is

$$E[(x - E[x])^n] \equiv \int_{-\infty}^{\infty} (x - E[x])^n p(x) dx \tag{A.27}$$

again assuming existence.

Substituting eqn A.1 into eqn A.25 yields $E[x] = \mu$, so x is an unbiased point estimate of the population parameter μ. Similarly, $E[\bar{x}] = \mu$. However, substituting eqn A.15 into eqn A.25 yields

$$E[s] \equiv \int_{-\infty}^{\infty} s p(s) ds = \sigma \left(\frac{2}{v}\right)^{1/2} \frac{\Gamma[(v + 1)/2]}{\Gamma(v/2)} \tag{A.28}$$

even though $E[s^2] = \sigma^2$. Since $E[s] \neq \sigma$, s is a *biased* point estimate of the population parameter σ. For convenience, the bias factor $c_4(v)$, which is solely a function of v, may be defined as

$$c_4(v) \equiv \frac{E[s]}{\sigma} = \left(\frac{2}{v}\right)^{1/2} \frac{\Gamma[(v + 1)/2]}{\Gamma(v/2)} \tag{A.29}$$

Table A.5 gives values of $c_4(v)$ as a function of v and shows that $c_4(v)$ is always less than unity. Therefore, s is a *negatively biased* point estimate of σ, and, as expected,

Table A.5 Bias Factors $c_4(v)$ and $\sigma_s(v)/\sigma$

v	$c_4(v)$	$\sigma_s(v)/\sigma$
1	0.797884560717817	0.602810275101658
2	0.886226925432232	0.463251375215371
3	0.921317731904314	0.388810541110566
4	0.939985602969366	0.341214106112743
5	0.951532861932930	0.307547090153243
6	0.959368788686538	0.282155147559150
7	0.965030456135815	0.262137785773627
8	0.969310699703934	0.245838905463455
9	0.972659274112909	0.232236811212496
10	0.975350077137712	0.220663152853980
11	0.977559351848255	0.210660185165636
12	0.979405604308569	0.201902605849866
13	0.980971436750611	0.194152106039162
14	0.982316177158397	0.187229613290509
15	0.983483531612152	0.180997632712942
16	0.984506405468635	0.175348617306288
17	0.985410043805315	0.170197078610673
18	0.986214136857812	0.165474095440346
19	0.986934267522618	0.161123404847323
20	0.987582928824418	0.157098563629920
21	0.988170253314359	0.153360850495280
22	0.988704545232774	0.149877690921812
23	0.989192674957496	0.146621457537545
24	0.989640375584896	0.143568544647445
25	0.990052468840277	0.140698645847332
26	0.990433039208967	0.137994184092264
27	0.990785569621407	0.135439857612092
28	0.991113048241783	0.133022274844784
29	0.991418053292603	0.130729658476970
30	0.991702821009622	0.128551603652222
31	0.991969300515434	0.126478879007207
32	0.992219198457462	0.124503261854583
33	0.992454015570442	0.122617400796561
34	0.992675076817746	0.120814700533018
35	0.992883556389471	0.119089224749327
40	0.993770137125334	0.111449156829001
45	0.994460302450977	0.105112829136650
50	0.995012810705528	0.099747213153478
60	0.995842193881336	0.091095141942958
70	0.996435062409525	0.084363300083190
80	0.996879958842223	0.078932551325345
90	0.997226133666984	0.074431433760201
100	0.997503163955958	0.070621794779329
200	0.998750786126600	0.049968662294465

(continued)

Table A.5   (*Continued*)

v	$c_4(v)$	$\sigma_s(v)/\sigma$
300	0.999167015334574	0.040807786847590
400	0.999375195922671	0.035344274989366
500	0.999500125312328	0.031614862021534
600	0.999583420319499	0.028861493765420
800	0.999687548904172	0.024996091013784
1000	0.999750031289128	0.022357883115095
1E+04	0.999975000314615	0.007070979124927
1E+05	0.999997500098973	0.002236022317652
1E+06	0.999999750384662	0.000706562534044

$c_4(v) \rightarrow 1$ as $v \rightarrow \infty$. Note that $E[s]$ may be approximated as the mean of a *large* number of *i.i.d.* s variates. In this case, $E[s] \cong \bar{s}$ and therefore an estimate of σ, denoted by $\hat{\sigma}$, is

$$\hat{\sigma} = \frac{\bar{s}}{c_4(v)} \cong \frac{E[s]}{c_4(v)} = \sigma \tag{A.30}$$

It is almost surely true that $\hat{\sigma}$ is an accurate and negligibly biased estimate of σ, provided that $\bar{s}$ is the mean of a *large* number of *i.i.d.* s variates.

However, if only a *single* s value is available, then an immediate problem arises because $s/c_4(v)$ is χ distributed. Hence, if $N \gg 1$ *i.i.d.* s variates are each divided by the *constant* $c_4(v)$, the N resulting $s/c_4(v)$ variates are still χ distributed; if the variate u is defined as ks, where $k > 0$ is a constant and s is distributed as in eqn A.15, then the PDF of u is

$$p(u) = \frac{v^{v/2} u^{v-1} e^{-vu^2/2(k\sigma)^2}}{(k\sigma)^v 2^{(v-2)/2} \Gamma(v/2)} U(u) \tag{A.31}$$

Thus, s and $s/c_4(v)$ are distributed as per eqn A.31, with $k \equiv 1$ and $k \equiv 1/c_4(v)$, respectively.

To illustrate this critically important point, consider the simulation model shown in Fig. A.6. As noted in Fig. A.6, 10^4 simulations are performed, with 10^4 steps per simulation. In each simulation step, 5 *i.i.d.* Gaussian variates are generated and their sample standard deviation, s, is computed. Each s variate is binned into the upper histogram block. They are also each divided by $c_4(4)$ and binned into the lower histogram block. In addition, for *each* simulation, the 10^4 generated s variates are averaged, producing 10^4 $\bar{s}$ variates, and these are saved. Thus, two histogram blocks contain 10^8 binned variates each and the 10^4 $\bar{s}$ variates are separately binned, afterward, into a third histogram.

The results are shown in Figs A.7 and A.8. From Fig. A.7, it is clear that the histograms are fitted extremely well by eqn A.31, with appropriate k values. Since $\sigma \equiv 1$, $E[s] = c_4(4) \cong 0.93999$. The sample mean of the $10^8 s$ variates was 0.94008,

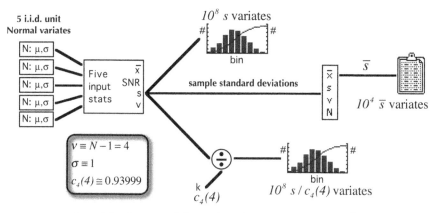

number of simulations = 10^4; number of steps/simulation = 10^4

Figure A.6 Simulation model demonstrating the statistics of s and related variates.

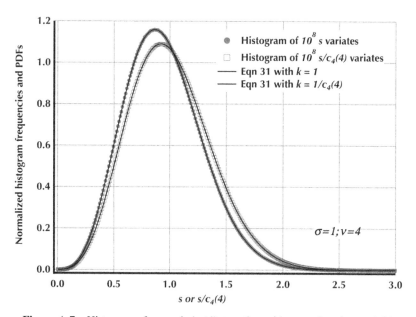

Figure A.7 Histograms for s and $s/c_4(4)$, together with curve fits via eqn A.31.

while the sample mean of the $10^8 s/c_4(4)$ variates was 1.0001. Therefore, $E[s]$ and $\overline{s}$ differ by less than 0.01%. Note, too, that the distributions are relatively wide: a random s variate is not necessarily close to $E[s]$.

As demonstrated by the data in Fig. A.8, the histogram of $\overline{s}$ variates is approximately Gaussian distributed, thanks to the hidden hand of the central limit

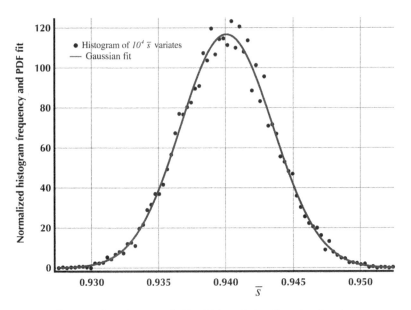

Figure A.8 Histogram for $\bar{s}$, with best fitting Gaussian function.

theorem. It is also approximately 100 times narrower than the histograms in Fig. A.7 and the Gaussian fit function is centered at 0.94008 ± 0.00003. Hence, $\bar{s}_{fit}/c_4(4) = 1.0001\sigma \cong \sigma$.

The above results clearly demonstrate that the following scenarios are *inequivalent* and only the *second* one is valid insofar as usage with confidence intervals, decision levels, and detection limits is concerned:

- Generate $N \gg 1$ independent s variates, divide each s variate by $c_4(v)$, use the N resulting $s/c_4(v)$ variates in N independent tests, and average the N test results.
- Generate $N \gg 1$ independent s variates, calculate $\bar{s}$ as their mean, then perform a *single* test, using $\bar{s}/c_4(v)$ as the estimate of σ.

The uncritical assumption that these two scenarios would yield equivalent results, as a mere consequence of N being large, is precisely where Currie [5], and the various standards organizations, went awry. In particular, never estimate $E[s]$ by any single s variate.

A.10 AN ADDITIONAL USEFUL RESULT

The population standard deviation of the sample standard deviation is simply the square root of the second central moment of s, denoted by $\sigma_s(v)$, and is given by

$$\sigma_s(v) \equiv \sqrt{E[(s - E[s])^2]} = \sigma[1 - c_4^2(v)]^{1/2} \tag{A.32}$$

Equation A.32 facilitates quantitative examination of the *relative precision* of s. Table A.5 gives values of $\sigma_s(v)/\sigma$ as a function of v and shows that $\sigma_s(v)/\sigma$ is relatively large for low degrees of freedom. As expected, $\sigma_s(v)/\sigma \to 0$ as $v \to \infty$, but this is a slow decline.

REFERENCES

1. E.S. Keeping, *Introduction to Statistical Inference*, Dover, Mineola, New York, ©1995.

2. A.B. Carlson, *Introduction to Communication Systems*, 2nd Ed., McGraw-Hill, New York, ©1975.

3. N.L. Johnson, S. Kotz, N. Balakrishnan, *Continuous Univariate Distributions, Volume 2*, 2nd Ed., John Wiley & Sons, New York, ©1995.

4. D.B. Owen, "A survey of properties and applications of the noncentral *t*-distribution", *Technometrics* **10** (1968) 445–478.

5. L.A. Currie, "Detection: international update, and some emerging di-lemmas involving calibration, the blank, and multiple detection decisions", *Chemom. Intell. Lab. Syst.* **37** (1997) 151–181.

APPENDIX B

AN EXTREMELY SHORT LIGHTSTONE® SIMULATION TUTORIAL

B.1 INTRODUCTION

Computer simulation is an indispensable tool in the quantitative analysis of detection limits. In the past, failure to perform even rudimentary computer simulations of model systems has allowed significant mistakes to go unrecognized and therefore uncorrected for decades. In contrast, computer simulations of model systems are deeply integrated throughout this text and they prove their worth over and over again. Even though it took 40 years for Currie's detection limit schema to be correctly instantiated [1–4], it would have taken even longer without computer simulations performed on carefully defined model systems.

Obviously, computer simulation requires a computer plus software suited to quantitatively evaluating the performance of posited model systems. As it happens, almost any computer, and operating system, is satisfactory. As for software, there are many viable possibilities, for example, Fortran or C++, so the choice of programming language or simulation software is at the user's discretion. But there are horses for courses.

The main simulation software used throughout the text is the author's freely available and fully commented *LightStone* libraries of add-on blocks for *Imagine That, Inc.'s ExtendSim*® (*v. 9.2*) simulation program. In two cases, *Wolfram Research's Mathematica*® (*v. 9*) was also used. All of the simulation executable files, that is, "programs," are freely available. Both *ExtendSim* and *Mathematica* are widely used commercial programs that have been readily available for more than 25 years. Each has freely available demonstration versions that may be used to perform their relevant simulations that appear in the text. These are available at http://www.extendsim.com and http://www.wolfram.com/mathematica. For the free *LightStone* libraries of add-on blocks, and many additional models, reference

Limits of Detection in Chemical Analysis, First Edition. Edward Voigtman.
© 2017 John Wiley & Sons, Inc. Published 2017 by John Wiley & Sons, Inc.
Companion Website: www.wiley.com/go/Voigtman/Limits_of_Detection_in_Chemical_Analysis

files, spreadsheets, videos, screencasts, and so on, go to either www.wiley.com/go/ Voigtman/Limits_of_Detection_in_Chemical_Analysis or https://umass.box.com/v/ LightStone.

B.2 WHY LIGHTSTONE AND EXTENDSIM?

The *LightStone* add-on libraries of simulation blocks require the use of *ExtendSim*, or the older *Extend*® program. *LightStone* augments the libraries of simulation blocks that are provided with *ExtendSim*, making it very easy to model sophisticated electronic and spectroscopic functionalities relevant to many modern chemical measurement systems. Similar to *ExtendSim*, *LightStone* is mature software: the first version, named *Voigt FX*, was developed in March 1990 and released several months later. Both *ExtendSim* and *LightStone* have become increasingly powerful over the years and feature in many of the author's publications since 1991 [1–24]. Those willing to learn more about their quantitative use in performing Monte Carlo simulations are advised to consult the references and the information at the URL given above. For brevity, simulations involving *LightStone*, and either *ExtendSim* or *Extend*, will simply be referred to as *LightStone* simulations.

B.2.1 Simulation Introduction

The *ExtendSim* program, in its "continuous" operation mode, is a time evolution engine. Users specify the following: start time, end time, *either* number of simulation steps *or* time between steps, and total number of simulations to perform. These will be given for each of the simulations discussed as follows. With every simulation step, a set of one or more calculations is performed, depending upon the specific simulation model. The simplest way to see how *LightStone* simulations work is by examining a few examples.

B.2.1.1 Simple Model 1 This simulation model, and the three that follow, all have arbitrary starting time = 0, arbitrary end time = 1, number of steps = 1 million, and number of simulations = 10. The histogram blocks all have 400 bins and differ only in their bin ranges, which will be clearly apparent. In Fig. B.1, the simulation model (that is, program) is designed to perform the following calculations for each of the 10 million total simulation steps:

1. Compute $r(x)$ for the user-specified values of x, β, and α.
2. Generate a unit normal random (really pseudorandom) variate.
3. Add $r(x)$ and the random variate together.
4. Place their sum into a histogram block, for construction of a histogram.
5. Calculate the sample mean and sample standard deviation after the last simulation ends.

The resulting histogram is shown in the figure, along with the obtained sample mean, $\bar{r}(x)$ and sample standard deviation, s. With $x = 3$, $\beta = 2$, and $\alpha = 1$, as shown,

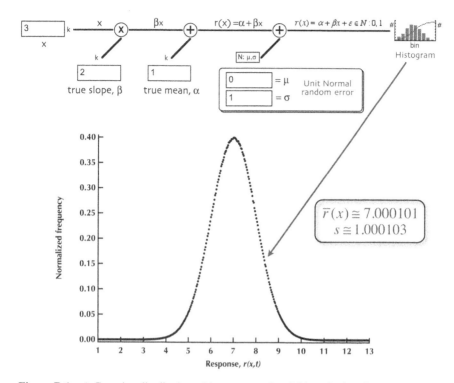

Figure B.1 A Gaussian distribution with a computed variable as its location "parameter."

$r(3) = 7$, which is then simply the constant population parameter of the normal variate. In other words, 10 million *i.i.d.* samples have been generated from N:7, 1, and binned into the 400 histogram bins that span the range from 1 to 13. The obtained sample test statistics are in excellent agreement with the expected values.

Note that the value of x, and therefore $r(x)$, stayed constant throughout all the simulations. More generally, however, it could have been replaced with a time-varying function, for example, a sine wave or temporal waveform. In such a case, the location "parameter" of the normal distribution would actually be a time-dependent variable, that is, $r(x(t))$.

B.2.1.2 Simple Model 2 This model differs from the first model by replacing the $r(x)$ function by an independent Gaussian variate. It may be viewed as a further extension of the possibility mentioned in the previous paragraph; the location "parameter" of a Gaussian distribution is, itself, another independent Gaussian variate. A *compound normal variate (CNV)* is a variate in which either (or both) of a Gaussian variate's parameters are, themselves, random variates [25, p. 163]. In Fig. B.2, we show model 2, with the indicated parameters.

The model in Fig. B.2 clearly meets the definition of a CNV. But it is also obviously just the *sum* of two independent Gaussian variates. Hence, it is a Gaussian variate

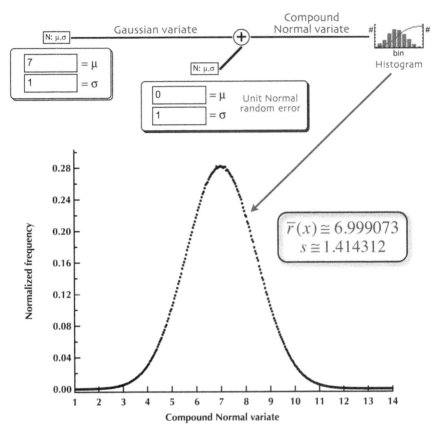

Figure B.2   A Gaussian distribution with an independent Gaussian variate as its location "parameter."

with population mean equal to the sum of the two individual population means, that is, expectation value equal to the sum of the two individual expectation values. Thus, $\mu_{sum} = 7 + 0 = 7$.

The variance of the sum is the sum of the variances: $\sigma^2_{sum} = 1^2 + 1^2 = 2$. Consequently, the CNV is distributed as $N{:}7, 2^{1/2}$. Note that the sample test statistics given in Fig. B.2 are in excellent agreement with $\mu_{sum} = 7$ and $\sigma_{sum} = 2^{1/2}$. This model is particularly relevant to the discussion in Chapter 8, where $N{:}\hat{\alpha}, \sigma_0$ is compound normal, and also normal, because $\hat{\alpha} \sim N{:}\alpha, \sigma_a$, where σ_a is the population standard error of the blank.

B.2.1.3   Simple Model 3   Sample standard deviations, s, are χ distributed random variates, as discussed in Appendix A. Model 3, shown in Fig. B.3, computes 10 million independent s variates, with three *i.i.d.* samples used for each standard deviation calculation. The histogram of s results is shown as follows, together with eqn A.15 (with $v \equiv 2$ and $\sigma \equiv 1$).

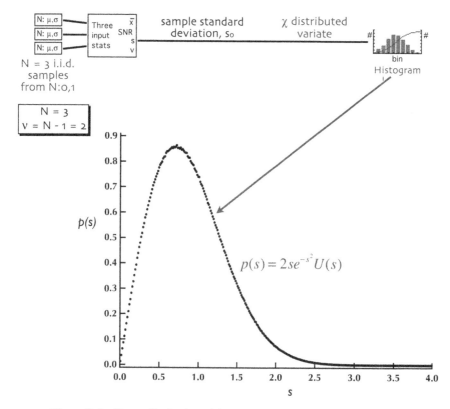

Figure B.3   The χ distribution of the sample standard deviation, for $v = 2$.

This distribution is quite asymmetric, tailing to the high side. Substituting $p(s)$ into eqn A.24 results in $E[s] = c_4(2) = \Gamma[1.5] = \pi^{1/2}/2 \cong 0.886227$, since $\sigma \equiv 1$. Thus, s is a negatively biased point estimate of σ, as is well known. The simulation estimate of $E[s]$, from the model in Fig. B.3, is 0.8863.

B.2.1.4   Simple Model 4   Model 4 is a hybrid of models 2 and 3: the χ distributed random variate from model 3 is used in place of the normal variate, that is, $N{:}7, 1$, at the left of model 2. The result is a CNV that is *not* normally distributed, as shown in Fig. B.4.

This compound variate is not Gaussian: it has a noticeable tail on the high side. Yet thanks to the central limit theorem, it is not highly non-Gaussian. The expectation value of the CNV should be 0.8862, that is, $E[s] + 0$. The model in Fig. B.4 gave 0.8861. Using eqn A.32, the population variance of s, denoted by $\sigma_s^2(v)$, is $\sigma_s^2(2) = \sigma^2[1 - c_4^2(2)] \cong 0.214602$, since $v = 2$. Therefore, $\sigma_{\text{CNV}} = (\sigma_s^2(2) + 1^2)^{1/2} \approx 1.102$, in good agreement with the value of 1.103 shown in Fig. B.4.

B.2.1.5   Simple Model 5   In many of the models in the text, either two variates need to be compared or a variate must be compared with a constant. A typical scenario

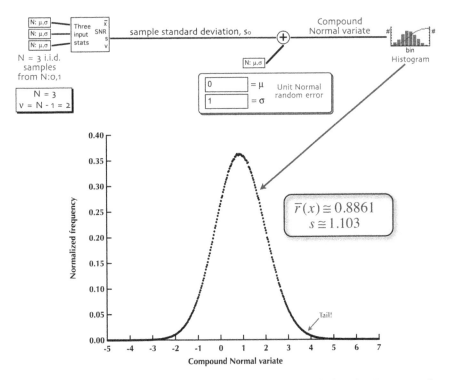

Figure B.4   A Gaussian distribution with a χ variate ($v = 2$) as its location "parameter."

involves millions of such pairwise comparisons, one comparison per simulation step. These comparisons are performed using a comparator block followed by an accumulator block and then a final block to record the results. An example illustrating comparison of a variate with a constant is shown in Fig. B.5.

In Fig. B.5, the distribution block at the left generates *i.i.d.* central *t* variates, with user-specified degrees of freedom, v. As shown, $v = 4$. With each simulation step, a new *t* variate is presented to the "+" input of the comparator. At the same time, a constant value, the critical *t* value, is presented to the comparator's "−" input. As in

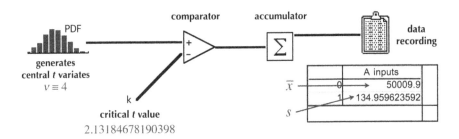

Figure B.5   A simulation model illustrating how the comparator and accumulator work.

Fig. B.5 in Appendix A, the critical t value in the model was chosen for $v = 4$ and 5% upper tail area of the central t distribution.

On any simulation step, if the "+" input is higher than the "−" input, the comparator's output will be 1 and this is added to a running sum in the accumulator block. If the "+" input is lower than the "−" input, the comparator's output will be 0, so the running sum in unchanged. Therefore, at the end of the simulation, the total sum is simply the number of times that t variates exceeded the critical t value, that is, these were the upper tail t variates. For a single simulation of 1 million steps, using the indicated critical t value, the expected number of upper tail t variates is 50 000, that is, 5% of 1 million. By running a set of 10 simulations, resetting the accumulator before each simulation, it is possible to determine an average number of upper tail t variates, $\bar{x}$, and a corresponding standard deviation, s. As shown in Fig. B.5, the obtained average probability of upper tail t variates was 5.001%, with a standard deviation of 0.0135%.

B.3   QUAM EXAMPLE A2

The following sample problem is directly based on Example A2 in QUAM [26, pp. 43–50, 112–114]. It is recommended that this problem be set aside, at least until the indicated QUAM pages have been read.

Consider the common laboratory practice of standardizing a freshly prepared solution of sodium hydroxide (NaOH) against primary titrimetric standard potassium hydrogen phthalate (KHP). The experimental procedure involves four steps:

1. Oven-dried KHP is weighed by difference.
2. The NaOH solution is prepared to roughly the desired concentration.
3. The KHP is dissolved and all of it is then titrated against the NaOH solution.
4. The molar concentration of NaOH is computed from eqn B.1, as follows:

$$c_{NaOH} = \frac{rep}{V_T} \times \frac{(m_{KHP+tare} - m_{tare}) \cdot P_{KHP}}{M_{KHP}} \times \frac{1000\,mL}{L} \qquad (B.1)$$

where c_{NaOH} = molar NaOH concentration (M), $m_{KHP+tare}$ = mass of KHP plus tare (g), m_{tare} = mass of tare (g), V_T = volume of NaOH solution used (mL), rep = repeatability, P_{KHP} = purity of KHP, M_{KHP} = molar mass of KHP (g/mole).

With the exception of the conversion factor, every quantity on the right-hand side of eqn B.1 has uncertainty; each is a random sample from some distribution, known or assumed. Therefore, the computed NaOH concentration has uncertainty, which may be estimated from the uncertainties in the constituent measured quantities. But the uncertainties are unequal: from largest to smallest, they are due to V_T, rep, net mass, P_{KHP} and M_{KHP} [26, p. 48, Fig. A2.9], and this is reflected in the way eqn B.1 is written. Thus, any proposed minimization of total uncertainty would start with the rep/V_T factor in eqn B.1.

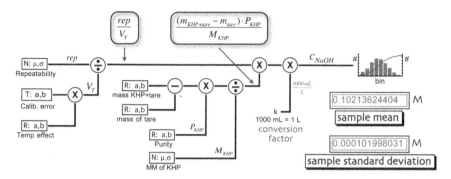

Figure B.6   A Monte Carlo simulation model used to estimate the uncertainty in c_{NaOH}.

It is easy to perform Monte Carlo computer simulations to estimate the uncertainty in c_{NaOH} because all of the needed numerical quantities and distribution assumptions are given in Example A2 [26, pp. 43–50, 112–114]. The simulation model is shown in Fig. B.6.

As shown in Fig. B.6, the mass, temperature effect, and purity values were assumed to be rectangularly (i.e., uniformly) distributed, the volumetric calibration error was assumed to be triangularly distributed, and the repeatability and KHP molar mass were assumed to be Gaussian distributed. Obviously, it is infeasible to compute the analytical distribution function of c_{NaOH} values. But the Monte Carlo simulation makes quick work of the problem, as shown in Fig. B.7.

From the histogram data in Fig. B.7, the 95% CI is 0.10214 ± 0.00020 M and the concentration uncertainty is 102 μM, rather than 87 μM [26, p. 114, Table E3.3]. The latter value is an error, due to inadvertent neglect of the repeatability and purity factors in eqn B.1, and is curious given that these factors had been identified as the second and fourth largest sources of uncertainty [26, p. 48, Fig. A2.9].

The same error also invalidated Fig. E3.3 in QUAM [26, p. 114], where the volumetric calibration error was assumed to be rectangularly distributed. The correct distribution, shown in Fig. B.8, is only slightly "flattened" in appearance, with uncertainty 122 μM, not 110 μM. From Fig. B.8, the 95% CI for the NaOH concentration is from 0.10190 to 0.10237 M and the mean concentration is 0.10214 M, as for the triangular distribution. Since the 95% CIs are almost the same, the volumetric calibration error's PDF is not critically important in this example.

B.4   MANY ADDITIONAL MODELS

One major advantage of the *LightStone* and *ExtendSim* models used in the text is that they are visually self-documenting: the "picture" is the program itself. In other words, the simulation models are essentially digital instantiations of modernized analog computer computations and block diagrams. This greatly facilitates understanding the purpose of the model, and, with a little experience, the models are

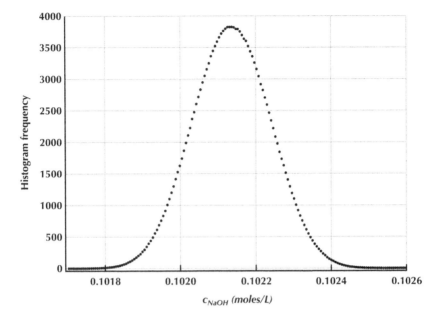

Figure B.7   Histogram of 10 million concentrations from the model in Fig. B.6.

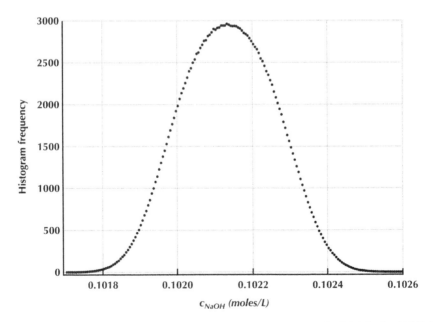

Figure B.8   The corrected 10 million concentrations histogram for Fig. E3.3 in QUAM [26, p. 114].

very easily understood. As well, the *LightStone* website provides many additional models, involving a number of topics, that may be freely downloaded and run. Some of these may be beneficial in understanding the simulations in the text or they may be of interest for other reasons, for example, in teaching.

REFERENCES

1. E. Voigtman, "Limits of detection and decision. Part 1", *Spectrochim. Acta, B* **63** (2008) 115–128.

2. E. Voigtman, "Limits of detection and decision. Part 2", *Spectrochim. Acta, B* **63** (2008) 129–141.

3. E. Voigtman, "Limits of detection and decision. Part 3", *Spectrochim. Acta, B* **63** (2008) 142–153.

4. E. Voigtman, "Limits of detection and decision. Part 4", *Spectrochim. Acta, B* **63** (2008) 154–165.

5. E. Voigtman, "Gated peak integration versus peak detection in white noise", *Appl. Spectrosc.* **45** (1991) 237–241.

6. E. Voigtman, "Computer simulations in spectrometry", *Anal. Chim. Acta* **246** (1991) 9–22.

7. E. Voigtman, "Mueller calculus spectrometric simulations of thermal lensing", *Appl. Spectrosc.* **45** (1991) 890–899.

8. E. Voigtman, "Effect of source 1/f noise on optical polarimeter performance", *Anal. Chem.* **64** (1992) 2590–2595.

9. U. Kale, E. Voigtman, "Gated integration of transient signals in 1/f noise", *Appl. Spectrosc.*, **46** (1992) 1636–1643.

10. E. Voigtman, "Computer modeling of fluorescence dip spectroscopy with pulsed laser excitation", *Spectrochim. Acta, B* **47B** (1992) E1549–E1565.

11. E. Voigtman, "Spectrophotometric precision: a case by case simulation study", *Anal. Instrum.* **21** (1993) 43–62.

12. E. Voigtman, "Block diagram computer simulation of analytical instruments", *Anal. Chem.* **65** (1993) 1029A–1035A.

13. E. Voigtman, "Computer experimentation in and teaching of modern instrumental techniques using circular dichroism as an example", *Anal. Chim. Acta* **283** (1993) 559–572.

14. E. Voigtman, A.I. Yuzefovsky, R.G. Michel, "Stray light effects in Zeeman atomic absorption spectrometry", *Spectrochim. Acta, B* **49B** (1994) 1629–1641.

15. U. Kale, E. Voigtman, "Signal and noise analysis of non-modulated polarimeters using Mueller calculus simulations", *Analyst* **120** (1995) 325–330.

16. U. Kale, E. Voigtman, "Signal processing of transient atomic absorption signals", *Spectrochim. Acta, B*, **50B** (1995) E1531–E1541.

17. E. Voigtman, "Comparison of signal-to-noise ratios", *Anal. Chem.* **69** (1997) 226–234.

18. D. Montville, E. Voigtman, "Statistical properties of limit of detection test statistics", *Talanta* **59** (2003) 461–476.

19. E. Voigtman, "Comparison of signal-to-noise ratios. Part 2", *MATCH Commun. Math. Comput. Chem.* **60/2** (2008) 333–348.

20. E. Voigtman, K.T. Abraham, "Statistical behavior of ten million experimental detection limits", *Spectrochim. Acta, B* **66** (2011) 105–113.

21. E. Voigtman, K.T. Abraham, "True detection limits in an experimental linearly heteroscedastic system. Part 1", *Spectrochim. Acta, B* **66** (2011) 822–827.

22. E. Voigtman, K.T. Abraham, "True detection limits in an experimental linearly heteroscedastic system. Part 2", *Spectrochim. Acta, B* **66** (2011) 828–833.

23. J. Carlson, A. Wysoczanski, E. Voigtman, "Limits of quantitation – yet another suggestion", *Spectrochim. Acta, B* **96** (2014) 69–73.

24. A. Wysoczanski, E. Voigtman, "Receiver operating characteristic-curve limits of detection", *Spectrochim. Acta, B*, **100** (2014) 70–77.

25. N.L. Johnson, S. Kotz, N. Balakrishnan, *Continuous Univariate Distributions, Volume 1*, 2nd Ed., John Wiley and Sons, NY, ©1994.

26. EURACHEM/CITAC Guide, "Quantifying Uncertainty in Analytical Measurement", 3rd Ed., 2012. Known as "QUAM" from the upper-case letters in the title.

APPENDIX C

BLANK SUBTRACTION AND THE $\eta^{1/2}$ FACTOR

C.1 INITIAL ASSUMPTIONS AND NOTATION

The fundamental assumptions, notations, and general theory of Currie method decision levels and detection limits are as previously described in considerable detail, beginning in Chapter 7. This is supplemented with information in the appendices, particularly Appendix A.

C.2 THE BEST CASE SCENARIO

From Fig. 7.2, a future blank is simply a random sample from $N{:}\alpha, \sigma_0$. With $p \equiv 0.05$, a future blank has only 5% probability of exceeding a *constant* response domain decision level, R_C, defined as $\alpha + z_p\sigma_0$. The difference between R_C and the center of the blank's probability density function (PDF), α, is a *constant*: $R_C - \alpha = z_p\sigma_0$. This difference is the net response and, indeed, the *constant* net response domain decision level, Y_C, is defined as $R_C - \alpha$. Hence, $Y_C = z_p\sigma_0$, and the offset parameter, α, has been harmlessly removed; it is usually the case that α does not carry any fundamentally significant information about analyte content. Note that subtracting α only shifts the zero point of the $r(x)$ axis. Transformation to the content domain follows by dividing Y_C by β, if it is known, or by an unbiased estimate, $\hat{\beta}$, if β is unknown. This yields $X_C = Y_C/\beta$ or $x_C = Y_C/\hat{\beta}$, respectively.

C.2.1 What Is the Problem?

Subtraction of α from a response yields a net response, as per eqn 3.3. However, if α is unknown, then an unbiased estimate, $\hat{\alpha}$, is used to convert responses to net responses.

Limits of Detection in Chemical Analysis, First Edition. Edward Voigtman.
© 2017 John Wiley & Sons, Inc. Published 2017 by John Wiley & Sons, Inc.
Companion Website: www.wiley.com/go/Voigtman/Limits_of_Detection_in_Chemical_Analysis

This is as per eqn 3.4, repeated as follows:

$$y(x) \equiv r(x) - \hat{\alpha} \tag{C.1}$$

Since $\hat{\alpha}$ is unbiased, its expectation value, $E[\hat{\alpha}]$, equals α. Hence, from eqn C.1,

$$E[y(x)] = E[r(x) - \hat{\alpha}] = E[r(x)] - E[\hat{\alpha}] = E[\alpha + \beta x] - \alpha = \beta x \tag{C.2}$$

since $E[\alpha + \beta x] = E[\alpha] + E[\beta x] = \alpha + \beta x$.

Thus, a future net blank, defined as a future blank $- \hat{\alpha}$, would be distributed about 0, since $x = 0 = X_0$ for any blank. If $\hat{\alpha}$ is normally distributed, a condition usually met in practice, then $y(x)$ is normally distributed as well, and this is also true, as a special case, for the future net blank. This result follows from the fact that eqn C.1 is the difference between two independent Gaussian variates, and any linear combination of independent Gaussian variates is Gaussian. However, it would *not* be true that $y(x) \sim N{:}\beta x, \sigma_0$, because subtraction of $\hat{\alpha}$ adds error, that is, noise. Similarly, it would *not* be true that a future net blank was distributed as $N{:}0, \sigma_0$.

C.2.2   What Are the Gaussian Distributions of $y(x)$ and a Future Net Blank?

It was assumed above that $\hat{\alpha}$ was a normally distributed, unbiased estimate of α. Hence,

$$\hat{\alpha} \sim N{:}\alpha, \sigma_{\hat{\alpha}} \tag{C.3}$$

where $\sigma_{\hat{\alpha}}$ is the population standard deviation of the Gaussian PDF of $\hat{\alpha}$. As noted above,

$$r_{\text{future blank}} \sim N{:}\alpha, \sigma_0 \tag{C.4}$$

so that

$$y_{\text{future net blank}} \equiv (r_{\text{future blank}} - \hat{\alpha}) \sim N{:}0, \sigma_d \tag{C.5}$$

where the *population variance of the difference*, σ_d^2, is the *pooled population variance*:

$$\sigma_d^2 \equiv \sigma_0^2 + \sigma_{\hat{\alpha}}^2 \tag{C.6}$$

C.2.3   A Small Detail

A slight generalization occurs if it was desired to compare the mean of M future *i.i.d.* blank replicates, $\bar{r}_{\text{future blanks}}$, with a decision level. In this case,

$$\bar{r}_{\text{future blanks}} \sim N{:}\alpha, \sigma_0/M^{1/2} \tag{C.7}$$

so that eqn C.6 generalizes to

$$\sigma_d^2 = \frac{\sigma_0^2}{M} + \sigma_{\hat{\alpha}}^2 \tag{C.8}$$

C.3 HOW MAY $\hat{\alpha}$ BE ESTIMATED?

There are several commonly used methods to obtain an experimental value for $\hat{\alpha}$. Assuming the noise is homoscedastic additive Gaussian white noise (AGWN), both produce estimates that are unbiased and normally distributed, and they are discussed as Method 1 and Method 2 in the following.

Method 1: Calibration Curves Processed with Ordinary Least Squares (OLS)

In this method, a value for $\hat{\alpha}$ is obtained by preparing an experimental calibration curve from N standards, performing OLS, and then taking $\hat{\alpha}$ as the intercept, a, of the OLS calibration curve. In this case, $\sigma_{\hat{\alpha}} \equiv \sigma_a$:

$$\sigma_{\hat{\alpha}}^2 \equiv \sigma_a^2 = \sigma_0^2 \left(\frac{1}{N} + \frac{\overline{X}^2}{S_{XX}} \right) \tag{C.9}$$

where σ_a is the *population standard error* of the intercept, $\overline{X}$ is the mean of the N standards (for $N > 2$), and S_{XX} is the sum of the squared differences: $S_{XX} \equiv \sum_{i=1}^{N} (X_i - \overline{X})^2$, with X_i being the ith standard. Degenerate standards are allowed, that is, multiple standards may have the same values. The number of degrees of freedom, v, is $N - 2$. Note that the *sample standard error about regression, s_r,* is a point estimate of σ_0, also with $v \equiv N - 2$.

Method 2: Replicate Measurements of An Analytical Blank or Standard

In this method, $\hat{\alpha}$ is taken as the sample mean of M_0 *i.i.d.* replicate measurements of *any* standard or the blank. In this case,

$$\sigma_{\hat{\alpha}}^2 \equiv \frac{\sigma_0^2}{M_0} \tag{C.10}$$

with $v \equiv M_0 - 1$. Then, the sample standard deviation, s_0, is the point estimate of σ_0, also with $v \equiv M_0 - 1$. The possibility of using any standard is due to the assumption of homoscedasticity.

In practice, this assumption must be checked by performing an F test on the experimental results from the two standards having the largest and smallest sample variances. The null hypothesis is that the relevant population variances are the same, that is, that the noise is homoscedastic. Hence "passing" the F test means the null hypothesis is not rejected. Even so, it is prudent to pick a standard in the general vicinity of the limit of detection.

C.4 DEFINITION OF η

From the above, $\sigma_{\hat{a}}^2 \propto \sigma_0^2$ and, hence, $\sigma_d^2 \propto \sigma_0^2$. It is then convenient to define a proportionality factor, η, as

$$\sigma_d^2 \equiv \eta \sigma_0^2 \tag{C.11}$$

so that $\eta^{1/2}$ is simply the scale factor that relates σ_d and σ_0:

$$\sigma_d = \eta^{1/2} \sigma_0 \tag{C.12}$$

The $\eta^{1/2}$ factor is called the "blank subtraction" factor [1–4], for obvious reasons. It may alternatively be called the "noise pooling" or "blank referencing" factor. Note that if α were to be estimated via a method that produced a non-Gaussian distributed $\hat{\alpha}$, then $\eta^{1/2}$ would not be defined and either eqn C.6 or eqn C.8 would be required. For the moment, continue assuming AGWN, with either Method 1 or Method 2 used to estimate $\hat{\alpha}$. Then, if eqn C.9 is substituted into eqn C.8,

$$\eta^{1/2} \equiv \left(\frac{1}{M} + \frac{1}{N} + \frac{\overline{X}^2}{S_{XX}} \right)^{1/2} \tag{C.13}$$

and if eqn C.10 is substituted into eqn C.8,

$$\eta^{1/2} \equiv \left[\frac{1}{M} + \frac{1}{M_0} \right]^{1/2} \tag{C.14}$$

More complicated variants of $\eta^{1/2}$ are possible, for example, if weighted least squares is used [3].

C.5 IMPORTANCE OF η

The $\eta^{1/2}$ factor is necessary whenever α is unknown and must be estimated. With the canonical definition of $M \equiv 1$, $\eta^{1/2} \geq 1$, with equality obtaining *only if* α is used for blank subtraction. As a practical matter, $\eta^{1/2}$ may be only slightly greater than unity if v is sufficiently high and $M \equiv 1$. In any event, note that $\eta^{1/2}$ is errorless, since it depends only on experimental design protocol, that is, how α is to be estimated. This means that $\eta^{1/2}$ may be computed *after* an experiment is designed, but *before* it is performed. If necessary, it may be modified or refined subsequently, given that experiments do not always go as planned.

Finally, note that "traditional" detection limits, of the form LOD $\equiv k s_{\text{blank}}/$slope, *neither specify nor control* the probability of false positives because k is really a product of $\eta^{1/2}$ and a relevant critical z or t value. But $\eta^{1/2}$ depends on the experimental design protocol.

As an example, suppose Method 1 was used, to determine a detection limit, and then a *subsequent* detection limit experiment was performed using either a different number of standards or with one (or more) standards different than those in the first experiment. This would change the value of $\eta^{1/2}$, perhaps by only a small amount. Nevertheless, if k was a *defined* constant, for example, 3, the *effective* critical value (z or t) would be changed, thereby giving *incorrect specification and control* of the probability of false positives. This is one reason why "traditional" detection limits are not recommended.

C.6 NONNORMAL ESTIMATES OF α

If $\hat{\alpha}$ is not normally distributed, then either eqn C.6 or C.8 still applies, but it becomes necessary to know the PDF of $y_{\text{future net blank}}$, or have a usable histogram approximation of it. This might happen if $\hat{\alpha}$ were estimated in unorthodox fashion. It will certainly happen if the noise is non-Gaussian, for example, uniformly distributed, binary, triangular, Poisson or Laplacian. In either case, the total variance is insufficient to the task of specifying a decision level, either y_C or Y_C.

To illustrate the issues posed by non-Gaussian noise, consider additive uniformly distributed white noise (AUWN). The uniform distribution is denoted by $U{:}a, b$, where a and b are the lower and upper range limits, respectively. It is also called the rectangular distribution [5], denoted by $R{:}a, b$. The population mean, μ, is simply $(a+b)/2$, while the population standard deviation, σ_0, is $(b-a)/2\sqrt{3}$ [5, p. 137]. Suppose the blank's true value, α, is unity and the noise is zero mean AUWN, with $\sigma_0 \equiv 0.1$. Then the noise on the blank is in the range $\alpha \pm 3^{1/2}\sigma_0$.

C.6.1 Both Parameters Known

If σ_0 and α are known, then

$$R_C = \alpha + (1 - 2p)3^{1/2}\sigma_0 \equiv \alpha + Y_C \tag{C.15}$$

and

$$R_D = R_C + (1 - 2q)3^{1/2}\sigma_0 \equiv \alpha + Y_D \tag{C.16}$$

with $M \equiv 1$ and both p and $q < 0.5$. The content domain expressions follow immediately: $X_C = Y_C/\beta$, $x_C = Y_C/b$, $X_D = Y_D/\beta$, and $x_D = Y_D/b$. If $p = q$, $Y_D = 2Y_C$. The expressions in eqns C.15 and C.16 are tested as shown in Fig. C.1.

As shown in the data tables in Fig. C.1, eqns C.15 and C.16 work correctly; the obtained probabilities of false positives and false negatives are statistically equivalent to p and q, respectively.

C.6.2 If Only σ_0 Is Known

If σ_0 is known, but α is unknown, a simple computer simulation is used to estimate Y_D. First, α is arbitrarily defined as a dummy value, such as 0 or 1. An unbiased estimate of α, $\hat{\alpha}$, may be obtained by averaging N *i.i.d.* blank variates. The simulation, with $\sigma_0 = 0.1$, $\alpha = 1$ and $N = 4$, is shown in Fig. C.2.

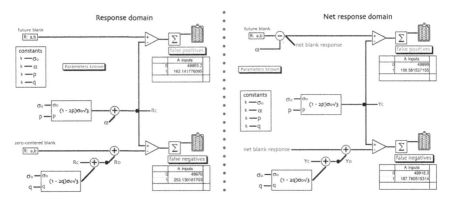

Figure C.1   Simulation models testing R_C, R_D, Y_C, and Y_D when both σ_0 and α are known.

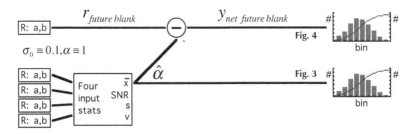

Figure C.2   Simulation program that generates histograms of 10 million $\hat{\alpha}$ and $y_{\text{future net blank}}$ variates.

The histogram of 10 million $\hat{\alpha}$ variates is shown in Fig. C.3. The histogram is approximately Gaussian near the center, but definitely not Gaussian in the tails, due to the finite range of the uniform distribution. It is, in fact, an approximation of the underlying Bates distribution [6] and its shape varies with v. As v increases, the histograms and Bates PDFs become narrower, more symmetric and more like Gaussians, thanks to the central limit theorem.

Concurrently with generation of the $\hat{\alpha}$ variates, 10 million $y_{\text{future net blank}}$ variates, as per eqn C.5, are generated and binned into a histogram. This is shown in Fig. C.4. The distribution is centered on zero, as it must be, and it is neither uniform nor Gaussian. It was acceptable to arbitrarily assume $\alpha \equiv 1$ because the same distribution of differences would have been obtained regardless of the assumed α value.

For every value of v, there is a unique PDF for $y_{\text{future net blank}}$ and Y_C must be estimated empirically as the value that separates the desired upper tail area, p, from everything below. There is no usable analog of the central t distribution, so there is no use of critical t values. There is also no use of critical z values and $\eta^{1/2}$ is not usefully definable. From the CDF of the histogram, the Y_C value is found to be approximately 0.1793, and, since $p = q$, $Y_D = 2Y_C$. In the response domain, $r_C = \hat{\alpha} + Y_C$ and $r_D = r_C + (1 - 2q)3^{1/2}\sigma_0$.

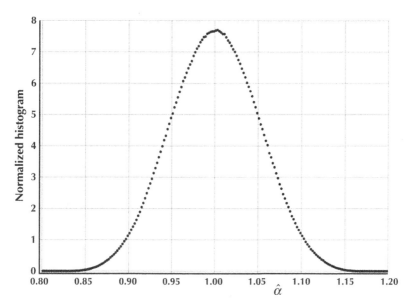

Figure C.3 Histogram of 10 million $\hat{\alpha}$ variates, each a sample mean of 4 *i.i.d.* uniformly distributed variates.

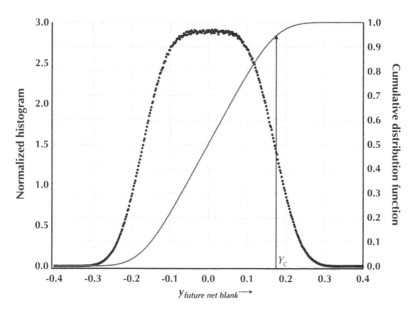

Figure C.4 Histogram of 10 million $y_{future\ net\ blank}$ variates, the CDF of the histogram, and Y_C (arrow).

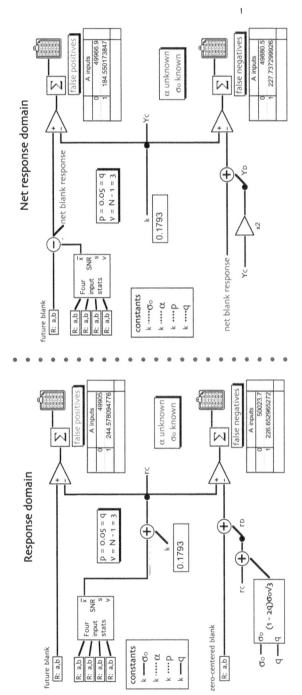

Figure C.5 Simulation models testing r_C, r_D, Y_C, and Y_D when only σ_0 is known.

The simulation models that test r_C, r_D, and the estimates of Y_C and Y_D, are shown in Fig. C.5. As shown in the data tables in Fig. C.5, the obtained probabilities of false positives and false negatives are statistically equivalent to p and q, respectively.

C.6.3   If Only α Is Known

If α is known and σ_0 is unknown, then the following procedure may be used to find $\hat{\sigma}_0$, the estimate of σ_0. First, obtain N *i.i.d.* blank replicates and sort them from r_{low} to r_{high}. Next, compute the *uniformly minimum variance unbiased* (UMVU) *estimators* [7]:

$$\hat{a} = (Nr_{low} - r_{high})/N - 1 \tag{C.17}$$

and

$$\hat{b} = (Nr_{high} - r_{low})/N - 1 \tag{C.18}$$

Then

$$\hat{\sigma}_0 = (\hat{b} - \hat{a})/2\sqrt{3} \tag{C.19}$$

In principle, α and $\hat{\sigma}_0$ may then be used in eqns C.15 and C.16. However, the simulation models in Fig. C.1 demonstrate that the uncertainty in $\hat{\sigma}_0$ must be less than 0.1% in order to obtain probabilities of false positives and false negatives that are statistically equivalent to p and q, respectively. This effectively means that N must be large.

To test the assertion that N must be large, 101 *i.i.d.* blank replicates were obtained and processed to yield $\hat{a} \cong 0.825253$, $\hat{b} \cong 1.173245$, and $\hat{\sigma}_0 \cong 0.100457$, with $\alpha = 1$ and $\sigma_0 = 0.1$ being the true values. In the response domain, using α and $\hat{\sigma}_0$ gave poor results: $4.793 \pm 0.020\%$ false positives and $4.811 \pm 0.026\%$ false negatives. In the net response domain, the results were similar: $4.789 \pm 0.017\%$ false positives and $4.798 \pm 0.029\%$ false negatives. Thus, even though $\hat{\sigma}_0 \cong \sigma_0$, $N = 101$ was too small, suggesting the receiver operating characteristic (ROC) method (see Chapter 6) as a possible alternative.

C.6.4   If Neither Parameter Is Known

This is a combination of the two previous scenarios. Aside from the ROC method, one possibility is to obtain $N \gg 100$ *i.i.d.* experimental blank replicates, compute $\hat{a}$, $\hat{b}$, $\hat{\sigma}_0$, and $\hat{\alpha} = (\hat{a} + \hat{b})/2$, and then substitute $\hat{\alpha}$ and $\hat{\sigma}_0$ into eqns C.15 and C.16. Obviously, this will not work unless N is large.

REFERENCES

1. E. Voigtman, "Limits of detection and decision. Part 1", *Spectrochim. Acta, B* **63** (2008) 115–128.

2. E. Voigtman, "Limits of detection and decision. Part 2", *Spectrochim. Acta, B* **63** (2008) 129–141.

3. E. Voigtman, "Limits of detection and decision. Part 3", *Spectrochim. Acta, B* **63** (2008) 142–153.

4. E. Voigtman, "Limits of detection and decision. Part 4", *Spectrochim. Acta, B* **63** (2008) 154–165.

5. M. Evans, N. Hastings, B. Peacock, *Statistical Distributions*, 2nd Ed., John Wiley and Sons, NY, ©1993.

6. G.E. Bates, "Joint distributions of time intervals for the occurrence of successive accidents in a generalized Polya urn scheme", *Ann. Math. Stat.* **26** (1955) 705–720.

7. R.C.H. Cheng, N.A.K. Amin, "Estimating parameters in continuous univariate distributions with a shifted origin", *J. Royal Stat. Soc. Ser. B* **45** (1983) 394–403.

APPENDIX D

PROBABILITY DENSITY FUNCTIONS FOR DETECTION LIMITS

D.1 THE MODEL CMS

The model chemical measurement system (CMS) is as given in Chapter 7. It is further assumed that σ_0, β, and α are unknown, as in Chapter 14, so ordinary least squares (OLS) is performed on N data pairs of the form X_i, r_i, with only one r_i measurement per X_i standard, but no restriction on the number of times a standard may be repeated. Thus, the OLS calibration curve is $\hat{r}(x) = a + bx$ and all population parameters, sample test statistics, and associated quantities, for example, S_{XX}, are exactly as given in Table 10.1.

D.2 HOW THE NONCENTRAL t DISTRIBUTION IS INVOLVED

Performing OLS produces a t_{slope} variate, as discussed in several chapters, for example, Chapter 14. By definition, t_{slope} is the quotient of a Gaussian variate, b, and the χ variate, s_b. Thus,

$$t_{\text{slope}} \equiv \frac{b}{s_b} = \frac{(b/\sigma_b)}{(s_b/\sigma_b)} \tag{D.1}$$

Then, since $b \sim N{:}\beta, \sigma_b$ and $\delta \equiv \beta/\sigma_b$, $(b/\sigma_b) \sim N{:}\beta/\sigma_b, 1$, that is, $(b/\sigma_b) \sim N{:}\delta, 1$. But

$$\frac{s_b}{\sigma_b} = \frac{(s_0/S_{XX}^{1/2})}{(\sigma_0/S_{XX}^{1/2})} = \frac{s_0}{\sigma_0} \sim \sqrt{\chi^2/\nu} \tag{D.2}$$

Consequently [1, p. 509], t_{slope} is distributed in accordance with the noncentral t distribution:

$$t_{\text{slope}} \sim t(u|\nu, \delta) = t(u|N-2, \beta/\sigma_b) \tag{D.3}$$

Limits of Detection in Chemical Analysis, First Edition. Edward Voigtman.
© 2017 John Wiley & Sons, Inc. Published 2017 by John Wiley & Sons, Inc.
Companion Website: www.wiley.com/go/Voigtman/Limits_of_Detection_in_Chemical_Analysis

where the noncentral t distribution is defined in eqn A.20. Note that t_{slope} is the sample test statistic point estimate of δ, and it is positively biased, as per eqn A.21.

Now consider the development of the content domain Currie decision level expression, x_C, given as eqn 14.6:

$$x_C \equiv \frac{t_p \eta^{1/2} s_0}{b} \equiv \frac{t_p \eta^{1/2} s_r}{b} \equiv \frac{t_p \eta^{1/2} s_b S_{XX}^{1/2}}{b} \equiv \frac{t_p \eta^{1/2} S_{XX}^{1/2}}{t_{\text{slope}}} \equiv \frac{c}{t_{\text{slope}}} \qquad (\text{D.4})$$

where $c \equiv t_p \eta^{1/2} S_{XX}^{1/2}$ is a positive constant. Thus, x_C is distributed as the scaled reciprocal of a noncentral t variate. Note that x_D, given by eqn 14.7, functionally differs from x_C only in having a factor of $(t_p + t_q)$ in place of t_p. Hence, x_D is also distributed as the scaled reciprocal of t_{slope}.

D.3   THE PDFs OF x_C AND x_D

Suppose $x \sim p_x(x)$, $y = c/x$, and c is a positive constant. Then $y \sim p_y(y) = (c/y^2) \times p_x(c/y)$ [2]. Applying this result to eqn D.4 yields the *modified noncentral t distribution* of x_C:

$$x_C \sim (t_p \eta^{1/2} S_{XX}^{1/2} / x_C^2) t(t_p \eta^{1/2} S_{XX}^{1/2} / x_C | N - 2, \delta) \qquad (\text{D.5})$$

Similarly, the PDF of x_D is also a modified noncentral t distribution:

$$x_D \sim ((t_p + t_q) \eta^{1/2} S_{XX}^{1/2} / x_D^2) t((t_p + t_q) \eta^{1/2} S_{XX}^{1/2} / x_D | N - 2, \delta) \qquad (\text{D.6})$$

D.4   IMPORTANT COROLLARY RESULTS

First, note that

$$\frac{\sigma_0}{\beta} = \frac{\sigma_r}{\beta} = \frac{\sigma_b S_{XX}^{1/2}}{\beta} = \frac{S_{XX}^{1/2}}{\delta} = \frac{N^{1/2} \sigma_X}{\delta} \qquad (\text{D.7})$$

where σ_X is the population standard deviation of the N *errorless* calibration standards. The $S_{XX}^{1/2}$ factor, equal to $N^{1/2} \sigma_X$, is purely a function of the calibration design, while σ_0 and β are parameters of the CMS. By substitution of eqn D.7 into the X_C and X_D expressions in Chapter 8, it is seen that $X_C = z_p \eta^{1/2} S_{XX}^{1/2} / \delta$ and $X_D = (z_p + z_q) \eta^{1/2} S_{XX}^{1/2} / \delta$. As always, the $\eta^{1/2}$ factor is unity if $M \equiv 1$ and α is used for blank subtraction. Therefore, any accurate estimate of δ immediately allows for accurate estimation of *both X_D and X_C*, since the other factors are errorless constants that are known in advance of the detection limit determinations [3, 4].

Second, as amplitude signal-to-noise ratios (SNRs), and for the operative calibration design value of $S_{XX}^{1/2}$, δ is the theoretical SNR of the CMS and t_{slope} is the corresponding experimental SNR that estimates δ. *Ceteris paribus*, δ and X_D are inversely proportional:

$$X_D = \frac{(z_p + z_q)\eta^{1/2}S_{XX}^{1/2}}{\delta} = \frac{(z_p + z_q)\eta^{1/2}N^{1/2}\sigma_X}{\delta} = \frac{\text{Composite constant}}{\delta} \propto \frac{1}{\delta} \quad (D.8)$$

and similarly for X_C.

Lastly, if δ is *known*, then the $\eta^{1/2}$ factor might potentially be omitted; as per Appendix C, it is ultimately just a technical factor needed when $M > 1$ or α is unknown and must be estimated.

REFERENCES

1. N.L. Johnson, S. Kotz, N. Balakrishnan, *Continuous Univariate Distributions, Volume 2*, 2nd Ed., John Wiley and Sons, New York, ©1995.

2. A. Papoulis, *Probability, Random Variables, and Stochastic Processes*, 2nd Ed., McGraw-Hill, New York, ©1984.

3. E. Voigtman, "Limits of detection and decision. Part 1", *Spectrochim. Acta, B* **63** (2008) 115–128.

4. E. Voigtman, "Limits of detection and decision. Part 2", *Spectrochim. Acta, B* **63** (2008) 129–141.

APPENDIX E

THE HUBAUX AND Vos METHOD

E.1 INTRODUCTION

In 1970, Hubaux and Vos published their method for determining experimental content domain limits of detection [1]. For a model system comprised of a univariate chemical measurement system (CMS), having linear response function and additive Gaussian white noise (AGWN), their method is statistically valid and has been discussed in detail by others, for example, Coleman and Vanatta [2–5]. However, for the assumed model system, the method has numerous disadvantages and needless restrictions relative to the correct instantiation [6, 7] of Currie's 1968 schema, as discussed below.

E.2 THEORY

To understand how the Hubaux and Vos method works, consider Fig. E.1. In Fig. E.1, the ordinary least squares (OLS)-determined calibration curve is shown, along with its flanking prediction interval hyperbola bands. For $\beta > 0$, as assumed in Fig. E.1, the upper prediction band depends on p, while the lower prediction band depends on q [2–5]. Typically both p and q are defined as 0.05, and $s_0 = s_r$, since it is assumed that each standard is measured just once, but degenerate standards are allowed. Thus, $\hat{\sigma}_0 = s_0 = s_r$.

The content domain detection limit, x_D, is the x value directly below where the horizontal arrow from the response domain decision level, r_C, intersects the lower hyperbola. In the net response domain, $y(x) \equiv r(x) - a = bx$. Therefore, y_C may be used in place of r_C to determine x_D, as should be obvious from Fig. E.1. The full set of relevant equations is shown in Fig. E.2.

There are no theoretical expressions on the left side of Fig. E.2 because they simply do not exist in the method. On the right side of Fig. E.2, only the r_C, y_C, and x_D expressions are relevant and they are not treated as estimates of theoretical values. Although it is possible to define x_C, r_D, and y_D in the customary way, that is,

Limits of Detection in Chemical Analysis, First Edition. Edward Voigtman.
© 2017 John Wiley & Sons, Inc. Published 2017 by John Wiley & Sons, Inc.
Companion Website: www.wiley.com/go/Voigtman/Limits_of_Detection_in_Chemical_Analysis

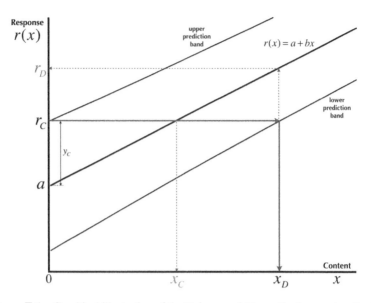

Figure E.1   Graphical illustration of the Hubaux and Vos method, assuming $\beta > 0$.

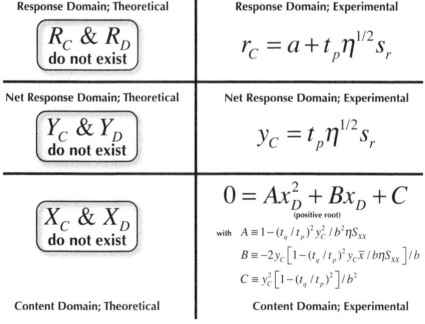

$\eta^{1/2} = 1$ *if ideal blank subtraction & $M \equiv 1$ future blank measurement*

Figure E.2   The equations for the Hubaux and Vos method. If $p = q$, $t_p = t_q$ and $x_D = -B/A$.

$x_C \equiv y_C/b$, $r_D \equiv a + bx_D$, and $y_D \equiv x_Db$, these are either irrelevant in the method or the cause of bias when paired with x_D, r_C, or y_C, respectively [6, 7]. Consequently, decision and detection *cannot* take place in any single domain: the method absolutely requires use of an experimental calibration curve, obtained via regression, to connect the content domain to the other domains. This is a significant disadvantage; in Chapter 6, it was shown that there is no absolute necessity of having a calibration curve. Indeed, all that is required is that the response function be monotonic.

Another serious issue is that, when σ_0 and β are unknown, as is usually the case, they must be estimated by $\hat{\sigma}_0$ and $\hat{\beta}$, respectively. In the Hubaux and Vos method, $\hat{\sigma}_0 \equiv s_r$ and $\hat{\beta} \equiv b$. Then, as a matter of long established and repeatedly validated statistics,

$$x_C = t_p\eta^{1/2}\hat{\sigma}_0/\hat{\beta} = t_p\eta^{1/2}s_r/b = y_C/b \tag{E.1}$$

This is simply the traditional detection limit expression, with $k \equiv t_p\eta^{1/2}$, doing double duty as both the decision level and the detection limit. Yet, in the Hubaux and Vos method, x_C has no place, despite its incontestable validity as a decision level.

E.3   COMPARISON OF THE TWO METHODS

A comparison of the two methods is presented in Table E.1. From Table E.1, it is evident that the Hubaux and Vos method is inferior in every row. On the plus side, their method has a solitary advantage: its prediction interval hyperbolas, which result from the OLS slope and intercept being negatively correlated, typically result in slightly lower detection limits, for example, 2% or so [8]. This improvement is insignificant compared to the typically broad widths of x_D distributions, for example, Fig. 15.9. Furthermore, if the prediction interval hyperbolas in Fig. E.1 were replaced by straight lines, having the same ordinate axis intercepts as the hyperbolas they replaced and parallel to the regression line, then the method would transform into the correct method [6, 7], with equations exactly as shown in Fig. 16.3. This is effectively what happens when $p = q$, and x_D is set equal to $2x_C$, as a simplifying "approximation" in the Hubaux and Vos method.

E.4   A COMPREHENSIVE EXAMPLE

Vogelgesang and Hädrich, in an exceptionally thorough paper, provide a comprehensive example calculation [8, Table 6] that nicely illustrates this. They implicitly used the Hubaux and Vos method to compute, in our notation, r_C, y_C, x_C, and two separate estimates of x_D, that is,

$$x_D = (t_p + t_q)\eta^{1/2}s_r/b \tag{E.2}$$

and the Hubaux and Vos x_D. With $p \equiv 0.05 \equiv q$, $r_C = 3922.50$, $y_C = 2167.93$, $x_C = 8.7_3$ mg/kg, and $x_D = 2x_C = 17.4_5$ µg/kg. Using iteration, Vogelgesang and Hädrich found the Hubaux and Vos x_D estimate to be about 17.2 µg/kg, and remarked

Table E.1   Method Comparison

Detection Limit Method	Voigtman [6, 7]	Hubaux and Vos [1]
Compatible with independent blanks?	Simple	No, must use regression intercept
Requires experimental calibration curve and OLS?	No	Yes, with prediction interval hyperbolas
Usable purely in *ONE* domain (response, net response, or content)?	Yes	No
Mathematical complexity?	Least possible	More complicated
Heteroscedastic extension?	Relatively easy	Via computer
Analytical detection limit PDFs known?	Yes, for simplest cases	No, only via computer simulation
Facilitates accurate estimation of X_D?	Yes, with bootstrapping	X_D does not exist
Allows response and net response domain detection limits?	Yes: R_D, r_D, Y_D, and y_D	No: x_D only
Allows (or tolerates) use of content domain decision level?	Yes: both X_C and x_C	No: response or net response domain only

that "This resulted in ID values that were up to 2% smaller than $2 \times$ DTC. Since this difference is quite small for the practitioner's purpose, we set: ID $= 2 \times$ DTC" [8, p. 252]. In the quotation, the notational equivalencies are ID $= x_D$ and DTC $= x_C$. As for the Hubaux and Vos x_D estimate they found, it is easily obtained by solving the x_D equation in Fig. E.2, which reduces to $x_D = -B/A$ since $t_p = t_q$. The result is $x_D = 17.2_3$ µg/kg, in agreement with what they reported.

E.5   CONCLUSION

In view of the above, the Hubaux and Vos method should not be used with the assumed model system. It is strongly recommended that Currie's schema, as correctly instantiated and demonstrated throughout this text, be used instead.

REFERENCES

1. A. Hubaux, G. Vos, "Decision and detection limits for linear calibration curves", *Anal. Chem.* **42** (1970) 849–855.
2. D. Coleman, L. Vanatta, "Part 28 – statistically derived detection limits", *Am. Lab.* **39**(20) (2007), 24–27.
3. D. Coleman, L. Vanatta, "Part 29 – statistically derived detection limits (continued)", *Am. Lab.* **40**(3) (2008) 44–46.

4. D. Coleman, L. Vanatta, "Part 30 – statistically derived detection limits (concluded)", *Am. Lab.* **40**(12) (2008), 34–37.

5. D. Coleman, L. Vanatta, "Part 34 – detection-limit summary", *Am. Lab.* **41**(6) (2009) 50–52.

6. E. Voigtman, "Limits of detection and decision. Part 1", *Spectrochim. Acta, B* **63** (2008) 115–128.

7. E. Voigtman, "Limits of detection and decision. Part 2", *Spectrochim. Acta, B* **63** (2008) 129–141.

8. J. Vogelgesang, J. Hädrich, "Limits of detection, identification and determination: a statistical approach for practitioners", *Accred. Qual. Assur.* **3** (1998) 242–255.

BIBLIOGRAPHY

This bibliography does not include references at the end of the Preface or at the ends of the chapters and appendices.

PAPERS

1. A.L. Wilson, "The performance characteristics of analytical methods – I", *Talanta* **17** (1970) 21–29.

2. A.L. Wilson, "The performance characteristics of analytical methods – II", *Talanta* **17** (1970) 31–44.

3. A.L. Wilson, "The performance characteristics of analytical methods – III", *Talanta* **20** (1973) 725–732.

4. A.L. Wilson, "The performance characteristics of analytical methods – IV", *Talanta* **21** (1974) 1109–1121.

5. C. Liteanu, E. Hopîrtean, I.O. Popescu, "Detection limit of ion-sensitive membrane-electrodes: the electrodic function in the nonlinear domain", *Anal. Chem.* **48** (1976) 2013–2019.

6. J.S. Garden, D.G. Mitchell, W.N. Mills, "Nonconstant variance regression techniques for calibration curve-based analysis", *Anal. Chem.* **52** (1980) 2310–2315.

7. D.F. Vysochanskij, Y.I. Petunin, "Justification of the 3σ rule for unimodal distributions", *Theory Probab. Math. Stat.* **21** (1980) 24–36.

8. M.H. Feinberg, "Calibration and confidence interval: the minimum allowable concentration", *J. Chemom.* **3** (1988) 103–113.

9. F.C. Garner, G.L. Robertson, "Evaluation of detection limit estimators", *Chemom. Intell. Lab. Syst.* **3** (1988) 53–59.

10. W.G. de Ruig, R.W. Stephany, G. Dijkstra, "Criteria for the detection of analytes in test samples" *J. Assoc. Off. Anal. Chem.* **72** (1989) 487–490.

Limits of Detection in Chemical Analysis, First Edition. Edward Voigtman.
© 2017 John Wiley & Sons, Inc. Published 2017 by John Wiley & Sons, Inc.
Companion Website: www.wiley.com/go/Voigtman/Limits_of_Detection_in_Chemical_Analysis

11. D. Lambert, B. Peterson, I. Terpenning, "Nondetects, detection limits and the probability of detection", *J. Am. Stat. Assoc.* **86** (1991) 266–277.

12. L.A. Currie, "In pursuit of accuracy: nomenclature, assumptions, and standards", *Pure Appl. Chem.* **64** (1992) 455–472.

13. M.C. Ortiz, J. Arcos, J.V. Juarros, J. López-Palacios, L.A. Sarabia, "Robust procedure for calibration and calculation of the detection limit of trimipramine by adsorptive stripping voltammetry at a carbon paste electrode", *Anal. Chem.* **65** (1993) 678–682.

14. K.S. Booksh, B.R. Kowalski, "Theory of analytical chemistry", *Anal. Chem.* **66** (1994) 782A–791A.

15. N. Cressie, "Limits of detection", *Chemom. Intell. Lab. Syst.* **22** (1994) 161–163.

16. L. Sarabia, M.C. Ortiz, "DETARCHI: a program for detection limits with specified assurance probabilities and characteristic curves of detection", *Tr. Anal. Chem.* **13** (1994) 1–6.

17. R. Ferrús, M.R. Egea, "Limit of discrimination, limit of detection and sensitivity in analytical systems", *Anal. Chim. Acta* **287** (1994) 119–145.

18. D.A. Armbruster, M.D. Tillman, L.M. Hubbs, "Limit of detection (LOD)/limit of quantitation (LOQ): comparison of the empirical and the statistical methods exemplified with GC-MS assays of abused drugs", *Clin. Chem. Rev.* **40** (1994) 1233–1238.

19. I. Kuselman, A. Shenhar, "Design of experiments for the determination of the detection limit in chemical analysis", *Anal. Chim. Acta* **306** (1995) 301–305.

20. T.L. Rucker, "Methodologies for the practical determination and use of method detection limits", *J. Radioanal. Nucl. Chem.* **192** (1995) 345–350.

21. M. Thompson, R. Wood, for IUPAC, "Harmonized guidelines for internal quality control in analytical chemistry laboratories", *Pure Appl. Chem.* **67** (1995) 649–666. IUPAC ©1995.

22. L.A. Currie, for IUPAC, "Nomenclature in evaluation of analytical methods including detection and quantification capabilities", *Pure Appl. Chem.* **67** (1995) 1699–1723. IUPAC ©1995.

23. M.E. Zorn, R.D. Gibbons, W.C. Sonzogni, "Weighted least-squares approach to calculating limits of detection and quantification by modeling variability as a function of concentration", *Anal. Chem.* **69** (1997) 3069–3075.

24. K. Danzer, L.A. Currie, for IUPAC, "Guidelines for calibration in analytical chemistry", *Pure Appl. Chem.* **70** (1998) 993–1014. IUPAC ©1998.

25. J. Vial, A. Jardy, "Experimental comparison of the different approaches to estimate LOD and LOQ of an HPLC method", *Anal. Chem.* **71** (1999) 2672–2677.

26. A. Herrero, S. Zamponi, R. Marassi, P. Conti, M.C. Ortiz, L.A. Sarabia, "Determination of the capability of detection of a hyphenated method: application to spectroelectrochemistry", *Chemom. Intell. Lab. Syst.* **61** (2002) 63–74.

27. D.T. O'Neill, E.A. Rochette, P.J. Ramsey, "Method detection limit determination and application of a convenient headspace analysis method for methyl tert-butyl ether in water", *Anal. Chem.* **74** (2002) 5907–5911.

28. M.C. Ortiz, L.A. Sarabia, A. Herrero, M. S. Sánchez, M.B. Sanz, M.E. Rueda, D. Giménez, M.E. Meléndez, "Capability of detection of an analytical method evaluating false positive and false negative (ISO 11843) with partial least squares", *Chemom. Intell. Lab. Syst.* **69** (2003) 21–33.

29. V. Thomsen, D. Schatzlein, D. Mercuro, "Limits of detection in spectroscopy", *Spectroscopy* **18(12)** (2003) 112–114.

30. R.J.C. Brown, R.E. Yardley, A.S. Brown, P.E. Edwards, "Analytical methodologies with very low blank levels: implications for practical and empirical evaluations of the limit of detection", *Anal. Lett.* **39** (2006) 1229–1241.

31. D. Coleman, L. Vannata, "Detection limits: editorial comments and introduction", *Am. Lab.* **39**(**12**) (2007) 24–25.

32. C.H. Proctor, "A simple definition of detection limit", *J. Agric. Biol. Environ. Stat.*, **13** (2008) 1–23.

33. N.M. Faber. "The limit of detection is not the analyte level for deciding between 'detected' and 'not detected'", *Accred. Qual. Assur.* **13** (2008) 277–278.

34. I. Janiga, J. Mocak, I. Garaj, "Comparison of minimum detectable concentration with the IUPAC detection limit", *Meas. Sci. Rev.* **8**(**5**) (2008) 108–110.

35. M.C. Ortiz, L.A. Sarabia, M.S. Sánchez, "Tutorial on evaluation of type I and type II errors in chemical analyses: from the analytical detection to authentication of products and process control", *Anal. Chim. Acta* **674** (2010) 123–142.

36. A. Shrivastava, V.B. Gupta, "Methods for the determination of limit of detection and limit of quantitation of the analytical methods", *Chron. Young Sci.* **2** (2011) 21–25, doi: 10.4103/2229-5186.79345.

37. M. Thompson, S.L.R. Ellison, "Dark uncertainty", *Accred. Qual. Assur.* **16** (2011) 483–487.

38. I. Lavagnini, D. Badocco, P. Pastore, F. Magno, "Theil-Sen nonparametric regression technique on univariate calibration, inverse regression and detection limits", *Talanta* **87** (2011) 180–188.

39. L.V. Rajaković, D.D. Marković, V.N. Rajaković-Ognjanović, D.Z. Antanasijević, "Review: the approaches for estimation of limit of detection for ICP-MS analysis of arsenic", *Talanta* **102** (2012) 79–87.

40. M. Thompson, "What exactly is uncertainty?", *Accred. Qual. Assur.* **17** (2012) 93–94.

41. E. Besalú, "The connection between inverse and classical calibration", *Talanta* **116** (2013) 45–49.

42. J.M. Andrade, M.G. Estévez-Pérez, "Statistical comparison of the slopes of two regression lines: a tutorial", *Anal. Chim. Acta* **838** (2014) 1–12.

43. E. Bernal, "Limit of detection and limit of quantification determination in gas chromatography", *Adv. Gas Chromatogr.*, 2014, doi: http://dx.doi.org/10.5772/57341.

44. M.H. Ramsey, S.L.R. Ellison, "Uncertainty factor: an alternative way to express measurement uncertainty in chemical measurement", *Accred. Qual. Assur,* **20** (2015) 153–155.

45. M. Alvarez-Prieto, J. Jiménez-Chacón, "What do we do chemical analysis for?", *Accred. Qual. Assur.* **20** (2015) 139–146.

46. D. Badocco, I. Lavagnini, A. Mondin, G. Favaro, P. Pastore, "Definition of the limit of quantification in the presence of instrumental and non-instrumental errors. Comparison among various definitions applied to the calibration of zinc by inductively coupled plasma-mass spectrometry", *Spectrochim. Acta, B* **114** (2015) 81–86.

47. H. Evard, A. Kruve, I. Leito, "Tutorial on estimating the limit of detection using LC-MS analysis, Part I: theoretical review", *Anal. Chim. Acta*, **942** (2016) 23–29.

48. H. Evard, A. Kruve, I. Leito, "Tutorial on estimating the limit of detection using LC-MS analysis, Part II: practical aspects", *Anal. Chim. Acta*, **942** (2016) 40–49.

BOOKS, BOOK CHAPTERS AND QUASI-OFFICIAL GUIDANCE

1. C. Liteanu, I. Rica, *Statistical Theory and Methodology of Trace Analysis*, Ellis Horwood, Chichester, UK ©1980.

2. D.T.E. Hunt, A.L. Wilson, *The Chemical Analysis of Water: General Principles and Techniques*, 2nd Ed., RSC Books, ©1986. Chapters 1–4.

3. J.C. Miller, J.N. Miller, *Statistics and Chemometrics for Analytical Chemistry*, 6th Ed., Pearson Education Canada, Canada ©2010.

4. R.D. Gibbons, D.E. Coleman, *Statistical Methods for Detection and Quantification of Environmental Contamination*, John Wiley & Sons, NY ©2001.

5. H. van der Voet, "Detection limits. Encyclopedia of environmetrics", Vol. 1, pp. 504–515. A.H. El-Shaarawi, W.W. Piegorsch (eds.), Wiley, Chichester, UK, ©2002.

6. AMC-RSC, "Recommendations for the definition, estimation and use of the detection limit" *Analyst* **112** (1987) 199–204.

7. AMC-RSC, "Measurement of near zero concentration: recording and reporting results that fall close to or below the detection limit" *Analyst* **126** (2001) 256–259.

8. AMC-RSC, "Measurement uncertainty evaluation for a non-negative measurand: an alternative to limit of detection", *Accred. Qual. Assur.* **13** (2008) 29–32.

GLOSSARY OF ORGANIZATION AND AGENCY ACRONYMS

AMC-RSC Analytical Methods Committee of the Royal Society of Chemistry
BIPM Bureau International des Poids et Mesures
CITAC Co-operation on International Traceability in Analytical Chemistry
CLSI Clinical Standards Laboratory Institute
DIN Deutsches Institut für Normung
DHS Department of Homeland Security (USA)
DoD Department of Defense (USA)
DoE Department of Energy (USA)
EPA Environmental Protection Agency (USA)
FDA Food and Drug Administration (USA)
GUM Guide to the Expression of Uncertainty in Measurement
IEC International Electrotechnical Commission
IFCC International Federation of Clinical Chemistry and Laboratory Medicine
ISO International Standards Organization
IUPAC International Union of Pure and Applied Chemistry
IUPAP International Union of Pure and Applied Physics
JCGM Joint Committee for Guides in Metrology
MARLAP Multi-Agency Radiological Laboratory Analytical Protocols
NBS National Bureau of Standards (USA, now NIST)
NCCLS National Committee for Clinical Laboratory Standards
NIST National Institutes of Standards and Technology (USA)
NRC Nuclear Regulatory Commission (USA)

Limits of Detection in Chemical Analysis, First Edition. Edward Voigtman.
© 2017 John Wiley & Sons, Inc. Published 2017 by John Wiley & Sons, Inc.
Companion Website: www.wiley.com/go/Voigtman/Limits_of_Detection_in_Chemical_Analysis

OIML   Organisation Internationale de Métrologie Légale
QUAM   Quantifying Uncertainty in Analytical Measurements
USGS   United States Geological Survey
VIM   International Vocabulary of Basic and General Terms in Metrology

INDEX

Limits of Detection in Chemical Analysis, First Edition. Edward Voigtman.
© 2017 John Wiley & Sons, Inc. Published 2017 by John Wiley & Sons, Inc.
Companion Website: www.wiley.com/go/Voigtman/Limits_of_Detection_in_Chemical_Analysis

CHEMICAL ANALYSIS

A SERIES OF MONOGRAPHS ON ANALYTICAL CHEMISTRY AND ITS APPLICATIONS

Series Editor
MARK F. VITHA

Editorial Board
Stephen C. Jacobson, Stephen G. Weber